Vasodilators in Chronic Heart Failure

Edited by
H. Just and W.-D. Bussmann

With Contributions from:

F. Burkart · W.-D. Bussmann
J.N. Cohn · H. Drexler
P. Hanrath · L. Horbach
H. Just · F. Kersting
H. Kirchheim · W. Klein
V.A.W. Kreye · J. Manthey
D.G. Mathey · Th. Meinertz
Ph. Morand · R.A. Morrison
M. Oeff · M. Schlepper
W. Schulz · U. Sigwart
K. Starke · B.E. Strauer
S.H. Taylor · G.A. Turini
R. Zelis

With 124 Figures and 17 Tables

Springer-Verlag
Berlin Heidelberg NewYork 1983

Prof. Dr. H. Just
Medizinische Klinik III
Kardiologie
Hugstetter Str. 55
7800 Freiburg im Breisgau

Prof. Dr. W.-D. Bussmann
Zentrum der Inneren Medizin
Abteilung für Kardiologie
Klinikum der J.-W.-Goethe-Universität
Theodor-Stern-Kai 7
6000 Frankfurt/Main 70

ISBN-13:978-3-642-68607-8 e-ISBN-13:978-3-642-68605-4
DOI: 10.1007/978-3-642-68605-4

Softcover reprint of the hardcover 1st edition 1983

Typesetting: Schreibsatz-Service Weihrauch, Würzburg

2119/3321-543210

Contents

List of Contributors

O. Bertel, Division of Cardiology, Department of Internal Medicine, University Hospital, Basel, Switzerland

H.R. Brunner, University Hospital Centre, Lausanne, Switzerland

F. Burkart, Division of Cardiology, Department of Internal Medicine, University Hospital, Basel, Switzerland

W.-D. Bussmann, Department of Cardiology, Center of Internal Medicine, University Clinic, Frankfurt/Main, Federal Republic of Germany

J. N. Cohn, Cardiovascular Division, Department of Medicine, University of Minnesota Medical School, Minneapolis, Minnesota, USA

R. Dietz, Third Department of Internal Medicine – Cardiology, University Hospital, Heidelberg, Federal Republic of Germany

H. Drexler, Third Department of Internal Medicine – Cardiology, University Medical Clinic, Freiburg im Breisgau, Federal Republic of Germany

St. F. Flaim, Division of Cardiology, Departments of Medicine and Physiology, The Pennsylvania State University College of Medicine, Hershey, Pennsylvania, USA

H.-L. Fung, Department of Pharmaceutics, State University of New York at Buffalo, Amherst, New York, USA

M. Grbic, University Hospital Centre, Lausanne, Switzerland

E. Hackenthal, Third Department of Internal Medicine – Cardiology, University Hospital, Heidelberg, Federal Republic of Germany

P. Hanrath, Department of Cardiology, University Hospital, Hamburg, Federal Republic of Germany

H.J. Hermann, Department of Cardiology, Center of Internal Medicine, University Clinic, Frankfurt/Main, Federal Republic of Germany

D. Höhmann, Department of Pharmaceutics, State University of New York at Buffalo, Amherst, New York, USA

L. Horbach, Institute of Medical Statistics and Documentation, University of Erlangen-Nürnberg, Federal Republic of Germany

M. Irmer, Third Department of Internal Medicine – Cardiology, University Medical Clinic, Freiburg im Breisgau, Federal Republic of Germany

E. Jähnchen, Department of Pharmacology, University of Mainz, Mainz, Federal Republic of Germany

H. Jahrmärker, Medical Clinic "Innenstadt", University of Munich, Munich, Federal Republic of Germany

H. Just, Third Department of Internal Medicine – Cardiology, University Medical Clinic, Freiburg im Breisgau, Federal Republic of Germany

N.Y. Ke, Third Department of Internal Medicine – Cardiology, University Hospital, Heidelberg, Federal Republic of Germany

F. Kersting, Evangelical Foundation of St. Martin, Joh.-Müller-Str. 7, Koblenz, Federal Republic of Germany

H. Kirchheim, First Physiological Institute, Heidelberg University, Heidelberg, Federal Republic of Germany

R. Kirsten, Department of Pharmaceutics, State University of New York at Buffalo, Amherst, New York, USA

W. Klein, Department of Cardiology, University Hospital, Graz, Austria

G. Kober, Department of Cardiology, Center of Internal Medicine, University Clinic, Frankfurt/Main, Federal Republic of Germany

W. Kramer, Kerckhoff Clinic, Bad Nauheim, Federal Republic of Germany

P. Kremer, Department of Cardiology, University Hospital, Hamburg, Federal Republic of Germany

V.A.W. Kreye, Second Department of Physiology, University of Heidelberg, Heidelberg, Federal Republic of Germany

W. Kübler, Third Department of Internal Medicine – Cardiology, University Hospital, Heidelberg, Federal Republic of Germany

G. Lavigne, Cardiology Service, University Hospital Centre, Nice, France

H. Leinberger, Third Department of Internal Medicine – Cardiology, University Hospital, Heidelberg, Federal Republic of Germany

T.B. Levine, Cardiovascular Division, Department of Medicine, University of Minnesota Medical School, Minneapolis, Minnesota, USA

H. Löllgen, Third Department of Internal Medicine – Cardiology, University Medical Clinic, Freiburg im Breisgau, Federal Republic of Germany

J. Manthey, Third Department of Internal Medicine – Cardiology, University Hospital Heidelberg, Federal Republic of Germany

D.G. Mathey, Department of Cardiology, University Hospital, Hamburg, Federal Republic of Germany

Th. Meinertz, Second Medical University Clinic and Polyclinic, Mainz, Federal Republic of Germany

Ph. Morand, Cardiology Service, University Hospital Centre, Nice, France

W. Motz, Department of Medicine, University Hospital, Munich, Federal Republic of Germany

M. Oeff, Department of Cardiology and Pneumology, Steglitz Clinic, Berlin (West)

G. Pierpont, Cardiovascular Division, Department of Medicine, University of Minnesota Medical School, Minneapolis, Minnesota, USA

M. Schlepper, Kerckhoff Clinic, Bad Nauheim, Federal Republic of Germany

P. Schölmerich, Second Medical University Clinic and Polyclinic, Mainz, Federal Republic of Germany

A. Schömig, Third Department of Internal Medicine – Cardiology, University Hospital, Heidelberg, Federal Republic of Germany

H. Scholz, University of Hannover Medical School, Institute of Pharmacology and Toxicology, Hannover, Federal Republic of Germany

R. Schröder, Department of Internal Medicine – Cardiology, Steglitz Clinic, Berlin (West)

W. Schulz, Department of Cardiology, Center of Internal Medicine, University Clinic, Frankfurt/Main, Federal Republic of Germany

F. Schwarz, Third Department of Internal Medicine – Cardiology, University Hospital, Heidelberg, Federal Republic of Germany

U. Sigwart, University Hospital Centre, Lausanne, Switzerland

K. Starke, Pharmacological Institute, Freiburg im Breisgau, Federal Republic of Germany

B.E. Strauer, Department of Medicine, University Hospital, Munich, Federal Republic of Germany

S.H. Taylor, Department of Cardiovascular Studies, University of Leeds, Great Britain

J. Thormann, Kerckhoff Clinic, Bad Nauheim, Federal Republic of Germany

G.A. Turini, University Hospital Centre, Lausanne, Switzerland

R. Zelis, Division of Cardiology, Departments of Medicine and Physiology, The Pennsylvania State University College of Medicine, Hershey, Pennsylvania, USA

Acknowledgments

We hope the symposium of the working group has helped to shed light on some of the unanswered questions in the field of vasodilator treatment of heart failure. At the same time new questions have arisen, as was to be expected from a fruitful scientific exchange of thought.

Our thanks go to CIBA-GEIGY AG, especially Dr. W. Neis, whose generous support made possible both the symposium and the publication of the proceedings. We are also indebted to Springer-Verlag (Dr. Graf-Baumann) for their help and expert work with the preparation and production of this book.

H. Just
W.-D. Bussmann

Introduction

The therapeutic principle of pharmacological vasodilation in congestive heart failure is based upon established physiological and pathophysiological mechanisms and has been proved to be effective and useful. In acute heart failure this method of treatment has assumed importance as the first measure to be taken. In chronic heart failure a sizeable increase in cardiac output can be achieved with a number of vasodilating drugs. This effect can be maintained over several weeks or months.

The adjustments of the heart and the circulation to the arteriolar and/or venous unloading effect of vasodilatation in the situation of acute and chronic heart failure are well understood. However, long-term adaptations of the organism to the vasodilating intervention have not yet fully been described. In many instances of chronic administration of vasodilators, attenuation of the response has been observed. The mechanisms responsible have not been fully elucidated, nor has there been a satisfactory demonstration of improved exercise performance or of life expectancy under this mode of treatment in the patient with chronic heart failure.

In an attempt better to understand the pathophysiological mechanisms of chronic congestive heart failure as they apply to long-term treatment with vasodilators, the working group on vasodilator therapy of the Deutsche Gesellschaft für Herz- und Kreislaufforschung has assembled a group of physiologists, pharmacologists, and clinicians. The Proceedings of their Symposium held in Hinterzarten in June 1981 are presented here.

The working group has in two previous symposia presented and discussed the current status and results of vasodilator treatment with particular reference to vasodilating substances currently in use and under investigation.[1]

Vasodilators currently in use may be divided into two groups:

1. Those having a direct action with nonspecific relaxant effect upon smooth muscle (hydralazine, cardralazine, nitrates, calcium antagonists such as nifedipine, dipyridamole, and others).
2. Vasodilators with receptor-mediated vascular smooth muscle relaxation such as α-blocking agents (phentolamine, phenoxybenzamine, prazosin), as well as converting enzyme inhibitors (teprotide, captopril).

A grouping of vasodilator drugs according to their principal site of action, that is, the arterioles or the venous capacitance bed, is found to be useful in acute heart failure (pulmonary edema). In chronic heart failure, however, the substances acting predom-

---

[1] See H. Just and W.-D. Bussmann, Therapie der Herzinsuffizienz mit Vasodilatantien, Edition Medizin, Weinheim 1981

inantly on the venous system, such as the nitrates, have not been proved to be reliably effective during long-term administration. It has not satisfactorily been established if other more venously acting agents, such as molsidomine, might have an advantageous effect.

Whatever the mechanism of action may be, the resultant venous and/or arteriolar dilatation evokes a cardiovascular response which can well be defined and predicted by means of known principles of cardiovascular physiology. A fall in systemic arteriolar resistance reduces afterload and thereby leads to an increased stroke volume as long as cardiac filling is sufficient. In congestive heart failure, stroke volume may be augmented by 60% or more, since ventricular filling is usually adequate. The falling resistance is not necessarily accompanied by a reduction of aortic pressure. The secondary increase in stroke volume is usually sufficient to maintain pressure. If, however, increased cardiac output cannot be effected, than a fall in blood pressure may ensue. It is for this very reason that with the induction of vasodilator treatment a very close observation of the patient is required, particularly if his systemic blood pressure is low to begin with.

Venular dilatation reduces preload, that is, the filling potential of the left and/or right ventricle. This may be beneficial in congestive heart failure because a reduction of ventricular dimensions can be achieved. This in turn allows the ventricle to assume a more favorable geometric configuration, and allows a more efficient performance of the contractile element.

The basic principle underlying vasodilator effects on the failing heart would seem primarily to be a reduction in ventricular dimensions, brought about either by reduced aortic impedance, and therefore facilitation of ventricular emptying, or by reduced filling. Improved geometry of contraction allows the contractile element to shorten to a greater extent with less expenditure of force.

Arteriolar vasodilatation has particular beneficial effects if there are regurgitant lesions of the left heart (aortic insufficiency, mitral incompetence, or ventricular aneurysm). Since the regurgitant volume is directly determined by the ratio of the systemic and the regurgitant resistance, lowering of the former will reduce the regurgitant volume, resulting in a proportional increase of forward stroke volume. In cases of mitral regurgitation due to ventricular dilatation or papillary muscle dysfunction, a complete cessation of the regurgitation may be achieved.

Unfavorable effects of vasodilator treatment are to be expected as soon as either the filling pressure is inadequate or the sympathetic regulations are blocked. Maintenance of the aortic pressure after arteriolar vasodilatation and lowering of systemic resistance is achieved through the baroreceptor reflex. In the presence of β-blockade this response may be attenuated to such a degree that dangerous hypotension may result, due to inadequate response of heart rate and stroke volume increase. Therefore, certain precautions will have to be observed with the induction and perpetuation of vasodilator treatment. This also applies to the effect of diuretic agents. With diuresis and subsequent contraction of blood volume, even in patients with heart failure, a critical reduction of ventricular filling potential may occur. In such cases, lowering of systemic resistance may not be adequately responded to by an increase in stroke volume and cardiac output. Here again, a possibly dangerous fall in blood pressure may ensue.

Chronic vasodilatation will inevitably induce changes in the fluid exchange mechanisms at the capillary level in the peripheral circulation. An increase in interstitial, extravascular fluid volume may occur. The accumulation of edema may by itself inhibit vascular dilatation. The processes related to altered fluid exchange between the vascular and the interstitial compartments have not been fully clarified. The fact that fluid retention does occur with chronic vasodilator treatment does, however, seems to be well established. Other mechanisms, such as changes in vascular compliance, are also awaiting clarification.

Aside from these problems related to pathophysiology, a number of very basic clinical questions have still not been answered satisfactorily. Although cardiac output can be increased, and although this increase in cardiac pumping can be maintained over longer periods of time under vasodilator treatment, it is not certain to what extent the concomitant increase in tissue perfusion is to the benefit of the patient. Subjective well-being does improve, as has been shown in many studies. The increase in working capacity, however, has not regularly been demonstrated. On the other hand, an improvement in performance, or the potential for such an improvement, will to a large extent depend upon the basic condition of the patient, and primarily upon the severity of the heart failure or the underlying ailment. In the patients studied so far this has generally been so severe that judgment of physical performance would be a difficult problem. Studies on the long-term effect of vasodilating agents in chronic congestive heart failure have so far provided only scarce evidence that, apart from subjective improvement, objectively measurable improvements can be achieved. Only the most recent long-term studies are attempting to tackle this problem. Up to now an improvement in life expectancy has not been demonstrated, although this may reasonably be expected from the rather favorable changes in the pathophysiological condition of the heart and the circulation. Vasodilation favorably combines with inotropic interventions in the failing heart. Improvement of geometry of chamber contraction through the unloading effect of the vasodilator by itself leads to augmented myocardial contractile performance. The effectiveness of inotropic drugs, be it digitalis glycosides or sympathomimetic amines, is greatly enhanced by the prior administration of vasodilators. Several combinations of vasodilators and inotropic agents have been tested clinically, e.g., hydralazine and prenalterol. During recent years, a number of drugs combining vasodilating and inotropic actions (AR-L 115 BS, amrinone) have been synthesized. Last but not least, substances such as aminophylline and related drugs of the methylxanthine group, which have been in use for many years, are beginning to be re-evaluated with a particular view to the combination of vasodilating and inotropic properties.

The evaluation of effects of treatment in a complex situation such as that prevailing in chronic congestive heart failure in a group of very heterogeneous underlying ailments of the heart is, of necessity, a difficult question. Further knowledge can only be gained if the methods of biostatistics are applied. It is well recognized that statistical planning of long-term clinical studies in this group of patients is particularly difficult due to reduced life expectancy and an extremely variable natural course of the disease, fraught with numerous complications. If we are to arrive at a sound basis for the indication of vasodilator treatment, we will have no choice but to discard studies not compliant with the necessary requirements and to strive for careful statistical planning and execution of future research.

I. Physiology

Vasoconstrictor Mechanisms and the Effects of Nitrates*

Robert Zelis, Stephen F. Flaim

Studies on the peripheral circulations in congestive heart failure have indicated that vasomotor tone is increased (1–3). Certainly, vasoconstriction is the hallmark of advanced biventricular failure; however, vasoconstriction is not always present, all the time, in all patients, and in all regional vascular beds. The state of the circulation in heart failure will depend on the stimulus for heart failure and its severity, and where in the time course of congestive failure the patient or animal model is studied. In addition, the state of the circulation is also dependent upon endogenous (exercise) or exogenous stress (environmental temperature). These factors work to alter the state of the circulation via normal physiological mechanisms (reflexes, hormones) operating normally or in excess, or by nonphysiologial mechanisms peculiar to heart failure. In general, the vasoconstriction of far-advanced congestive heart failure can be related to four primary factors: (a) increased neurogenic sympathetic vasoconstrictor tone; (b) increased circulating levels of norepinephrine; (c) increased circulating levels of angiotensin II; and (d) mechanical factors impeding vascular relaxation (increased vascular sodium content, increased interstitial pressure). This brief review will consider how these factors mediate the development of vasoconstriction in heart failure, how they affect the distribution of blood flow to the arterial beds at rest and during exercise, the consequences of this blood flow redistribution, and the role these factors play in determining venous capacitance. Lastly, the effects of nitrate administration on this complex system will be considered. More thorough reviews of cardiac and circulatory compensatory mechanisms in heart failure are published elsewhere (1–4).

Mechanisms of Vasoconstriction

There are a variety of cardiac compensatory mechanisms that come into play to help sustain the failing heart (1, 4). In general, these mechanisms function to try to return systemic cardiac output to normal (tachycardia, ventricular hypertrophy and dilatation). There is a second group of vasomotor mechanisms that also come into play to redistribute blood flow between circulations and without organs to maximize "circulatory efficiency." In the advanced phase of heart failure, intense vasoconstriction

* This work was supported in part by grants from the National Heart, Lung, and Blood Institute, HL 23836 and HL 25186

is employed to maintain blood pressure at a normal level so that blood flow to the heart and brain is preserved.

With the acute onset of heart failure, cardiac output is reduced and a sympathetically mediated vasoconstrictor response occurs, probably mediated by arterial baroreceptors. Sympathetic neurogenic vasomotor tone is enhanced and adrenal catecholamine release is stimulated (5). The increased circulating norepinephrine levels, however, probably originate primarily from washout of released neurotransmittor from peripheral vascular sites. Flow is reduced most in those circulations with large populations of α-receptors, the cutaneous, splanchnic, and renal vascular beds (5). With renal vasoconstriction and stimulation of renal sympathetic nerves, there is an intrarenal redistribution of flow favoring sodium retention, and the renin-angiotensin-aldosterone cascade is initiated, leading to plasma volume expansion (6, 7).

However, the initial response of the circulations to an abrupt heart failure stimulus can be modulated by activation of certain reflex arcs. For example, atrial distension, or regional left ventricular deformation (as might occur with a functional aneurysm secondary to acute myocardial infarction) may stimulate the afferent limb of a vasodepressor response (8). We have called this reaction "endogenous impedance reduction" and it can be demonstrated in animal models and humans.

After the acute stress a phase of cardiocirculatory compensation may follow. This may result from plasma volume expansion, an enhancement of stroke volume, ventricular hypertrophy, an improvement in global pump performance, and a stabilization of heart rate at a higher level. The factors responsible for vasoconstriction during the acute phase abate (increased circulating norepinephrine and angiotensin, and increased sympathetic neurogenic tone) and a normal distribution of a normal cardiac output under basal conditions may result (6). However, in the compensated phase of chronic heart failure, the cardiac response to exercise is usually abnormal, and the inadequate cardiac output must again be redistributed (9).

A normal individual performing moderate exercise employs 70%–80% of the cardiac output in exercising skeletal muscle (Fig. 1A) (10). Vasoconstriction may occur in the renal bed; however, with moderate exercise, renal flow does not decrease in the face of an increased arterial pressure (Fig. 1B). Normal individuals constrict their cutaneous vessels at the onset of exercise; however, as the thermal load from the exercising muscles increases, a marked cutaneous vasodilatation occurs. Concomitant with the cutaneous vasodilatation is a splanchnic vasoconstriction. The inverse relationship between splanchnic and skin blood flow is the normal response of the body in handling environmental extremes of temperature. Coronary flow increases commensurate with the increased workload of the heart, and cerebral blood flow is maintained at normal levels (Fig. 1B).

The response of the patient with moderate compensated heart failure is different. Cardiac output fails to rise normally with exercise and proportionately less flow is delivered to the exercising skeletal muscles (Fig. 1A) (9, 11). A marked cutaneous, renal, and splanchnic vasoconstriction occurs (Fig. 1C). The failure of skin vessels to dilate may result in slight increases in core temperature (12). The renal response is important in that it again activates those mechanisms responsible for salt and water retention. It is this salt and water retention that plays a major role in increasing "vascular stiffness" and in increasing interstitial fluid pressure. These factors appear

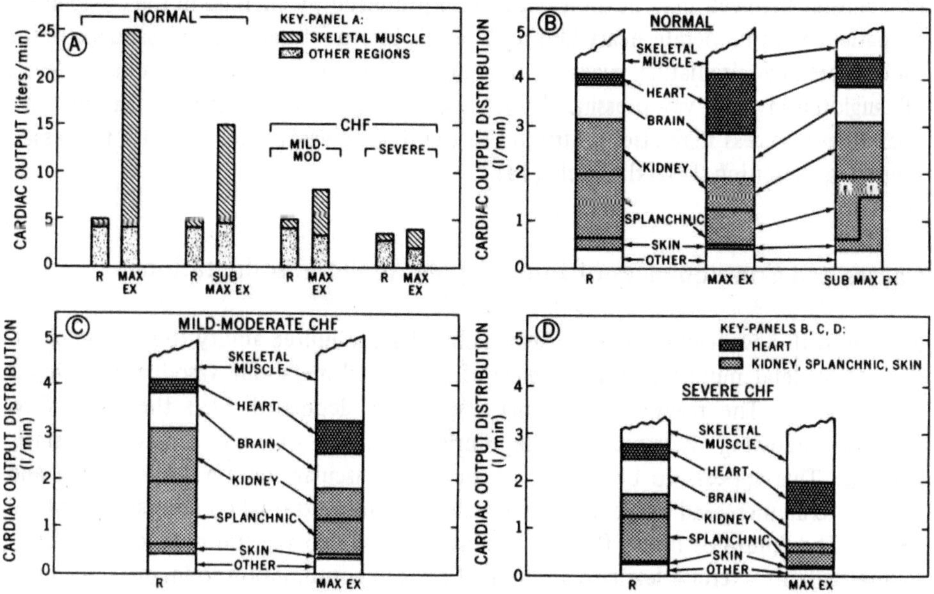

Fig. 1A–D. Regional distribution of the cardiac output at rest (*R*) and during exercise (*EX*) in normal subjects and patients with congestive heart failure (*CHF*). These data are estimates from limited quantitative data in humans and supplemented by directional changes in regional blood flow from animal studies. Total cardiac output and its distribution to skeletal muscle (*cross-hatched*) and all other regions (*light stipple*) is seen in **A**. The distribution of blood flow to the circulations exclusive of skeletal muscle is seen in **B, C,** and **D.** Attention is called to blood flow to the heart (*double cross hatch*) and to the circulations rich in α-receptors (kidney, skin, splanchnic) (*heavy stipple*) by the pattern of shading (*key in* **D**). *MAX EX*, maximum exercise; *SUB MAX EX*, submaximal exercise at 50%–60% of maximum; *E*, the distribution of cardiac output to splanchnic and cutaneous circulations early during submaximal exercise in normals; *L*, the late response. Zelis and Flaim (2)

to play a major role in limiting blood flow to exercising skeletal muscle (Table 1, Figs. 1 and 2). This will be considered further below.

The patient with severe heart failure who has symptoms at rest has a reduced cardiac output which is redistributed away from the cutaneous, renal, and splanchnic beds with relative preservation of coronary and cerebral flow (Fig. 1D) (13). The

Table 1. Congestive heart failure: vasoconstrictor mechanisms. *SNS*, sympathetic nerve stimulation; *NE*, norepinephrine; *AT*, angiotensin; *MECH*, mechanical factors. Zelis and Flaim (2)

Arterial			Venous	
Rest	Ex		Rest	Ex
++	++++	SNS	0	++
+	++	NE	+	+?
++	?++	AT	?	?
+0	++++++	MECH	+++	++++

mechanisms involved may be similar to those produced by exercise in the compensated patient with moderate heart failure. There is increased neurogenic vasoconstrictor tone, increased circulating levels of norepinephrine, and increased circulating levels of angiotensin and vasopressin. These neurohumoral vasoconstrictor mechanisms operating in excess may also contribute to the limited vasodilator capacity for vessels supplying exercising skeletal muscle (14).

Causes and Consequences of Impaired Metabolic Vasodilatation

Patients with advanced congestive heart failure have a limited ability to increase blood flow to skeletal muscle in the presence of a maximal metabolic vasodilator stimulus (15) (Fig. 2). The reactive hyperemia response is depressed, as is the blood flow response following dynamic exercise, during dynamic exercise, and during static exercise. This appears to be related to increased vascular sodium content (16) and perhaps to an increased tissue pressure which may cause external compression of blood vessels. The failure of blood flow to increase normally to skeletal muscle when heart failure patients exercise leads to a failure of oxygen consumption to increase normally (11). This leads to an earlier dependence on anaerobic metabolism, a lactic acidemia, the early appearance of fatigue, and slowed recovery from exercise. It has been suggested that this "increased vascular stiffness" is most important in limiting blood flow to exercising muscle at the conductance vessel level (17). When blood flow is mechan-

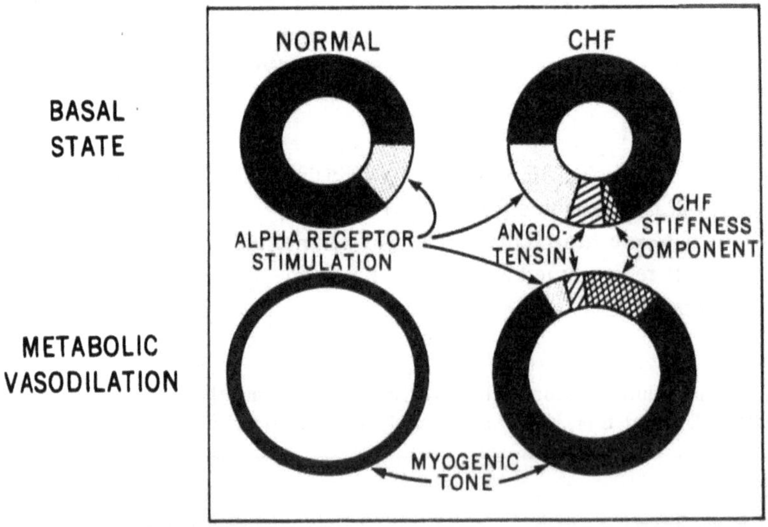

Fig. 2. Determinants of vasomotor tone in the *basal state* and under conditions of *metabolic vasodilatation* in normal subjects and in patients with congestive heart failure (*CHF*). The lumina of the blood vessels are drawn to scale from data derived from the study of blood flow in human limbs. Blood vessels in CHF are more constricted than normal ones. The estimated relative contribution of α-receptor stimulation, angiotensin, and the stiffness factor to this vasoconstriction in CHF is portrayed. Zelis and Flaim (2)

ically restricted at the large vessel level in an exercising skeletal muscle preparation, a similar metabolic pattern develops (17). There is a reduction in oxygen consumption of exercising skeletal muscle proportionate to the reduction in blood flow. This produces skeletal muscle ischemia.

The major consequence of relative skeletal muscle ischemia during exercise in heart failure is the stimulation of somatic afferent nerves in skeletal muscle which, when activated, initiate a reflex arc, resulting in an increased sympathetic tone (18). Thus relative skeletal muscle ischemia during exercise may be one of the factors that contributes significantly to the excessive sympathoadrenal response so characteristic of the exercising heart failure patient.

Although part of the excessive sympathoadrenal response to exercise in heart failure may be ascribed to relative skeletal muscle ischemia, part of it may be related to failure of normal baroreceptor regulation of the autonomic response to exercise (19, 20). Normally, during exercise, the systemic arterial baroreceptors are activated by the increase in arterial pressure. Similarly, atrial baroreceptors are activated by the increased venous return leading to increased stretch of these chambers. In heart failure, both of these reflexes function abnormally (19, 20). Thus the normal inhibition of the sympathetic vasoconstrictor response to exercise that is usually modulated by baroreflexes does not occur. The net result is an exaggerated sympathoadrenal response to exercise.

During the early phase of heart failure, when the renin-angiotensin system is first activated, the systemic vasoconstrictor response to circulating angiotensin is important in maintaining a normal blood pressure (Table 1) (6). However, the role of angiotensin in modifying blood flow to exercising skeletal muscle has not been well defined. The role of α-receptor stimulation in exercising skeletal muscle, however, has been defined (17).

In a canine exercising gracilis muscle, the infusion of low doses of norepinephrine intraarterially also reduces blood flow. However, oxygen consumption is maintained at a normal level (17). It seems that norepinephrine produces a change in the intramuscular distribution of flow within skeletal muscle, distributing flow away from inactive skeletal muscle fibers to active fibers. Thus it appears that the normal activation of sympathetic nervous system during exercise plays a role in maintaining circulatory "efficiency" in a normal individual, and, if increased but slightly, may be of significant benefit to the heart failure patient. However, as more norepinephrine is delivered to the blood vessels of exercising skeletal muscle, flow falls further and, at that point, the normal increase in oxygen consumption is impaired. This is best explained by an effect of norepinephrine at the conductance vessel level, whereas the response to the low-dose norepinephrine selectively constricts nonnutritional vessels, the response to high-dose norepinephrine is nonselective, and total flow to exercising muscle falls (2). This leads to further skeletal muscle ischemia and a heightening of the excessive activation of the sympathetic nervous system. It is clear that this is a self-reinforcing cycle leading to further impairment of nutritional flow and further activation of the sympathetic nervous system. At this late stage of heart failure, the excessive activation of the sympathetic nervous system further restricts flow to those organs most susceptible: gut, skin, and kidney. Abnormalities may then be noted in renal, liver, and cutaneous function.

The Venous Beds in Heart Failure

The four vasoconstrictor mechanisms operative on the arterial side of the circulation also play a role in the regulation of the volume of blood in the capacitance vessels in heart failure (Table 1, Fig. 3) (2, 21). However, the relative extent to which each is employed varies.

It is convenient to consider three venous beds as being the major repositories of systemic blood volume (2, 22). The veins draining the skin are thick-walled, contain considerable vascular smooth muscle, are highly innervated, and respond vigorously to neurohumoral vasoconstrictor stimuli. They subserve the function of body temperature regulation. A reduction in core temperature causes a sympathetic neurogenic venoconstrictor response in skin veins and a diversion of blood from these vessels through communicating veins into the large central limb veins that run parallel to arteries. Core temperature is preserved by countercurrent heat exchange. The veins draining skeletal muscle are thin-walled, have minimal vascular smooth muscle, are

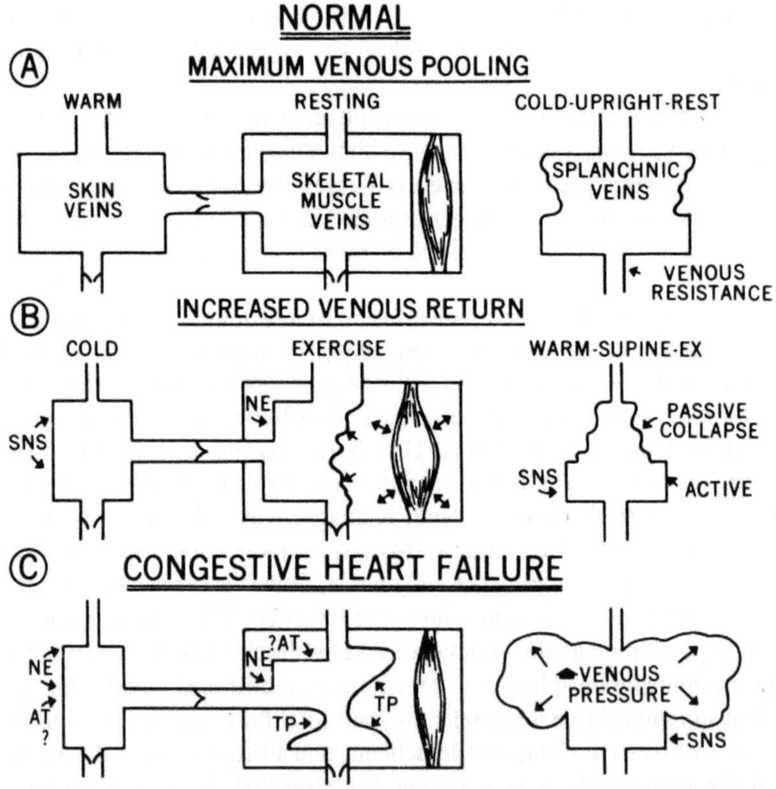

Fig. 3A–C. Regulation of the venous volume in the three major venous beds during maximum venous pooling (**A**), increased venous return (**B**) in normal subjects, and the factors altering venous volume in patients with congestive heart failure (**C**). *SNS*, neurogenic sympathetic venomotor tone; *NE*, norepinephrine; *EX*, exercise; *AT*, angiotensin; *TP*, increased tissue pressure. Zelis and Flaim (2)

not innervated, and respond poorly to humorally delivered vasoconstrictor agents. The return of blood from these vessels to the heart is dependent almost entirely on external compression from contracting skeletal muscle and respiratory efforts. Centripetal flow is favored by competent venous valves, which also prevent reflux of blood into the cutaneous veins. The third venous bed, which has been poorly studied in humans, is that of the splanchnic circulation (23). The volume of blood in these vessels is controlled by both active and passive elements. In a person who is warm or supine, the restriction of arterial inflow can cause a passive collapse of veins at low venous pressures. During exercise and at higher venous pressures, the activation of sympathetic nerves to the splanchnic veins can also actively reduce splanchnic blood volume. In an upright person at rest or in the cold, the splanchnic veins fill passively and a reduction in sympathetic venomotor tone may also contribute to venous pooling (Fig. 3).

What is known about the state of these three venous beds in heart failure? Very little is known about splanchnic blood volume, since this is difficult to study in humans. However, the veins in the limbs (skin and muscle) have undergone extensive examination (21, 24). In a comfortable patient with compensated heart failure, even advanced heart failure, sympathetic neurogenic venomotor tone is low (2). Humorally delivered venoconstrictor agents (norepinephrine, epinephrine) may play a greater role in reducing the volume of skin veins. More important, however, are local factors (21). Much of the increased "venous tone" in heart failure can be ascribed to external compression of the veins (presumably from an increased interstitial fluid pressure) rather than to an increase in vascular smooth muscle tension. Maximal venodilatation induced by intraarterial injections of phentolamine and multiple large doses of sodium nitrate fail to increase venous volume in heart failure patients to normal. However, a substantial increase in limb venous volume has been noted following diuresis (21).

If the limb venous volume in heart failure is not reduced, the venous volume of the splanchnic vessels must be increased to accomodate the expanded plasma volume. Presumably, this is a passive increase in venous volume mediated by a high venous pressure (Fig. 3). An interrelationship of some of these factors can be appreciated if one considers the mechanisms by which venous pressure increases markedly when heart failure patients exercise. Since some of the increased venous pressure can be eliminated by ganglionic blockade, it has been suggested that active venoconstriction was the mechanism by which venous pressure rose during exercise (24). Alternative explanations are possible. First, the exercising skeletal muscle is expelling greater volumes of blood from its venous bed. If the skin veins fail to dilate with the thermal stress of exercise, much of the venous volume expelled from skeletal muscle may contribute to raising the pressure in an already tense venous system. Ganglionic blockade may work by dilating arterial inflow into the splanchnic vessels. Some of the volume expelled from the veins in skeletal muscle may be taken up by veins in the gut by passive venous pooling. Equally likely is the possibility that ganglionic blockade reduces sympathetic neurogenic venoconstriction in the splanchnic system, which may be an important contributor to splanchnic venous volume at high venous pressures.

Effects of Nitrates on the Circulation in Heart Failure

Although the nitrates are classified as venodilators, the preferential effect on veins is relative (25). Nitrates can be shown to relax both arteries and veins, but the amount of venodilatation and arteriolar dilatation that occurs is a function of the status of the individual or animal model studied and is definitely modulated by baroreceptor reflexes. After nitroglycerin administration to normal subjects, venodilatation predominates. A direct arteriolar dilatation is present but is masked by a reflex arteriolar constriction (26, 27). Pooling of blood peripherally reduces stretch on cardiac low-pressure mechanoreceptors, which leads to an increase in sympathetic vasomotor tone. The net result is an increase in systemic arterial resistance. If the reflex vasoconstriction is blocked, an arteriolar dilatation is clearly evident. This reflex is more marked in an upright subject or a subject in whom plasma volume is contracted (25).

The response of heart failure patients to nitrates produces a different type of interaction. In the setting of increased systemic arterial resistance in heart failure, a nitrate-induced arteriolar dilatation is easier to demonstrate (28). Although they are not α-blockers, nitrates antagonize the high basal sympathetic tone and an arteriolar dilatation results. In addition, the venodilator effects of nitrates can significantly reduce pulmonary capillary pressure and relieve dyspnea. This can lead to a reflex withdrawal of sympathetic tone and contribute to the net peripheral vasodilatation.

The amount of systemically administered nitrate reaching the peripheral resistance vessels is small in comparison to what can be delivered by direct intraarterial injections. The latter have been used to look at maximal vascular responsiveness to nitrates. When sodium nitrate was injected intraarterially in graded amounts it produced greater arteriolar dilatation in normal subjects than in heart failure patients (15). This was despite the fact that the nitrate reaching the resistance vessels in heart failure arrived there in a higher concentration, since basal blood flow was lower. These data suggest that the heart failure "vascular stiffness factor" also impairs the arteriolar dilator response to vasodilator drugs. This concept goes unnoticed when nitrates are administered systemically, since the reflex changes in the circulation are greater and the amount of nitrate reaching the resistance vessel in considerably less.

Venous reactivity to intraarterially administered sodium nitrate produced an opposite pattern of response (21). There was a proportionally greater venodilatation in the patients with heart failure than in normal controls. This is probably a reflection of the fact that the normal individual resting in a comfortably warm environment possesses little intrinsic venomotor tone measured in the limbs. In addition, the concentration of nitrate reaching the limb veins is higher in the heart failure patient for the reasons stated above. These problems preclude any definitive statement regarding the comparative responsiveness of veins to nitrates between normals and patients with heart failure.

One can make a definitive statement about the maximum venous volume that can be achieved with nitrate administration. When high doses of sodium nitrate were repeatedly administered intraarterially, the maximal venodilator potential of limb veins was seen in normals and heart failure patients (21). It was noted above that a "mechanical factor" seemed most important in preventing filling of the limb veins by extrinsic compression in heart failure. We suggested that this "mechanical factor" was an increase in tissue pressure or subclinical edema. That this may be true was suggested by the recent report that nitrates were ineffective in lowering ventricular filling pressure in acute heart failure accompanied by massive peripheral edema (29). In this extreme case, it appeared that ascites may also have contributed to extrinsic compression of splanchnic vessels. In acute heart failure in the absence of massive edema, nitrates did prove to be effective in reducing filling pressures, presumably by dilating truly constricted veins that were subject to minimal extravascular compression.

Nitroglycerin has minimal effects on exercise performance when administered acutely to patients with congestive heart failure. They do not exercise longer, nor are they symptomatically improved (30, 31). In general, acute studies have shown a moderate nitroglycerin-induced reduction in left ventricular filling pressure and a slight augmentation in cardiac output during exercise. Further, we demonstrated that the small increase in exercise cardiac output produced by nitroglycerin did not seem to go to exercising skeletal muscles to improve nutritional flow. During submaximal and maximal exercise, the increase in arterial lactate was similar, as was catecholamine release following acute administration of nitroglycerin (32). It seemed likely, then, that the small increase in cardiac output may have been diverted to other circulations. This was studied in a rat model of volume-overload heart failure. Exercise performance on a treadmill and the regional distribution of cardiac output was measured by means of the radionuclide-tagged microsphere technique (33).

Nitroglycerin administered to normal exercising rats had a deleterious effect on regional hemodynamics. This was probably mediated by an increase in sympathetic tone. During the prenitrate control periods, blood flow to the renal, splanchnic, and cutaneous circulations was relatively unchanged during exercise in the normal rat. After nitroglycerin, there was a small reduction in blood flow to the highly innervated circulations during exercise. By contrast, heart failure animals displayed marked vasoconstriction in the renal, splanchnic, and cutaneous circulations during exercise prior to nitrate administration. After nitrate, there was less of a reduction in flow, and the absolute level of flow was similar to that in the nitroglycerin-treated normal exercising animals. Thus nitroglycerin "normalized" the regional blood flow response in the heart failure animals (33). If the renal circulation was protected by nitrates during exercise, it is possible that much of the hormonal response and the stimulus for salt and water retention might be minimized. If this were to lead to a reduced vascular sodium content, then nutritional blood flow to exercising skeletal muscle might ultimately be improved. This should lead to an improvement in exercise tolerance with chronic nitrate administration, but not during acute administration. Such a situation has been observed clinically (30, 31). Whether or not the renal effects of the drug were the cause remains to be established.

Summary

In far-advanced heart failure, vasoconstriction can be demonstrated in a large number of the arterial and venous beds. In the arterial circulations, an increase in sympathetic neurogenic vasomotor tone, increased circulating norepinephrine, and increased circulating angiotensin contribute primarily to this constrictor response. Patients with moderate heart failure may have a normal distribution of regional arterial flow at rest, but an exaggerated vasoconstriction on exercise. Circulations to which blood flow is most reduced are the cutaneous, renal, and splanchnic vascular beds.

In addition, an increased sodium content of blood vessels in skeletal muscle limits their ability to dilate to the metabolic stimulus of exercise. Whereas increased α-receptor stimulation enhances the efficiency of blood flow to exercising skeletal muscle (diverting flow from inactive to active fibers) at the nutritional vessel level, the heart failure "vascular stiffness" factor has its most marked effect in limiting blood flow at the conductance vessel level. A decrease in flow through the conductance vessels leads to an impairment of skeletal muscle oxygen consumption to rise normally during exercise. Greatly reduced oxygen tension in these relatively ischemic exercising muscles stimulates somatic afferent nerves which further exaggerate the sympatho-adrenal response. The sympathetic vasoconstrictor response is further amplified by the failure of low- and high-pressure baroreceptors normally to modulate the sympathetically mediated vasoconstriction associated with exercise.

Although the veins in the extremities have a reduced capacitance in patients with heart failure, the mechanisms involved are quantitatively different from those operating in the arterial bed. Sympathetic neurogenic venomotor tone may be normal and circulating constrictor hormones may contribute more to the reduced capacitance of veins draining skin. Veins draining skeletal muscle appear to be compressed exogenously, probably by an increase in tissue pressure resulting from salt and water retention and plasma volume expansion. The major repository of the increased plasma volume of chronic heart failure is most likely the splanchnic bed which appears to fill passively due to an increased venous pressure.

The effects of nitrates on this system are complex. The acute administration of nitroglycerin to normals and subjects with mild heart failure can be detrimental in that vasoconstrictor reflexes are enhanced. When vasoconstrictor tone is high, such as in advanced heart failure, or during exercise in moderate heart failure, nitrates may improve flow selectively to certain circulations. A nitrate-induced improvement in renal flow may tip the balance away from activation of the renin-angiotensin-aldosterone system, leading to natriuresis, reduced vascular stiffness, and reduction in sympathetic nervous system activation.

Acknowledgments. The authors gratefully acknowledge the secretarial assistance of Mrs. Nina Gingerich.

References

1. Zelis R, Flaim SF, Liedtke AJ, Nellis SH (1981) Cardiocirculatory dynamics in the normal and failing heart. Annu Rev Physiol 43:455–476
2. Zelis R, Flaim SF (to be published) Alterations in vasomotor tone in congestive heart failure. Prog Cardiovasc Dis
3. Fishman AP (ed) (1978) Heart Failure. Hemisphere, Washington DC
4. Braunwald E, Ross J Jr, Sonnenblick EH (eds) (1967) Mechanisms of Contraction of the Normal and Failing Heart. Little, Brown, Boston Massachusetts

5. Higgins CB, Vatner SF, Franklin D, Braunwald E (1972) Effects of experimentally produced heart failure on the peripheral vascular response to severe exercise in conscious dogs. Circ Res 31:186–194
6. Watkins L Jr, Burton JA, Haber E, Cant JR, Smith FW, Barger AC (1976) The renin-angiotensin-aldosterone system in congestive heart failure in conscious dogs. J Clin Invest 57: 1606–1617
7. Schmid PG, Mayer HE, Mark AL, Heistad DD, Abboud FM (1977) Differences in the regulation of vascular resistance in guinea pigs with right and left heart failure. Circ Res 41:85–93
8. Ratz P, Ealy R, Flaim SF (1980) Effects of acute myocardial infarction on cardiocirculatory hemodynamics in conscious rat. Fed Proc 39:633
9. Flaim SF, Minteer WJ (1980) Ventricular volume overload alters cardiac output distribution in rats during exercise. J Appl Physiol 49:482–490
10. Astrand PO, Rodahl K (1970) Textbook of Work Physiology. McGraw Hill, New York
11. Longhurst J, Gifford W, Zelis R (1976) Impaired forearm oxygen consumption during static exercise in patients with congestive heart failure. Circulation 54:477–480
12. Zelis R, Mason DT, Braunwald E (1969) Partition of blood flow to the cutaneous and muscular beds of the forearm at rest and during leg exercise in normal subjects and in patients with heart failure. Circ Res 24:799–806
13. Wade OL, Bishop JM (1962) Cardiac Output and Regional Blood Flow. Blackwell Scientific, Oxford
14. Flaim SF, Crede W, Beech A, Nellis SH, Zelis R (1979) The effects of norepinephrine on active hyperemia in the canine gracilis muscle. Circ Res 44:660–666
15. Zelis R, Mason DT, Braunwald E (1968) A comparison of the effects of vasodilator stimuli on peripheral resistance vessels in normal subjects and in patients with congestive heart failure. J Clin Invest 47:960–970
16. Zelis R, Delea CS, Coleman HN, Mason DT (1970) Arterial sodium content in experimental congestive heart failure. Circulation 41:213–216
17. Nellis SH, Flaim SF, McCauley K, Zelis R (1980) Alpha-stimulation protects exercise increment in skeletal muscle oxygen consumption. Am J Physiol 238:H331–H339
18. Longhurst J, Zelis R (1979) Cardiovascular responses to local hindlimb hypoxemia: relation to the exercise reflex. Am J Physiol 237:H359–H365
19. Higgins CB, Vatner SF, Eckberg DL, Braunwald E (1972) Alterations in the baroreceptor reflex in conscious dogs with heart failure. J Clin Invest 51:715–724
20. Zucker IH, Earle AM, Gilmore JP (1977) The mechanism of adaptation of left atrial stretch receptors in dogs with chronic congestive heart failure. J Clin Invest 60:323–331
21. Zelis R (1974) The contribution of local factors to the elevated venous tone of congestive heart failure. J Clin Invest 54:219–224
22. Shepherd JT, Vanhoutte PM (1975) Veins and Their Control. Saunders, Philadelphia
23. Rowell L (1975) The splanchnic circulation. In: Zelis R (ed) The Peripheral Circulations. Grune & Stratton, New York
24. Wood JE (1962) The mechanism of the increased venous pressure with exercise in congestive heart failure. J Clin Invest 41:2020–2024
25. Zelis R, Flaim SF, Moskowitz RM, Nellis SH (1979) Editorial: how much can we expect from vasodilator therapy in congestive heart failure? Circulation 59:1092–1097
26. Vatner SF, Pagani M, Rutherford JD, Millard RW, Manders WT (1978) Effects of nitroglycerin on cardiac function and regional blood flow distribution in conscious dogs. Am J Physiol 3:H244–H252
27. Flaim SF, Zelis R (1978) Cardiac and regional circulatory dynamics in conscious volume overload rats: detrimental effects of nitroglycerin. Clin Res 26:599A
28. Goldberg S, Mann T, Grossman W (1978) Nitrate therapy of heart failure in valvular heart disease. Importance of resting level of peripheral vascular resistance in determining cardiac output response. Am J Med 65:161–166
29. Magrini F, Niarchos AP (1980) Ineffectiveness of sublingual nitroglycerin in acute left ventricular failure in the presence of massive peripheral edema. Am J Cardiol 45:841–847

30. Chatterjee K, Parmley WW (1980) Vasodilator therapy for chronic heart failure. Ann Rev Pharmacol Toxicol 20:475–512
31. Cohn JN (1980) Vasodilator therapy of congestive heart failure. Adv Intern Med 26:293–315
32. Moskowitz RM, Kinney EL, Zelis R (1979) Hemodynamic and metabolic responses to upright exercise in patients with congestive heart failure. Chest 76:640–646
33. Flaim SF, Weitzel RL, Zelis R (to be published) Mechanism of action of nitroglycerin during exercise in a rat model of heart failure: improvement of blood flow to the renal, splanchnic and cutaneous beds. Circ Res

Discussion

Bassenge, Freiburg: Can you give information on the actual wall thickness of major arterial branches or smaller arteries in the state of heart failure?

Zelis, Philadelphia: We have not measured wall thickness, but we have looked at the sodium content of the large vessels and have found it to be increased. Water content of the walls of the large arteries in also increased. Studies of the mechanical properties of these vessels are, however, not available.

Auditorium: Is the increased vessel stiffness in the state of heart failure reversible toward normal when heart failure is successfully treated? Are there experimental studies showing that reequilibration of a vessel is possible, for example, in an ordinary organ bath?

Zelis, Philadelphia: When we study vascular stiffness in an organ bath, one of the problems we encounter is to have an animal model with advanced and chronic heart failure. There is no clear evidence, to answer the question. It is my impression, however, that if one perfuses a vessel, it will probably lose its mechanical stiffness.

Brunner, Basel: It is difficult to understand how an increase in sodium content of the vascular wall might come about.

Zelis, Philadelphia: The increased sodium content of the vessel is probably related to sodium retention in heart failure through activation of the aldosterone mechanism. Sodium content is increased in all tissues, including the vascular wall. Our knowledge of the reversibility of this increased sodium content of the tissues, including that of the vascular wall, is still incomplete, however.

Stepanec, Basel: What is your recommendation for more specific local vasodilators for the kidney and for the skeletal muscle?

Zelis, Philadelphia: It is not at all clear whether a specific renal vasodilator would be of particular use in heart failure. We have decided to look at small animal models of heart failure where we can exercise the animals. We are using a rat heart failure model. The rats are studied conscious and awake. They are exercised on a treadmill, and we inject microspheres and study regional blood flow. One of the interesting results we have obtained with one vasodilator is that the intensive renal vasoconstriction in the exercising heart failure rat can be overcome. I believe that studying regional distribution of flow in heart failure at rest and under exercise is of the greatest importance, mainly flow to the kidneys and to the intestinal, splanchnic area. As far as pharmacological influence on the altered distribution of blood flow in heart failure is concerned, I do not think we are able to do anything to the exercising skeletal muscle. The reduction of nutritional flow to the muscle at rest in heart failure is a protective mechanism in a state of decreased cardiac output.

Taylor, Leeds: The hemoglobin-oxygen dissociation curve changes dramatically in exercising muscle. Of course, this mechanism will improve nutrition in a state of reduced oxygen availibility. What significance will the shunt connection have in various organs, especially in muscle, by deviating residual blood flow from the parenchyma?

Zelis, Philadelphia: There is no simple answer to this question. Sympathetic-mediated increase in vascular tone will generally decrease nutritional flow to the organs. Lessening of vasoconstriction will improve parenchymal flow. The significance of shunt connections cannot quantitatively be estimated. The increase in sympathetic tone with subsequent peripheral vasoconstriction is, of course, more prevalent in early failure and less pronounced in persisting states of failure. Here shunt vessels may play a different role, which can again not be defined quantitatively.

Schröder, Berlin: Shifts in the oxygen dissociation curve as a means of improving availibility of oxygen in the peripheral tissues is generally overestimated in its clinical significance. In heart failure hemoconcentration may occur. This factor may influence peripheral oxygen availibility. On the other hand, increased blood viscosity in the hemoconcentrated state will decrease fluidity of the blood, thereby increasing a net resistance of flow in the vascular periphery. I wonder if this factor, together with the swelling of the arterial wall, will not amount to what is being observed as increased stiffness.

Zelis, Philadelphia: Hemoconcentration may indeed play an important role in the pathogenesis of increased peripheral resistance. It adds to the effect of vasoconstriction, but is quantitatively far less important.

Functional and Metabolic Consequences of Chronic Alterations in Preload and Afterload in the Hypertrophied and Dilated Heart*

B.E. Strauer, W. Motz

Introduction

Chronic hypertrophic and dilative heart disease is one of the most common cardiac diseases in man (1). From the functional point of view, ventricular hypertrophy provides a basic mechanism that permits the heart to maintain normal cardiac pump function despite abnormal pressure, stress, or volume load (5–8, 13). However, progression of cardiac hypertrophy is a significant precursor of cardiac dilatation and failure and of coronary insufficiency. Therefore, the question has been raised whether specific regression of hypertrophy (14, 15) might be appropriate to prevent both the myocardial and coronary consequences of cardiac hypertrophy.

Accordingly, this chapter will focus on three topics:

1. The evaluation of the appropriateness of left ventricular hypertrophy and energetics in clinical hypertrophic heart disease
2. The influence of chronic alterations in preload and afterload on ventricular wall dynamics and function
3. The contractile and hemodynamic consequences of prevention and regression of cardiac hypertrophy by chronic, pharmacologically induced alterations in preload and afterload.

Methods

The clinical studies were performed in more than 900 patients including hypertensive, coronary, congestive, valvular, septal, and myocardial heart disease (9–13). Quantitative ventriculography was performed in all patients. Circumferential wall stress was calculated frame by frame from ventriculograms. Left ventricular mass and left ventricular mass to volume ratio were determined from end-diastolic chamber dimensions. Coronary blood flow was measured by the argon method, with gas chromatographic analysis of argon in the arterial and coronary sinus blood (9, 10).

---

* Supported by the Deutsche Forschungsgemeinschaft, Heidelberg, Federal Republic of Germany

The experimental studies were carried out on both normotensive (n = 42) and spontaneously hypertensive rats (n = 92) at various ages (20–80 weeks). Hemodynamics, including determinations of left ventricular pressures, cardiac output, end-diastolic volume, wall mass, and systolic wall stress, as well as papillary muscle mechanics in isotonic, isometric, and isotonic afterload contractions, were analyzed (2–4).

Results

Appropriateness of Left Ventricular Hypertrophy and Energetics

Left ventricular size as represented by the end-diastolic volume shows an inverse relationship with left ventricular function as evidenced by the ejection fraction. With increase in end-diastolic volume the ejection fraction decreases. The reason for this reciprocal behavior is to be found in the intrinsic inverse relationship between systolic wall stress (afterload) and function (Fig. 1). With increase in systolic wall stress the left ventricular function decreases. Doubling of stress leads to reduction in the ejection fraction by approximately 50%. Since systolic wall stress results from systolic pressure and from the mass to volume ratio, it is compatible with the ventricular afterload imposed on the left ventricular wall. It is therefore reasonable to assume that left ventricular size, that is end-diastolic volume, and systolic wall stress are important determinants of ventricular performance.

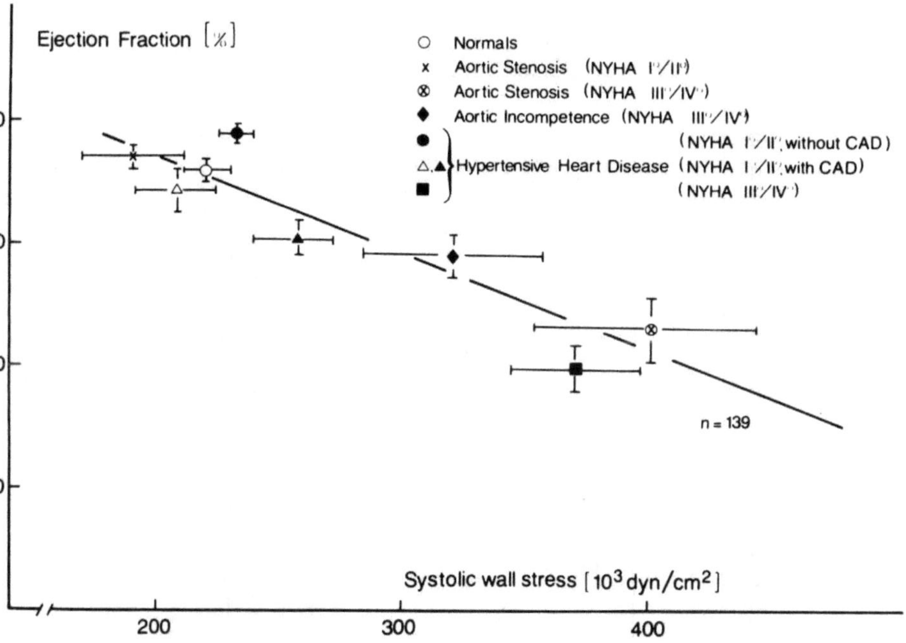

Fig. 1. Relationship between systolic wall stress and ejection fraction. Note the decrease in ejection fraction with increase in wall stress (afterload). *CAD*, coronary artery disease (9–12)

Table 1. Geometric and energetic values in clinical heart disease

	No.	Ejection fraction (%)	LV mass (g/m²)	LV mass to volume ratio (g/ml)	T_{syst} 10^3 dyn/cm²)	MVO_2 (ml/min · 100 g)	LVO_2 (ml/min)
N	12	72 ± 2	92 ± 6	1.21 ± 0.12	220 ± 9	7.98 ± 0.52	13.3 ± 2.1
CAD	36	52 ± 11	145 ± 22[a]	1.12 ± 0.16	236 ± 18	7.9 ± 0.39	20.6 ± 3.2
SVD	8	69 ± 8	84 ± 14	1.18 ± 0.21	206 ± 22	6.4 ± 0.6[a]	9.7 ± 0.91[a]
AS*	6	74 ± 4	145 ± 10[b]	2.1 ± 0.31	192 ± 23	8.1 ± 0.8	21.1 ± 2.3[b]
AS**	9	46 ± 11	190 ± 17[c]	1.01 ± 0.11	396 ± 95[b]	14.9 ± 1.6[c]	51.0 ± 6.9[c]
AI	12	58 ± 7	174 ± 22[c]	1.12 ± 0.13	329 ± 36[a]	14.2 ± 1.4[c]	45.4 ± 4.7[c]
HOC	12	78 ± 6	~228	~3.78	142 ± 52	8.6 ± 1.21	35.3 ± 3.6
EH	92	62 ± 6	152 ± 12[c]	1.52 ± 0.33	266 ± 18[b]	10.7 ± 0.38[c]	29.3 ± 4.1[c]
MI	20	63 ± 9	132 ± 11[a]	1.19 ± 0.09	267 ± 11[a]	9.82 ± 0.21[b]	21.6 ± 0.92
MV	44	62 ± 12	149 ± 9[b]	1.2 ± 0.1	248 ± 32[a]	9.22 ± 0.99[a]	24.9 ± 2.9[a]

[a] $P < 0.05$; [b] $p < 0.01$; [c] $p < 0.001$

Abbreviations: CAD, coronary artery disease; *SVD*, small coronary vessel disease (systemic immuncomplex vasculitis); *AS**, clinically compensated aortic stenosis (NYHA I–II), concentric hypertrophy; *AS***, clinically decompensated aortic stenosis (NYHA III–IV), LV dilatation; *AI*, compensated and decompensated aortic incompetence (NYHA II–IV); *HOC*, hypertrophic obstructive cardiomyopathy; *EH*, essential hypertension; *MI*, mitral incompetence; *MV*, combined mitral valve lesions; *LV*, left ventricular; T_{syst}, systolic wall stress; MVO_2, myocardial oxygen consumption; LVO_2, left ventricular oxygen consumption

Myocardial oxygen consumption per unit weight is quite different in clinical heart disease (9, 10). It is lowest in normotensive patients with coronary artery disease; it is normal in concentric and clinically compensated left-ventricular hypertrophy, even in extreme pressure load; and it is increased in dilated hearts with aortic valve disease (Table 1). There was no correlation between the oxygen consumption and ventricular function parameters, as cardiac index, isovolumic contractility indices, and ejection phase parameters. However, significant correlation was found between systolic wall stress and myocardial oxygen consumption (Fig. 2, Table 1). Patients with decompensated aortic valve disease showed above average degrees of this correlation. Extrapolation to zero stress resulted in an intercept of 3.28 ml/min · 100 g, a value that corresponds quite well with the oxygen consumption of the empty beating heart.

In hypertrophic heart disease an inverse, unlinear relationship exists between the mass to volume ratio and peak systolic wall stress (Fig. 3). Largest mass to volume ratio was found for hypertrophic obstructive cardiomyopathy, and lowest values were present for decompensated pressure and volume overload due to aortic valve disease. Concentric left ventricular hypertrophy due to essential hypertension and aortic stenosis was within this correlation, whereas normotensives were shifted to lower systolic stress at equal mass to volume ratio, that is to a lower isobaric relationship (9, 10).

With regard to these characteristics, three types of left ventricular hypertrophy may be classified:

1. Appropriate hypertrophy, which keeps systolic wall stress normal even at extreme pressure load, as a result of an appropriate increase in the mass to volume ratio parallel to pressure load

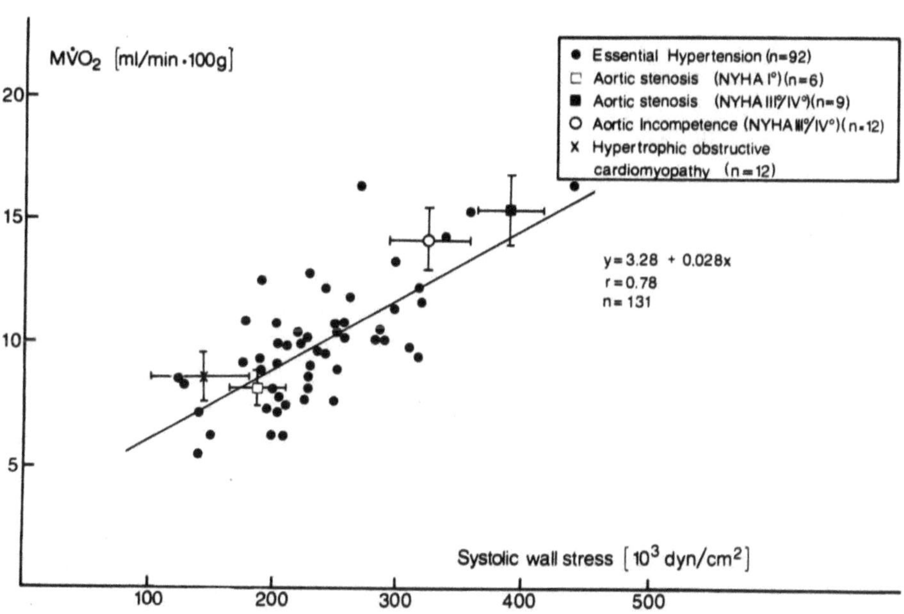

Fig. 2. Relationship between systolic wall stress and myocardial oxygen consumption (MVO_2) (9–12)

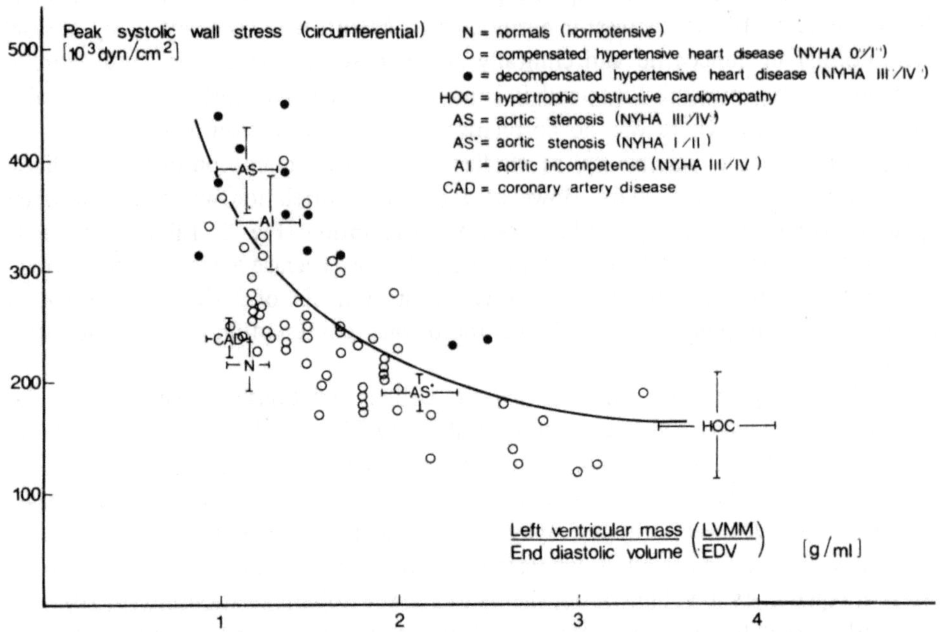

Fig. 3. Relationship between the mass to volume ratio and systolic wall stress. Note the inverse nonlinear dependence between the two variables. Note further that normotensive patients (*CAD*, *N*) are shifted to lower stress values at comparable mass to volume ratio (9–12)

2. Inappropriate (low-stress) hypertrophy, which is associated with marked increase in left ventricular mass out of proportion to intraventricular volume
3. Inappropriate (high-stress) hypertrophy, which is characterized by excess dilatation out of proportion to ventricular mass development.

Thus at least two forms of inappropriate hypertrophy may occur in chronic hypertrophic heart disease.

From the metabolic point of view, high-stress hypertrophy has increased oxygen consumption per mass unit at an impairment of left ventricular function. In contrast, low-stress hypertrophy may have normal or even decreased oxygen consumption per left ventricular mass unit at normal left ventricular function. This helps to explain, for example, the existence of normal, decreased, or increased oxygen consumption per left ventricular mass unit in chronic pressure or volume overload: despite large pressure load, the oxygen consumption may be normal or even decreased in aortic stenosis as long as heart size and systolic wall stress are normal or decreased respectively. On the other hand, the same pressure load may have high oxygen consumption when left ventricular dilatation occurs and when systolic wall stress increases.

Chronic Alterations in Preload and Afterload, Ventricular Wall Dynamics and Function

Reductions of both preload and afterload are therapeutically much more effective in dilated than in normally shaped hearts, as can be seen from the isobars of the mass to volume to stress relationship (Fig. 4).

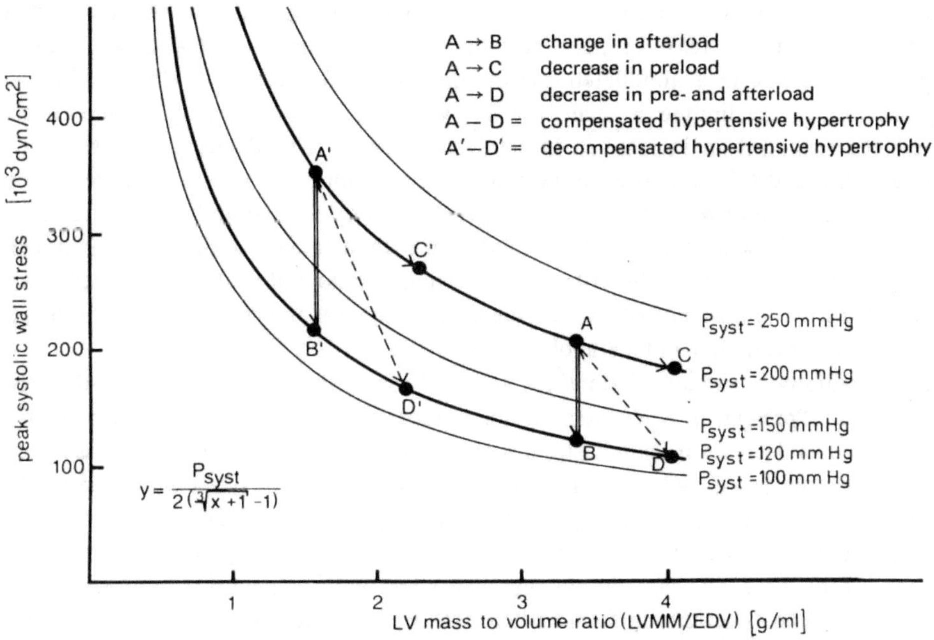

Fig. 4. Diagrammatic representation of the concept of the relation between mass to volume ratio (*abscissa*) and peak systolic wall stress (*ordinate*) for different isobaric conditions. Calculation of isobars was performed by means of the Laplace equation. The lack of parallelism and curvature of these isobaric relations implies that systolic pressure (P_{syst}) changes occurring at points with a high mass to volume ratio lead to fewer stress changes than the same systolic pressure variations at a low mass to volume ratio. An equal increase in systolic pressure, for example, from 120 to 200 mmHg (that is, from *B* to *A*), at a mass to volume ratio of 3.5 leads to a stress increase of only 80 units (10^3 dyn/cm^2). However, with the same increase in pressure (that is from *B'* to *A'*) at a mass to volume ratio of 1.5, systolic wall stress is considerably increased by 160 units. The same consequences are valid for therapeutically indiced pressure reductions (from *A'* to *B'* and from *A* to *B*, respectively). This means that the increase in systolic wall stress at a given rise in systolic pressure in hypertensive heart disease is greater in the dilated than in the nondilated heart. Conversely, systolic and metabolic unloading of the left ventricle at equal pressure reduction is greater or more effective in a dilated than in a nondilated heart. This relation between mass to volume ratio and stress emphasizes the importance of heart size and systolic wall stress on changes in stress, and hence in ventricular function and myocardial energy demand (9–12)

With an increase in systolic pressure (~ afterload) at constant end-diastolic volume, systolic wall stress increases in all ventricles and at each comparable level of mass to volume ratio. However, the degree of pressure-induced wall stress alteration closely depends on the individual isobaric condition as well as on the initial mass to volume ratio. An equal rise in pressure (for example, from 120 to 200 mmHg) (B → A), at a mass to volume ratio of 3.5 g/ml leads to a stress increase of only 70 · 10^3 dyn/cm^2, whereas the same rise in pressure (B' → A') at a mass to volume ratio of 1.5 g/ml is followed by a considerable increase in stress of 160 · 10^3 dyn/cm^2 (Fig. 4). The same calculations and consequences as to systolic stress, and hence MVO_2, are valid for therapeutically induced pressure reductions. This means that from a diagnostic and prognostic point of view, a rise in systolic pressure in a dilated hypertensive heart causes a greater increase in peak systolic wall stress and in MVO_2 than the same pressure increase in a nondilated hypertensive heart. Because both stress and metabolic reserve are limited in man, the left ventricular stress capacity is increasingly reduced with: (a) an increase in initial systolic stress, (b) a decrease in mass to volume ratio, and (c) an increase in systolic pressure. However, a therapeutic reduction in systolic pressure will lead to greater reduction in both stress and oxygen consumption in patients with left ventricular dilatation than in patients with concentric hypertrophy. Thus the relation between mass to volume ratio and peak systolic wall stress elucidates the importance of pressure-dependent changes in systolic stress, and hence in ventricular function and metabolic reserve.

Qualitatively similar relationships are valid for chronic alterations in end-diastolic volume (~ "preload"). At constant systolic pressure, i.e., in the course of each individual isobar, a decrease in end-diastolic volume and hence an increase in the mass to volume ratio are associated with larger decrease in systolic wall stress when ventricular dilatation is present. Accordingly, therapeutic reductions in preload are more effective with regard to left ventricular unloading in a dilated than in a nondilated heart. The most pronounced decreases in systolic wall stress with consecutive improvement in left ventricular function are found when both preload and afterload alterations occur (Fig. 4), i.e., when both blood pressure reduction and diminution in ventricular size are induced.

On the basis of the original values and isobar spectra, the functional and metabolic consequences may be derived for at least four types of pre- and afterload alteration:

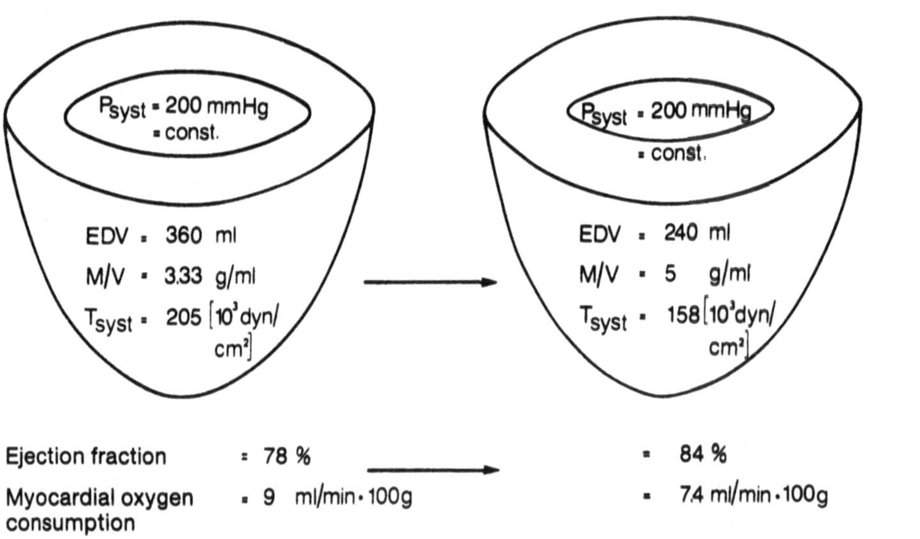

Fig. 5. Influence of preload reduction [end-diastolic volume (*EDV*) reduced from 360 ml to 240 ml] on mass to volume ratio (*M/V*), systolic wall stress (T_{syst}), ejection fraction, and myocardial oxygen consumption. Note that only minor changes in the ejection fraction and in myocardial oxygen consumption occur in compensated, concentrically hypertrophied heart. P_{syst}, systolic pressure

1. In compensated concentric hypertrophy (with large mass to volume ratio) halving of end-diastolic volume (e.g., during hemofiltration in renal hypertensive heart disease) at constant systolic pressure leads to considerable increase in the mass to volume ratio and decrease in systolic wall stress (Fig. 5); however, ventricular function and myocardial oxygen consumption only change by 7%–23% of that of controls.
2. In contrast, the same preload reduction in a dilated heart may enhance ventricular function, as evidenced by the left ventricular ejection fraction, by 82% and may lower myocardial oxygen consumption by 35% of that of controls (Fig. 6).
3. Reduction of systolic blood pressure in concentric hypertrophy, for instance, from 220 to 120 mmHg (e.g., in the course of treatment of hypertensive crisis), has little or no direct effect on the mass to volume ratio (Fig. 7). Despite marked decrease in systolic wall stress, left ventricular function – as the consequence of only small changes in the mass to volume to stress relationships – does not improve significantly, and myocardial oxygen consumption is lowered by only 33%.
4. In contrast, the same reduction of systolic blood pressure in a dilated heart with low ejection fraction leads to effective improvement in left ventricular function and normalization of myocardial energy demand by alterations of these parameters (ejection fraction, myocardial oxygen consumption) by 94% and 40% respectively (Fig. 8).

These data demonstrate that both preload and/or afterload reduction exert more beneficial effects with regard to left ventricular function and myocardial energetics in decompensated dilated hearts than in concentrically hypertrophied or nonhypertrophied ventricles of normal size.

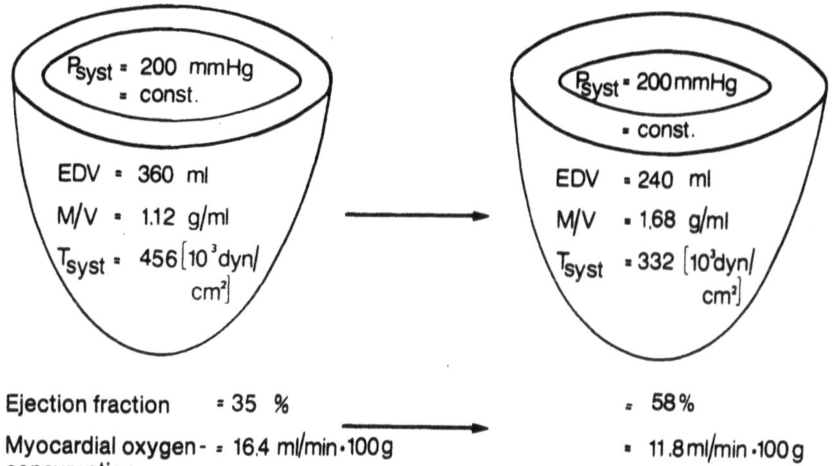

Fig. 6. Influence of preload reduction [end-diastolic volume (*EDV*) reduced from 360 ml to 240 ml] in dilated, decompensating heart. Note that the same preload reduction as in compensated hypertrophy (compare Fig. 5) leads to considerable increase in ventricular function and decrease in myocardial oxygen consumption. *M/V*, mass to volume ratio; P_{syst}, systolic pressure; T_{syst}, systolic wall stress

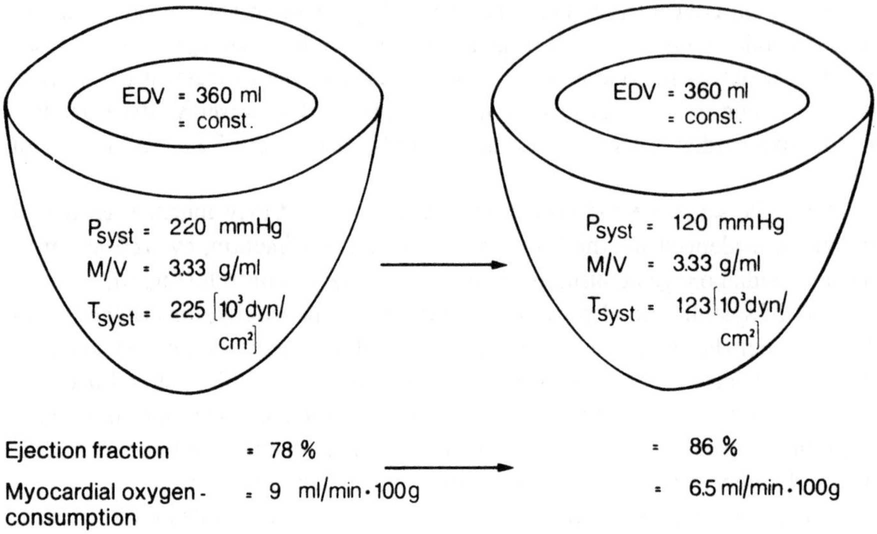

Fig. 7. Influence of systolic pressure reduction (from 220 to 120 mmHg) in the compensated, concentrically hypertrophied heart. Note that there are only minor changes in ventricular function and myocardial oxygen consumption. *EDV*, end-diastolic volume; *M/V* mass to volume ratio; P_{syst}, systolic pressure; T_{syst}, systolic wall stress

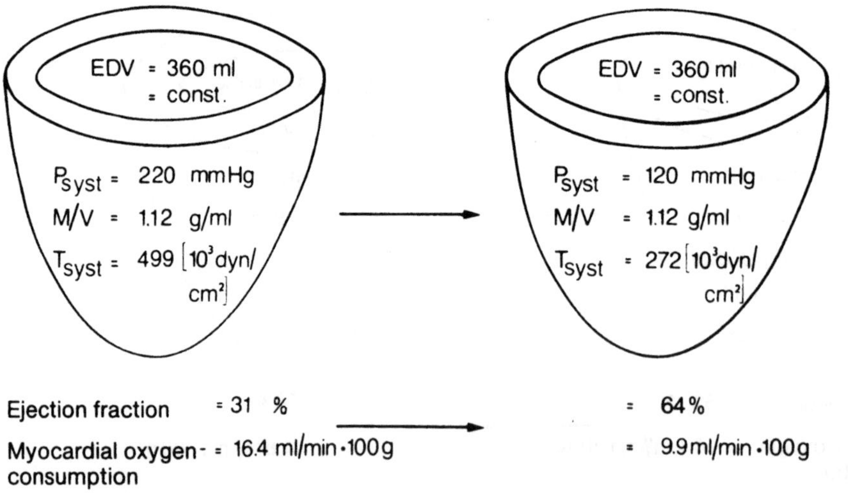

Fig. 8. Influence of systolic blood pressure reduction (from 220 to 120 mmHg) in a decompensated, dilated heart. Note that the same pressure reduction (compare Fig. 7) leads to considerable increase in ventricular function, as evidenced by the increase in left ventricular ejection fraction, as well as to significant decrease in myocardial energy demand. *EDV*, end-diastolic volume; *M/V*, mass to volume relationship; P_{syst}, systolic pressure; T_{syst}, systolic wall stress

Pharmacological Management of Regression of Cardiac Hypertrophy

The functional and metabolic consequences of progression of left ventricular hypertrophy raise the question whether specific regression of hypertrophy may be possible and may be also useful for ventricular function. Spontaneously hypertensive rats were therefore treated from the 4th–5th weeks of life over a period of 40 weeks with metoprolol, a β-receptor-blocking agent; with hydralazine, and arteriolar vasodilating drug; and with a combination of both:

Hydralazine	15 mg/kg/day
Metoprolol	50 mg/kg/day
Hydralazine + metoprolol	15 mg/kg/day hydralazine + 50 mg/kg/day metoprolol

By monotherapy with hydralazine as well as by the combined therapy with metoprolol and hydralazine, blood pressure was lowered to normotensive levels. However, blood pressure could not be lowered by β-receptor blockade alone. Systolic wall stress was only essentially reduced in the hydralazine group. The treatment with hydralazine resulted in a decrease in the left ventricular muscle mass by 17% (Table 2). The group treated in a combined manner showed a more extensive reduction by 26%. Corresponding changes resulted from left ventricular wall thickness and internal radius. The wall thickness in the hydralazine group decrease by 15%. There was also a significant decrease (27%) in the group on combined treatment.

Wall thickness to radius ratio decreased by 15% in the hydralazine group and by 29% in the animals on combined treatment. In contrast, the metoprolol group could not be normalized. Corresponding changes in the mass to volume ratio were also found for all groups.

The cardiac index was increased in the hydralazine group by 16%. It was reduced by 29% in the metoprolol group, mainly due to the decreased stoke volume, and to a lesser extent due to heart rate. Thus the group on combined treatment showed the most distinct decreases of left ventricular muscle mass, but did not significantly differ from untreated controls in cardiac index, stroke volume index, and heart rate. It therefore may be concluded that: (a) regression of left ventricular hypertrophy is possible; (b) regression may be modified by specific pharmacotherapy; and (c) the pumping ability of these treated hearts may be quite normal when compared with untreated hypertrophied hearts.

Despite normal ventricular function at rest, the decrease in mass to volume ratio could imply that these hearts might have more pronounced decrease in function under abnormal ventricular load. Accordingly, stepwise pressure load was induced by aortic clamping in treated and in nontreated rats. At an equal systolic pressure increase, wall stress was enhanced in the treated, i.e., the regressed hearts, parallel to decrease in mass to volume ratio. This means that the regressed hearts bear more wall stress or afterload imposed on the ventricular wall than do the hypertrophied hearts.

The relationship between pressure and function revealed significantly lower ejection fraction at comparable systolic pressure in the treated rats (Fig. 9). Moreover, the maximum pressure development, i.e., the abscissa intercept, was reduced. Thus the regressed hearts with lower mass to volume ratio were more sensitive to pressure load increase. From the therapeutic point of view, blood pressure control will have to be carefully performed if deterioration in function is to be avoided.

Table 2. Hemodynamic values in spontaneously hypertensive rats: controls (*C*), metoprolol-treated rats (*M*), hydralazine-treated rats (*H*) and those given combined treatment (*M + H*)

		BW (g)	LVW/BW (mg/g)	P_{syst} (mmHg)	T_{syst} (10^3 dyn/cm^2)	LVMM (mg)	LVMM/EDV (mg/µl)	*d* (mm)	*d/r*	CI (ml/min • kg)	SVI (ml)	EF (%)
C	*x*	304	3.08	181	171	892	3.76	2.62	0.69	191	0.54	66.6
(*n* = 25)	SD	31	0.29	18	26	61	0.47	0.18	0.07	31.3	0.15	6
H	*x*	288	2.67	129	140	738	2.94	2.24	0.58	220	0.56	65.5
(*n* = 8)	SD	22	0.24	19	26.5	101	0.45	0.12	0.06	36.6	0.11	8.9
	p (C/H)	N.S.	<0.001	< 0.001	< 0.01	< 0.01	<0.001	<0.001	<0.001	< 0.05	N.S.	N.S.
M	*x*	291	3.07	173	156	848	4.03	2.60	0.71	134	0.41	55.2
(*n* = 6)	SD	23	0.28	29.4	30.4	77	1.08	0.23	0.12	42.4	0.12	11.6
	p (C/M)	N.S.	N.S.	N.S.	N.S.	N.S.	N.S.	N.S.	N.S.	< 0.05	<0.05	< 0.05
M + H	*x*	279	2.48	124	169	660	2.31	1.92	0.48	188	0.57	56.9
	SD	30	0.17	17.3	43.4	120	0.57	0.18	0.08	63.8	0.18	10.8
	p (C/M+H)	< 0.05	<0.001	< 0.001	N.S.	< 0.001	<0.001	<0.001	<0.001	N.S.	N.S.	< 0.05
(*n* = 9)	*p* (H/M+H)	N.S.	<0.05	N.S.	N.S.	< 0.05	<0.05	<0.001	<0.001	N.S.	N.S.	< 0.05

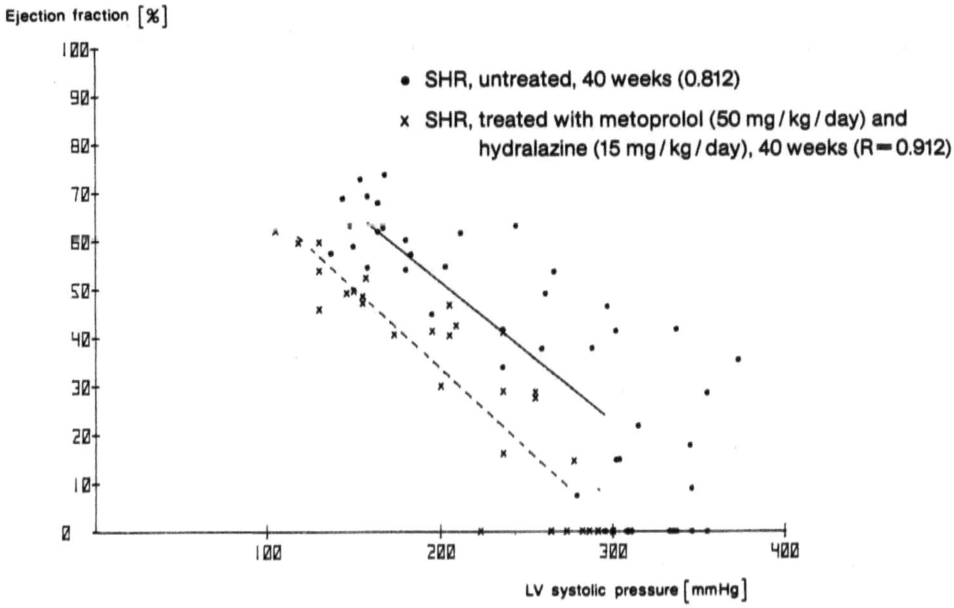

Fig. 9. Ejection fraction in untreated spontaneously hypertensive rats (*SHR*) and in treated SHR following aortic clamping. Note the lower ejection fraction, at comparable systolic pressure, in the treated group

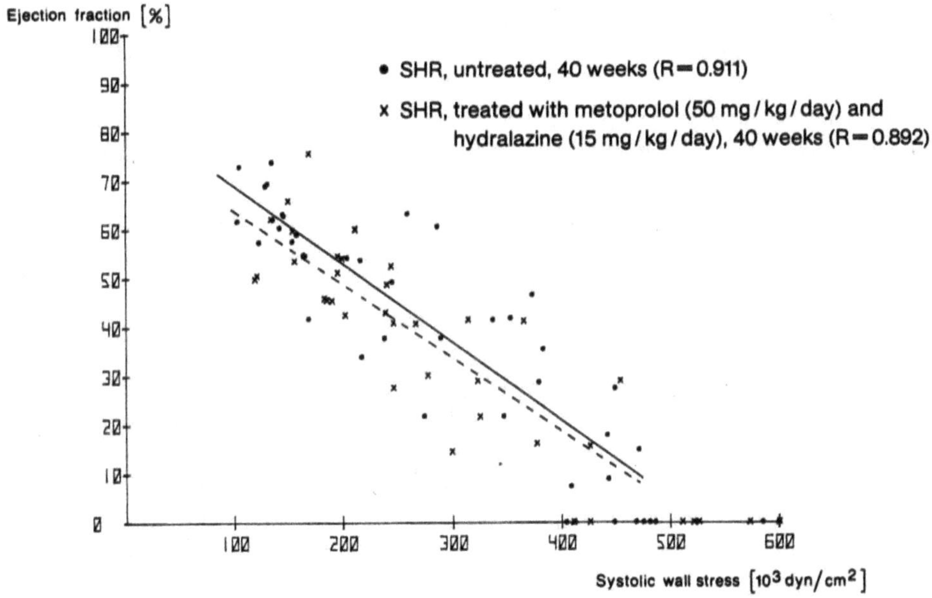

Fig. 10. Ejection fraction in untreated spontaneously hypertensive rats (*SHR*) and treated SHR with increasing of systolic wall stress (aortic clamping). Note the nearly identical regression lines for both experimental groups

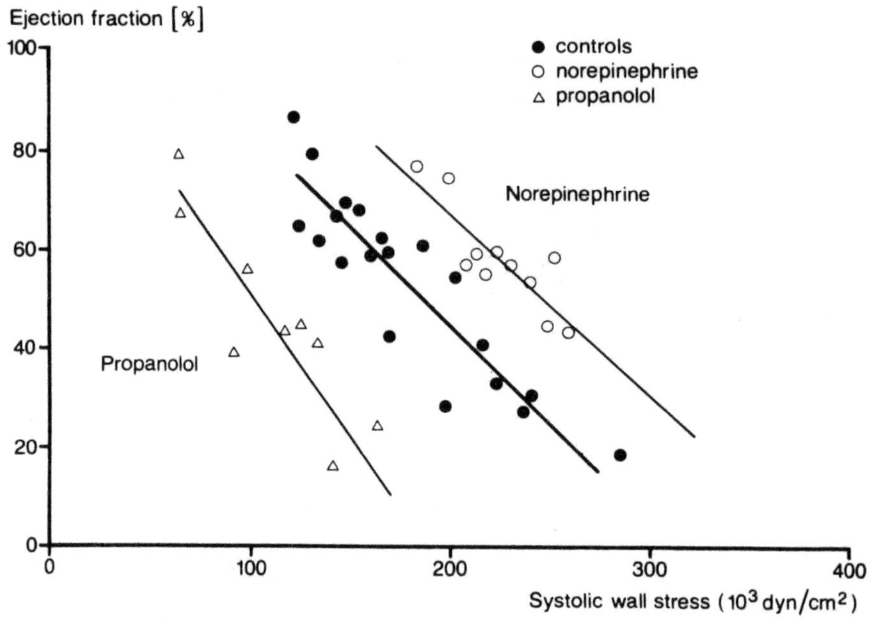

Fig. 11. Variations in the relationship between systolic wall stress and the ejection fraction following positive inotropic (norepinephrine) and negative inotropic interventions (propranolol)

In contrast to the pressure function relationship, nearly identical regressions were present for both groups of rats when considering systolic wall stress and function (Fig. 10). As in chronic clinical heart disease, the ejection fraction progressively decreased with an increase in systolic wall stress. The relationship between both variables therefore again elucidates the importance of wall stress for left ventricular function.

It may be noted that the relationship between stress and function may be modified by inotropic interventions (Fig. 11). Following noradrenaline, larger ejection fraction occurred at comparable wall stress, and following propranolol a downward shift to lower ejection fraction at comparable wall stress was found.

Regression of myocardial hypertrophy therefore has both desired and undesired effects. Favorable are the possible prevention of late cardiac failure, the reversibility of irregular left ventricular hypertrophy, and presumably the normalization of small coronary vessel hypertrophy. Undesirable and deleterious effects may be provoked if high blood pressure is not controlled. Thus increases in systolic wall stress and in myocardial oxygen consumption and an impairment in left ventricular function may occur. Regression of ventricular hypertrophy should therefore be attempted only if strict control of arterial blood pressure is guaranteed.

Summary

Chronic pressure overload results in smaller (mean) ventricular size than chronic volume overload. Left ventricular function is inversely related to both left ventricular volume (size) and systolic wall stress. Systolic wall stress represents the main determinant of the ventricular afterload. Systolic wall stress is directly correlated with myocardial oxygen consumption. Systolic stress is an important resultant of the appropriateness and degree of cardiac hypertrophy. In the course of chronic heart disease, at least two types of inappropriate hypertrophy may occur (low-stress hypertrophy, high-stress hypertrophy).

Preload and afterload are important determinants of left ventricular function. Alterations in preload are regularly associated with alterations in afterload (wall stress), and chronic alterations in afterload usually induce alterations in preload. For the clinical evaluation of the chronically diseased heart, end-diastolic volume and end-diastolic wall stress are a relevant equivalent of ventricular preload, whereas alterations of afterload may be quantified by the assessment of systolic wall stress. The functional and energetic consequences of chronic reduction of preload and afterload are more effective in a dilated heart with low ejection fraction than in a normally improvement in function and energetics is obtained by combined pre- and afterload reduction.

Regression of hypertrophy may be induced pharmacologically and depends on the specific kind of chronic antihypertensive treatment. Regression of hypertrophy may be useful for ventricular function and for the prevention of late cardiac failure; however, cautions control of blood pressure is necessary in order to avoid increases in systolic pressure and wall stress, which may both be deleterious for the regressed heart.

References

1. Alpert N (ed) (1971) Cardiac Hypertrophy. Academic Press, New York
2. Bürger S, Meinardus A, Strauer BE (1978) Hypertrophiegrad und Dynamik des linken Ventrikels bei der spontanen essentiellen Hypertonie der Ratte. Klin Wochenschr 56:207
3. Bürger S, Strauer BE (1981) Left ventricular hypertrophy in chronic pressure load due to essential hypertension. I. Left ventricular function, left ventricular geometry and wall stress. In: Strauer BE (ed) The heart in hypertension. Springer, Berlin Heidelberg New York, pp 13–36
4. Bürger S, Strauer BE (1981) Left ventricular hypertrophy inchronic pressure load due to essential hypertension. II. Contractility of the isolated left ventricular myocardium and left ventricular stiffness. In: Strauer BE (ed) The heart in hypertension. Springer, Berlin Heidelberg New York, pp 37–52
5. Ford LF (1976) Heart size. Circ Res 39:297
6. Hood WP (1971) Dynamics of hypertrophy in left ventricular wall of man. In: Alpert NR (ed) Cardiac hypertrophy. Academic Press, New York, p 445
7. Linzbach AJ (1960) Heart failure from the point of view of quantitative anatomy. Am J Cardiol 5:370
8. Linzbach AJ (1981) Structural adaption of the heart in hypertension and the physical consequences. In: Strauer BE (ed) The heart in hypertension. Springer, Berlin Heidelberg New York, pp 243–250

9. Strauer BE (1979) Myocardial oxygen consumption in chronic heart disease: role of wall stress, hypertrophy and coronary reserve. Am J Cardiol 44:730
10. Strauer BE (1979) Ventricular function and coronary hemodynamics in hypertensive heart disease. Am J Cardiol 44:999
11. Strauer BE (1979) Das Hochdruckherz. Springer, Berlin Heidelberg New York
12. Strauer BE (1980) Hypertensive Heart Disease. Springer, Berlin Heidelberg New York
13. Strauer BE (ed) (1981) The hart in Hypertension. Springer, Berlin Heidelberg New York
14. Strauer BE, Motz W, Bürger SB (to be published) Myocardial and metabolic consequences of development and regression of cardiac hypertrophy in chronic heart disease. In: Biology of myocardial hypertrophy and failure. Int Symp Burlington/Vermont 1981. Raven, New York
15. Tarazi RC, Sen S (1981) Reversal of cardiac hypertrophy by antihypertensive therapy. In: Strauer BE (ed) The heart in hypertension. Springer, Berlin Heidelberg New York, pp 74–88
16. Weber KT, Reichek N, Janicki JS, Shroff S (1981) The pressure overloaded heart: physiological and clinical correlates. In: Strauer BE (ed) The heart in hypertension. Springer, Berlin Heidelberg New York, pp 287–306

Discussion

Reindell, Freiburg: The determinants of contractile performance are preload and afterload, besides the sympathetic regulation with enhancement of contractility and of heart rate. Pre- and afterload may be important only in the diseased heart, since we know that a healthy heart is able to perform at exceedingly high workloads without increase in preload and afterload rises. Healthy hearts increase their performance by means of increased contractility. The opinion that an increase in contractility is always accompanied by an increase in preload is not in accordance with clinical experience. In cases of aortic stenosis with concentric hypertrophy, the filling pressure may be normal. We are therefore dealing with a situation of highly increased afterload, where an increase in preload to match the increased load is not needed. Finally, what is meant by an increase in preload? Is it merely increased residual and diastolic volume activating the Frank-Straub-Starling mechanism?

Strauer, München: Our clinical studies have led us to conclude that in the hypertrophied and dilated heart adaptation of ventricular function to situation of varying load is achieved with greater ease through variations of pre- and/or afterload as compared to isolated changes in contractility. In the dilated heart, reduction of afterload by 50% results in doubling of ventricular ejection fraction. With inotropic measures, increments in ejection fraction of this magnitude may not be achieved under clinical conditions.

The degree of hypertrophy can be expressed by the relation of wall thickness to chamber size and wall stress. In concentric hypertrophy as in hypertension, or in aortic stenosis without decompensation, a normal wall stress will be found. That means that under the conditions of increased pressure load in this situation, afterload may be normal or even reduced. A small, massively hypertrophied ventricle in aortic stenosis may also have normal wall stress, a normal or even reduced myocardial oxygen consumption, and normal ventricular function. On the other hand, a dilated hypertensive heart or a decompensated left ventricle in aortic stenosis will show elevated values for wall stress, increased myocardial oxygen consumption, and reduced ventricular function at the same pressure level, as a result of the increase in chamber size. The chamber size and geometry is therefore of great importance for ventricular performance and myocardial oxygen consumption.

Reindell, Freiburg: In concentric hypertrophy wall stress may be low. How then can the stimulus for hypertrophy or the mechanism of dilatation be explained? According to your arguments, increased wall stress cannot be the reason, neither can it be lack of oxygen, since oxygen consumption was shown to be normal. According to Linzbach, however, a deficient oxygen supply may result from increased fibre thickness even in situations with normal oxygen supply and normal total oxygen consumption.

Strauer, München: A great variety of biochemical, mechanical, and other factors may be responsible for the development of dilatation and decompensation. Arterial elongation, increased wall thickness of the small arteries and arterioles, reduced oxygen transport and increased distances from the capillaries to the myocardial fibers, alterations of calcium flux, and a whole host of other factors independent of wall stress will lead to deterioration of myocardial function and subsequent development of decompensation.

Bussmann, Frankfurt: Variations in afterload will regularly produce quantitatively similar variations in ventricular output. The magnitude of change in end-diastolic ventricular volumes, however, remains relatively small.

Just, Freiburg: On the subject of wall stress, we will probably have to differentiate between global wall stress, as calculated in the experimental or clinical situation, and the actual stress existing at the myocardial fiber. The complex architecture of the ventricle with different orientation of the muscle fibers in different layers of the ventricular wall will subject different fibers to different degrees of stress. Is then the determination of global wall stress a clinically relevant factor?

Strauer, München: There are in fact regional differences in wall stress, albeit quantitatively of limited significance. In animal experiments, as well as in clinical studies, measurements of wall stress were performed in different layers of the myocardial wall. From subendocardial to subepicardial layers a difference of approximately 15% could be measured. In view of the high wall stress attained in systole, such a difference is probably not of major importance.

Kirchheim, Heidelberg: How do you explain the high myocardial oxygen consumption per unit myocardium in hypertrophied states as compared to normal hearts? One finds increased values of myocardial oxygen consumption although wall tension has not significantly changed or is even lower.

Strauer, München: In concentric hypertrophied hearts wall stress is normal, but the velocity-related indices of contractility may be increased. Myocardial oxygen consumption is normal or only slightly elevated. In the case of a dilated heart, wall stress and myocardial oxygen consumption are increased. Wall stress and myocardial oxygen consumption show a direct linear relationship. Increased oxygen consumption is brought about with increased sympathetic stimulation and increased contractile performance as measured by velocity-related indices. It can, however, be stated that wall stress is always the more significant determinant of myocardial oxygen consumption; the contractility indices are in comparison less important. Alternatively, with constant wall stress, changes in myocardial oxygen consumption will be found when the contractile performance is changed through inotropic interventions. These influences are, however, quantitatively less important than changes in systolic wall stress.

The Sympathetic Nervous System in Chronic Heart Failure and Its Response to Vasodilator Therapy

Klaus Starke

When the pumping action of the heart fails, one response of the body is an increase in the firing rate of the sympathetic fibers. The main peripheral sensors that trigger this reflex adjustment are the arterial baroreceptors. Other sensors may contribute, such as mechanoreceptors at the venoatrial junctions, which, when stretched, enhance the sympathetic outflow to the sinus node, although sympathetic outflow to the kidneys is reduced. The increase in sympathetic tone may, on the one hand, improve the patient's circulatory function: for instance, released noradrenaline may shift the depressed myocardial length-active tension curve back toward normal; vasoconstriction may mobilize blood and maintain arterial pressure generation. On the other hand, the vascular effects of the sympathetic nervous system may be detrimental, in that they may increase ventricular filling pressure unduly and reduce cardiac output; in such cases, drugs that directly or indirectly antagonize the noradrenaline-induced vasoconstriction (α-adrenoceptor blocking agents and vasodilators sensu stricto, respectively) will be helpful.

In this chapter, three aspects of the sympathetic nervous system in heart failure are briefly discussed. First, some of the evidence for an increase, in sympathetic tone will be summarized. Secondly, it will be shown that the cardiac sympathetic neuroeffector system becomes intrinsically defective in heart failure. Thirdly, the response of the sympathetic system to vasodilatation will be compared in normal individuals and in patients with heart failure.

Increase in Sympathetic Tone

In patients with congestive heart failure, plasma noradrenaline levels are elevated both at rest and during exercise and increase with the severity of the disease (Chidsey et al. 1962; Thomas and Marks 1978). Although plasma noradrenaline levels do not necessarily reflect the activity of the sympathetic nervous system, it seems very likely that the change observed in heart failure is due to an increased overall sympathetic tone. The urinary excretion of noradrenaline is also enhanced (Chidsey et al. 1965).

It was early observed that blood vessels are constricted in heart failure. For instance, Wood et al. (1956) found that veins of patients with signs of heart failure were less distensible than veins of patients without cardiovascular disease. Although at that time the mechanism for the venoconstriction was not clarified, it seems very likely that an increased release of noradrenaline was responsible.

The sympathetic nervous system, and in particular its cardiac section, my help us to compensate for heart failure. In support of this view, (±)-propranolol has little effect on cardiovascular parameters in normal calves; in calves with congestive heart failure produced by ligation of the right pulmonary artery, however, it causes a marked further deterioration of hemodynamics, with a further rise in right ventricular end-diastolic pressure, a further fall of cardiac output, and a further widening of the arteriovenous oxygen difference (Vogel and Chidsey 1969). The animal experiment confirms that treatment with β-adrenoceptor antagonists is hazardous in patients with heart failure. Obviously, these patients depend on the (hyper)activity of the sympathetic nervous system for a tolerable ventricular function.

Defect in the Cardiac Sympathetic Neuroeffector Mechanism

It is surprising, therefore, that the cardiac sympathetic neuroeffector mechanism becomes intrinsically defective in chronic heart failure. Both in patients and in animals models, cardiac noradrenaline concentrations are reduced (Chidsey et al. 1963; Sole et al. 1975; Penttilä et al. 1977; Petch and Nayler 1979). The reduction is not due to dilution of the sympathetic axons by the hypertrophied muscle, since it is found even in nonhypertrophied areas, and since there is also a reduction when the levels are calculated per heart rather than per gram tissue. Probably in consequence of the reduced levels, the release of noradrenaline from cardiac sympathetic fibers may be decreased (Haneda et al. 1980). As a further consequence, myocardial responses to sympathetic nerve impulses are depressed, while responses to exogenous, injected noradrenaline may be normal (Covell et al. 1966).

What is the mechanism of the defect? An obvious possibility would be that sympathetic activity is enhanced to such an extent that the synthesis of noradrenaline does not keep pace with its utilization. This, however, is not a sufficient explanation. Tyrosine hydroxylase is the rate-limiting enzyme in the formation of noradrenaline. Increased activity of postganglionic sympathetic neurones leads to an adaptive increase in the in vitro tyrosine hydroxylase activity in the ganglia, as well as in the innervated tissues (Thoenen 1974). Yet in patients and in dogs with heart failure, tyrosine hydroxylase activity is reduced concomitantly with the decrease of noradrenaline (Pool et al. 1967; DeQuattro et al. 1973), suggesting that the neurones are not simply depleted but entirely degenerated. It has been reported that in one model of heart failure, the cardiomyopathic hamster, cardiac noradrenaline levels are reduced and tyrosine hydroxylase activity is increased; furthermore, there is a marked rise in cardiac dopamine (Sole et al. 1977). The authors suggest that, in the late stages of hamster cardiomyopathy, dopamine β-hydroxylase may become rate-limiting, and that this may be a consequence of the acceleration of neuronal firing. Although the observations are intriguing, the explanation proposed is difficult to reconcile with the fact that sympathetic hyperactivity produced by other means increases dopamine β-hydroxylase as well as tyrosine hydroxylase activity (Thoenen 1974). More work is obviously needed to clarify what happens with the cardiac sympathetic neurones when the heart fails.

The presynaptic defect, loss of noradrenaline, may not be the only one. The positive inotropic but not the positive chronotropic effect of injected noradrenaline or isoprenaline is diminished in dogs with heart failure (Newman 1977). Moreover, peripheral blood lymphocytes from patients with heart failure generate less cyclic AMP upon exposure to isoprenaline than do lymphocytes from healthy subjects (Thomas and Marks 1978). Thomas and Marks (1978) suggest that β-adrenoceptors may be desensitized in heart failure, and that this may blunt the response of the myocardium to catecholamines. On the other hand, radioligand binding studies indicate that the number of β-adrenoceptors (as well as α-adrenoceptors) is increased in guinea pigs with heart failure due to supravalvular aortic constriction (Karliner et al. 1980). Hence the nature of the postsynaptic β-adrenoceptor defect in the myocardium – if there is any defect – remains doubtful. Perhaps the number of β-adrenoceptors or the efficiency of the translation mechanism behind the receptors may differ in the various stages of failure.

Response to Vasodilators

Blood vessels can be dilated either by α-adrenoceptor antagonists (if they are constricted by α-agonists) or by vasodilators sensu strictu, i.e., drugs that interfere with some other link of the contraction machinery. In normal individuals, vasodilators invariably produce a reflex increase in sympathetic tone plus a withdrawal of vagal tone resulting, for instance, in tachycardia (Vatner et al. 1974). The increase in sympathetic activity tends to be smaller after prazosin than after phentolamine or hydralazine (Graham et al. 1980) one reason may be that prazosin, in contrast to phentolamine, causes little, if any, blockade of presynaptic α-adrenoceptors (see Starke 1981).

Little is known about the influence of vasodilators on the sympathetic nervous system in heart failure. Cohn et al. (1978) reported that, in hypertensive patients, heart rate and plasma noradrenaline levels increased as they do in normal individuals. In patients with heart failure, however, heart rate and (high)plasma noradrenaline levels decreased. Similarly, a decrease in plasma noradrenaline at least in some patients, and in particular those with advanced disease, is reported in the accompanying chapters by Klein and by Manthey. The reason for the anomalous response in the severely ill patients is not clear. Possibly their arterial blood pressure decreases less than the blood pressure of the less severely affected. Possibly, also, venodilatation markedly reduces the stretch of venoatrial mechanoreceptors in advanced failure, thus diminishing the firing rate of at least part of the sympathetic nervous system.

Conclusion

The sympathetic nervous system developed in the course of evolution because it assisted animals in adapting rapidly to temporary changes in their environment, not because it helped them to endure a chronic and perhaps progressive disease: hence, probably, its Janus-faced role in heart failure. It may be beneficial in some patients or some stages of the disease, but do more harm than good in others, in particular where its vascular fibers are concerned. The present brief discussion has shown that many aspects of its function in heart failure remain to be elucidated.

References

Chidsey CA, Braunwald E, Morrow AG (1965) Catecholamine excretion and cardiac stores of norepinephrine in congestive heart failure. Am J Med 39:442–451

Chidsey CA, Braunwald E, Morrow AG, Mason DT (1963) Myocardial norepinephrine concentration in man. N Engl J Med 269:653–658

Chidsey CA, Harrison DC, Braunwald E (1962) Augmentation of the plasma norepinephrine response to exercise in patients with congestive heart failure. N Engl J Med 267:650–654

Cohn JN, Taylor N, Vrobel T, Moskowitz R (1978) Contrasting effect of vasodilators on heart rate and plasma catecholamines in patients with hypertension and heart failure. Clin Res 26:547A

Covell JW, Chidsey CA, Braunwald E (1966) Reduction of the cardiac response to postganglionic sympathetic nerve stimulation in experimental heart failure. Circ Res 19:51–56

DeQuattro V, Nagatsu T, Mendez A, Verska J (1973) Determinants of cardiac noradrenaline depletion in human congestive failure. Cardiovasc Res 7:344–350

Graham RM, Stephenson WH, Pettinger WA (1980) Pharmacological evidence for a functional role of the prejunctional alpha-adrenoceptor in noradrenergic neurotransmission. Naunyn Schmiedebergs Arch Pharmacol 311:129–138

Haneda T, Miura Y, Arai T, Nakajima T, Miura T, Honna T, Kobayashi K, Sakuma H, Adachi M, Miyazawa K, Yoshinaga K, Takishima T (1980) Norepinephrine levels in the coronary sinus in patients with cardiovascular diseases at rest and during isometric handgrip exercise. Am Heart J 100:465–472

Karliner JS, Barnes P, Brown M, Dollery C (1980) Chronic heart failure in the guinea pig increases cardiac α_1- and β-adrenoceptors. Eur J Pharmacol 67:115–118

Newman WH (1977) A depressed response of left ventricular contractile force to isoproterenol and norepinephrine in dogs with congestive heart failure. Am Heart J 93:216–221

Penttilä O, Kyösola K, Partanen S, Merikallio E, Siltanen P (1977) Studies of auricular catecholamines by fluorescence histochemistry in various heart diseases of man. Virchows Arch [Pathol Anat] 373:279–292

Petch MC, Nayler WG (1979) Concentration of catecholamines in human cardiac muscle. Br Heart J 41:340–344

Pool PE, Covell JW, Levitt M, Gibb J, Braunwald E (1967) Reduction of cardiac tyrosine hydroxylase activity in experimental congestive heart failure. Circ Res 20:349–353

Sole MJ, Kamble AB, Hussain MN (1977) A possible change in the rate-limiting step for cardiac norepinephrine synthesis in the cardiomyopathic Syrian hamster. Circ Res 41:814–817

Sole MJ, Lo CM, Laird CW, Sonnenblick EH, Wurtman RJ (1975) Norepinephrine turnover in the heart and spleen of the cardiomyopathic Syrian hamster. Circ Res 37:855–862

Spann JF, Chidsey CA, Pool PE, Braunwald E (1965) Mechanism of norepinephrine depletion in experimental heart failure produced by aortic constriction in the guinea pig. Circ Res 17:312–321

Starke K (1981) α-Adrenoceptor subclassification. Rev Physiol Biochem Pharmacol 88:199–236

Thoenen H (1974) Trans-synaptic enzyme induction. Life Sci 14:223–235

Thomas JA, Marks BH (1978) Plasma norepinephrine in congestive heart failure. Am J Cardiol 41:233–243

Vatner SF, Higgins CB, Braunwald E (1974) Sympathetic and parasympathetic components of reflex tachycardia induced by hypotension in conscious dogs with and without heart failure. Cardiovasc Res 8:153–161

Vogel JHK, Chidsey CA (1969) Cardiac adrenergic activity in experimental heart failure assessed with beta receptor blockade. Am J Cardiol 24:198–208

Wood E, Litter J, Wilkins RW (1956) Peripheral venoconstriction in human congestive heart failure. Circulation 13:524–527

Discussion

Brunner, Basel: The data presented stemmed from dogs with right ventricular insufficiency due to damage to the tricuspid valve and pulmonary artery bading. In this supposedly purely right heart disease a decrease of the noradrenaline content of the myocardium of the left ventricle was observed. How can this be explained?

The sensitivity of the β-receptors mediating chronotropism may be variable, and may be dissociated from the sensitivity of the β-receptors mediating inotropic effects. We have had experience according to which β-receptors seem to increase in sensitivity in heart failure as the catecholamine content is decreased. This is the normal regulation which could be expected to occur. Is there a true difference between the sensitivity of β-receptors mediating chronotropism and receptors mediating inotropic effectiveness?

Starke, Freiburg: There is no ready answer to the questions. I shall try to attempt an interpretation on the basis of the information derived from the literature. It has repeatedly been shown that when left ventricular function is impaired initially, then right ventricular function will decline subsequently with a reduction of myocardial noradrenaline content. However, we do not really know whether this observation also holds in disease states with isolated impairment of right ventricular function. It must be admitted that we cannot say at present why left ventricular noradrenaline then declines. It is not even possible to say whether we are dealing with a simple depletion of noradrenaline or with a degeneration of the sympathetic fibers. I have the impression that the evidence is against a depletion of existing fibers, and in favor of a degenerative process affecting the nerve fibers. Why they degenerate, however, remains an open question. Workers in this field have been unable to offer an explanation. Even though we do not understand the basic pathogenetic mechanism, however, it can be stated confidently that noradrenaline levels in the myocardium decline everywhere in the heart, independently of where the primary lesion is located.

As to a dissociation of chronotropic and inotropic effects, an answer is not readily at hand. We do not even know whether there is a decrease or an increase in the sensitivity of the β-receptors or whether, as I mentioned earlier, something is wrong with the mechanism behind the receptors. There is evidence according to which in the guinea pig heart β-receptors and α_1-receptors may numerically be increased in congestive heart failure. On the other hand, however, all functional experiments show that noradrenaline never causes an enhanced effect in the failing heart. It may well be that the number of β-receptors increases, but nevertheless the functional response decreases.

Scholz, Hannover: Is it possible that chronotropic and inotropic effects are mediated through different types of β-receptors?

Starke, Freiburg: In the sinus node as well as in the working myocardium β-1- and β-2-receptors have been demonstrated. In both compartments the β-1-receptors predominate.

Cohn, Minneapolis: Some of our data which you mentioned in your paper deserve updating and extension. The response to vasodilators in patients with heart failure is very heterogeneous. As we extended our studies and looked at the specific drug responses, we found that a clear-cut distinction between normals and patients with heart failure did not hold up in large series. In fact, there are two subsets. One group of patients with congestive heart failure, responding to nitroprusside, who do not increase plasma catecholamine levels or plasma renin activity in response to body tilt or lower body negative pressure. There may be abnormalities on the efferent side of the reflex arc of the pressure mechanoreceptor systems. There is another subset of patients with heart failure who tend to maintain a normal response. These patients may present with higher than normal levels of plasma catecholamines in response to tilt or to nitroprusside.

It is interesting that those two groups of patients cannot easily be distinguished by hemodynamic or clinical parameters. We cannot be certain how to isolate the patient who has developed a severe abnormality of reflex control from patients who maintain normal reflex mechanisms.

Upon exercise, all these patients exhibit a marked increas of plasma catecholamines, thereby proving that the sympathetic system is able to respond, although the stimulus of falling arterial pressure cannot be responded to in a normal. We have to admit that we do not understand this selective abnormality of the baroreceptor reflex in certain patients with heart failure.

It is common clinical experience that patients with heart failure hardly ever develop orthostatic hypotension. Yet in the upright position these patients may neither increase plasma catecholamine nor plasma renin activity. None of the known vasoconstrictor mechanisms seems to be activated and yet the calculated systemic vascular resistance rises.

Auditorium: The situation of right heart failure with the decline of myocardial catecholamine content in the right and the left heart is not a unique situation at all. With pulmonary artery banding right heart failure may be seen, and at the same time depression of the myocardial function of the right and of the left heart. The reason for this remains obscure.

Depressed left ventricular function or reduced response of the left ventricular myocardium to catecholamines in patients with congestive heart failure or in animals with experimental congestive heart failure is well known. Although the mechanism is not well understood, in some of these patients availability of calcium may be the decisive point.

Kersting, Koblenz: In heart failure alterations of sympathetic nervous activity may be demonstrated both pre- and postsynaptically, as well as in the myocardial and the vascular muscle. Is there evidence that the central nervous system is also affected in congestive heart failure?

Starke, Freiburg: Up to now no data have been published, as far as I know. I would, however, anticipate central nervous system anomalies in controlling the sympathetic activity in congestive heart failure. There may very well be a disorder of the central integration of cardiovascular reflexes in congestive heart failure. Catecholamine turnover in the central nervous system may be abnormal in congestive heart failure, but systematic studies have not been published as yet.

Regulation of Renal Hemodynamics in Congestive Heart Failure*

Hartmut Kirchheim

The influence of the renal sympathetic nerves on kidney function has been a subject of controversy since Bernard (1859) observed that acute section of the splanchnic nerves in the anesthetized dog results in an increase in urine production by the ipsilateral kidney. This controversy centers around two questions:

1. Is the "denervation diuresis" exclusively hemodynamic in origin due to the increase of renal blood flow and glomerular filtration rate, or is there also an effect of the renal nerves on the tubular mechanism of reabsorption?
2. Is the phenomenon of denervation diuresis evident only under the artificial circumstances of anesthesia and surgical trauma and therefore not of functional significance?

Smith (1951) criticized Bernard's and a great number of subsequent similar findings, noting that denervation diuresis appears to be a release from enhanced vasoconstrictor tone engendered by anesthesia and traumatic operative procedures; the renal vasomotor pathways in man, and probably in other mammals, are normally at rest; like the sympathetic nervous system generally, they meet the emergencies of anesthesia, hemorrhage, oligemia, traumatic excitation, fear and shock so frequently created, among other ways, by unwary physiologists. Although I fully agree with Smith that experiments involving surgical trauma and anesthesia have severely hampered progress, especially in this field, his statement that denervation diuresis is purely hemodynamic in origin is according to most recent experiments probably no longer tenable. Sadowski et al. (1979) have studied the function of the innervated and the denervated kidney in unilaterally renal-denervated conscious and anesthetized dogs, before and after saline loading. Barbiturate anesthesia distinctly depressed hemodynamics and increased the functional difference between the denervated and innervated kidneys. In conscious moderately hydrated dogs the denervated kidney excreted slightly more sodium and water, while glomerular filtration rate and renal plasma flow were the same; after saline loading higher excretion was observed in the innervated side, probably by reflex inhibition of the renal sympathetic nerves. Other recent studies have clearly demonstrated that direct stimulation of renal sympathetic nerves increases renal tubular sodium reabsorption without alterations in glomerular filtration rate, renal

* This study was supported by the Deutsche Forschungsgemeinschaft, S.F.B. 90 (Kardiovaskuläres System), Heidelberg, Federal Republic of Germany

blood flow, or intrarenal distribution of blood flow (DiBona 1978; Gottschalk et al. 1981). Finally, renal sympathetic nerves will influence kidney function by stimulating renin release (Freeman and Davis 1977). Aside from the well-known systemic effects of the renin-angiotensin system, angiotensin II may function as an intrarenal antinatriuretic hormone (Melton and Frazier 1976). Thus renal nerve activity can affect both tubular transport of salt and water and the resistance of the renal blood vessels. The physiological conditions under which these mechanisms become operative remain to be elucidated.

The experiments from my laboratory that I am going to present are designed to separate nerve-mediated effects on renal function from nerve-mediated influences on renal vasomotor tone under physiological conditions in trained conscious dogs.

The simplified diagram in Fig. 1 summarizes the major inputs to the central nervous system that influence the activity of the renal sympathetic nerves, and shows their effect on the nephron. The systemic arterial baroreceptors and the myelinated and nonmyelinated vagal afferents from the atria and the ventricles exert an inhibitory effect on renal sympathetic nerve discharge, while the chemoreceptors and the cardiac sympathetic afferents activate the renal nerves. The cardiac sympathetic afferents discharge in response to myocardial stretch and may play an important role under pathophysiological conditions (Meckler et al. 1981). These "specific afferents" converge onto the neuron pools in the medullary reticular formation, which control

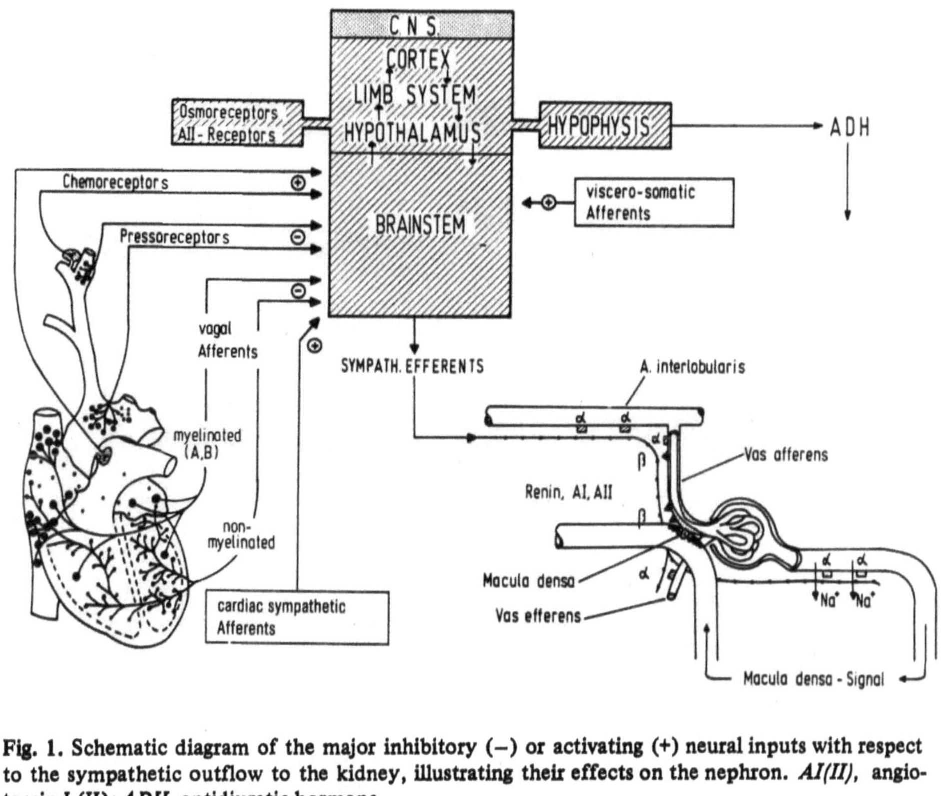

Fig. 1. Schematic diagram of the major inhibitory (–) or activating (+) neural inputs with respect to the sympathetic outflow to the kidney, illustrating their effects on the nephron. *AI(II)*, angiotensin I (II); *ADH*, antidiuretic hormone

the sympathetic outflow to the kidney with a great number of "nonspecific" viscerosomatic afferents from all parts of the body. These afferents, especially the C-fibers, have an activating influence on renal sympathetic nerve activity. This nonspecific input may very well explain a number of previous contradictory findings regarding the vasomotor response of the kidney to baroreceptor stimuli when the anesthetized animal is compared with the unanesthetized animal. The poorly controlled nonspecific activating input due to preparative trauma – for instance, in the anesthetized preparation – may elicit a high tonic sympathetic discharge of a nondifferentiated pattern causing an abnormally high generalized vasoconstrictor tone.

In the kidney the sympathetic nerves can induce an α-adrenergic vasoconstriction, mainly in the interglobular artery and the afferent arteriole. Humoral stimuli (e.g., isoproterenol) can induce a significant β-adrenergic dilatation, probably of the afferent arteriole, resulting in a 30%–40% increase of total renal blood flow (Hardt et al. 1971). Three groups of mechanisms can alter renin release: (a) two intrarenal "receptors," the renal "vascular receptor" and the macula densa; (b) the renal nerves; and (c) a number of humoral agents (not shown in Fig. 1), including epinephrine, norepinephrine, sodium and potassium ions, angiotensin II and ADH, estrogens, and adrenal steroids. The renal nerves can either directly influence the juxtaglomerular cells via a β-adrenergic receptor, or indirectly affect the vascular receptor by α-adrenergic vasoconstriction of the interlobular artery or the afferent arteriole. There is a second indirect effect on renin release: the effect of the renal nerves on proximal tubular

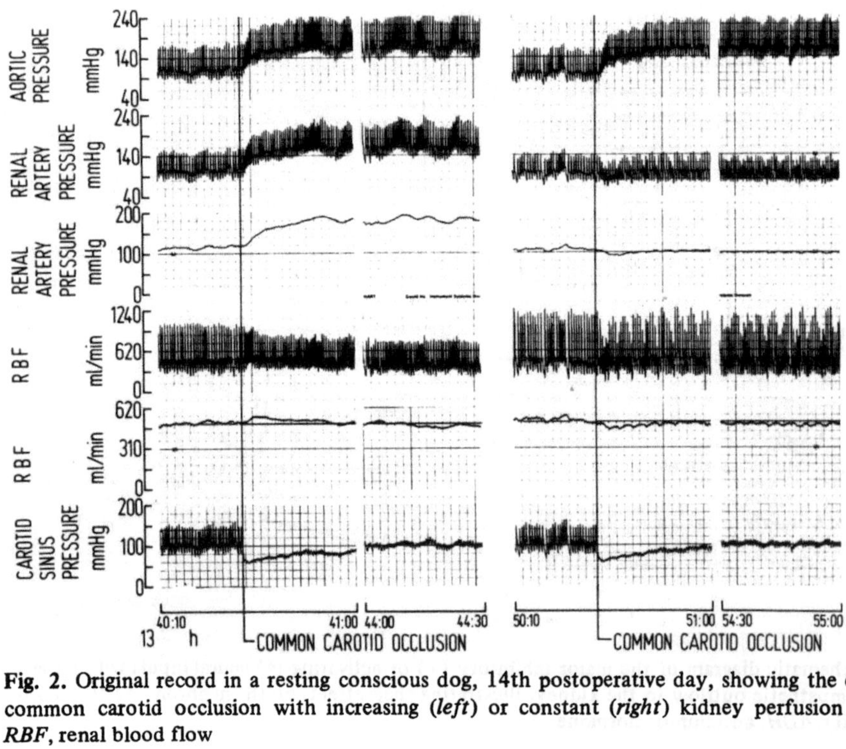

Fig. 2. Original record in a resting conscious dog, 14th postoperative day, showing the effects of common carotid occlusion with increasing (*left*) or constant (*right*) kidney perfusion pressure. *RBF*, renal blood flow

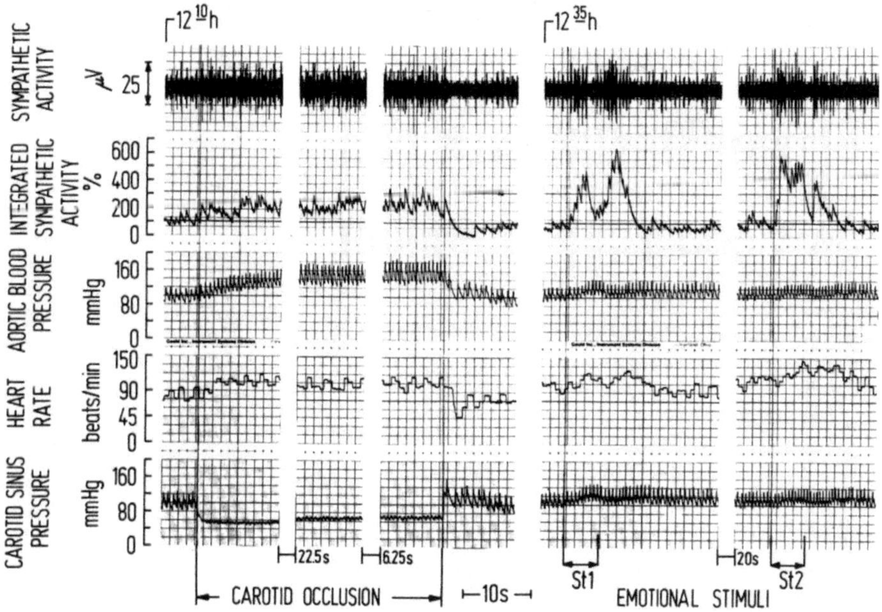

Fig. 3. Original record in a resting conscious dog, 3rd postoperative day. The effect on renal sympathetic nerve discharge of common carotid occlusion (*left*) is compared with auditory ("emotional") stimuli (*right*) tested 25 min later in the same animal. *St 1*, shouting; *St 2*, whistling (Gross and Kirchheim 1980)

reabsorption of sodium (probably mediated by α-adrenergic receptors) could influence the sodium load to the macula densa.

The experiments I am going to present were all performed in conscious chronically instrumented foxhounds on a normal sodium intake of 4.7 mmol/kg body wt./day. Time does not allow the presentation of details concerning the applied methods, which have previously been published elsewhere (Gross et al. 1979; Gross and Kirchheim 1980; Kirchheim et al. 1980; Gross et al. 1981a; Gross et al. 1981b).

In the original experiment shown in Fig. 2, renal blood flow was measured using an electromagnetic flowmeter. Renal perfusion pressure could be controlled independently of aortic blood pressure by use of a renal artery-cuff connected to an extracorporeal pneumatic control system. In the left-hand panel common carotid occlusion (implanted carotid artery-cuffs) induces a reflex rise of aortic and renal artery blood pressure by some 60 mmHg. This is due to a reduction of baroreceptor afferent activity (Kirchheim 1976; Gross et al. 1979). An increase in chemoreceptor activity is unlikely, since diastolic sinus pressure remained above the pressure level (50–60 mmHg) below which chemoreceptors discharge continuously. The right-hand panel shows a carotid occlusion experiment in the same dog performed 10 min later; in this experiment kidney perfusion pressure was kept constant during common carotid occlusion. The maximum change in renal blood flow during the steady state was less than 5% of control under both conditions. Thus the increase of renal vascular resistance observed with increasing perfusion pressure (left panel) must be attributed to autoregula-

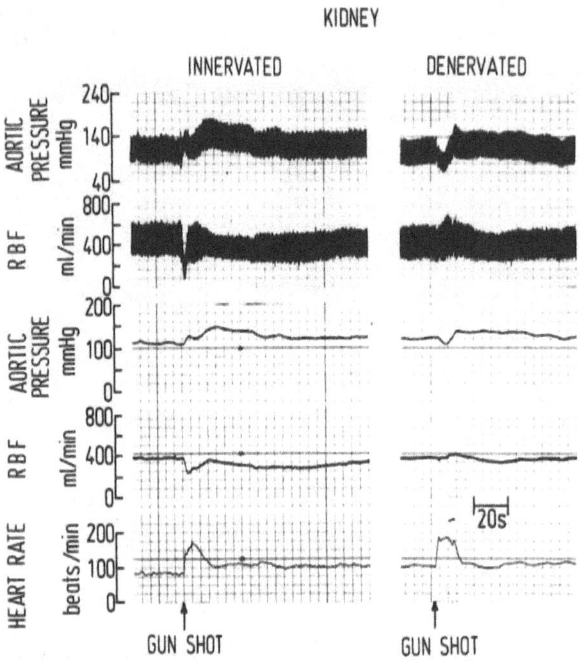

Fig. 4. Original record in conscious dogs, 3 weeks postoperative, showing the effect of excitement (gun shot) on blood pressure, renal blood flow (*RBF*), and heart rate. An intact dog (*left*) is compared to a dog with a surgically denervated kidney (*right*)

tion of renal blood flow, which was regularly present in the dogs with blood pressure down to 60 mmHg. Since we have observed identical results, i.e., no influence on renal blood flow with increasing or constant perfusion pressure in dogs after severance of the aortic nerves (Gross et al. 1979), the negative finding with respect to renal blood flow cannot be due to the attenuating influence exerted by the aortic nerves. However, we cannot exclude some possible attenuation by the "cardiopulmonary afferents," since the vagi were intact. In general this counterregulating effect cannot be great, since we observed an increase in systemic blood pressure of 50–60 mmHg. It should be mentioned in this context that common carotid occlusion with control of perfusion pressure induces a pronounced vasoconstriction in the muscle vascular bed under identical conditions in the conscious dog (Kirchheim et al. 1977). In other experiments on conscious dogs we were able to demonstrate that a nonhypotensive hemorrhage (maximum mean blood pressure drop 5 mmHg) elicited by a loss of 16% of total blood volume reduced central venous pressure by 2 mmHg and increased heart rate by 8–14 beats/min but caused no decrease in renal blood flow (Gross et al. 1979). The failure to observe significant changes in renal blood flow during carotid sinus hypotension raised the important question whether there is a resting sympathetic nerve discharge to the kidney of the conscious dog, and how it could be modulated.

Fig. 3 shows an original record taken on the 3rd postoperative day in a resting conscious dog. Without doubt there was a resting pulse-synchronous discharge in the renal sympathetic nerves. The left-hand panel demonstrates that it is increased during carot-

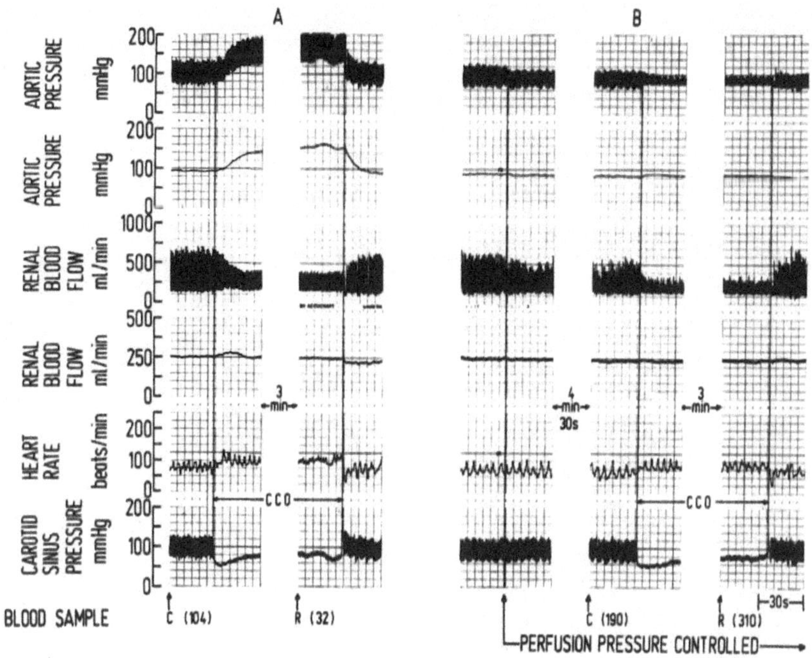

Fig. 5A, B. Original record in a conscious dog, 20th postoperative day, showing the effects of common carotid occlusion with increasing (*A*) and constant (*B*) kidney perfusion pressure. *CCO*, common carotid occlusion. Time of renal-venous and arterial blood-sampling is indicated (*C*, control; *R*, reflex). *Numbers in brackets* indicate renin release (ng/min) calculated from the respective blood sample and renal plasma flow. (Gross et al. 1981a)

id occlusion. It is also obvious from the record that the increase of carotid sinus pressure due to the opening of the carotid cuffs reduces renal sympathetic nerve discharge to the noise level of 2–4 s. When measuring renal blood flow under comparable experimental conditions we never observed a corresponding increase of renal blood flow when sinus pressure was suddenly increased. This indicates that the resting nervous discharge is obviously too small to induce a significant resting vasomotor tone. The right-hand panel shows that auditory stimuli (loud noises) cause significant transient bursts of sympathetic activation. Common carotid occlusion caused a steady state increase of average renal sympathetic nerve discharge by 62% of control. Excitement, on the other hand, was associated with bursts of activation of average sympathetic nerve discharge by 500% of control. In an experiment in which renal blood flow was measured, excitement of the dogs by firing a pistol regularly induced an initial transient decrease in renal blood flow by 30%–40% of control, followed by a smaller delayed fall. The example shown in the original record in Fig. 4 demonstrates that the initial transient vasoconstriction is no longer observed in the denervated kidney, while a slight secondary decrease, probably due to the release of adrenaline by the adrenal cortex, is still present. These experiments provide further evidence that preparatory procedures necessary for the chronic implantation of a flowmeter probe do not destroy the sympathetic nerve supply of the kidney.

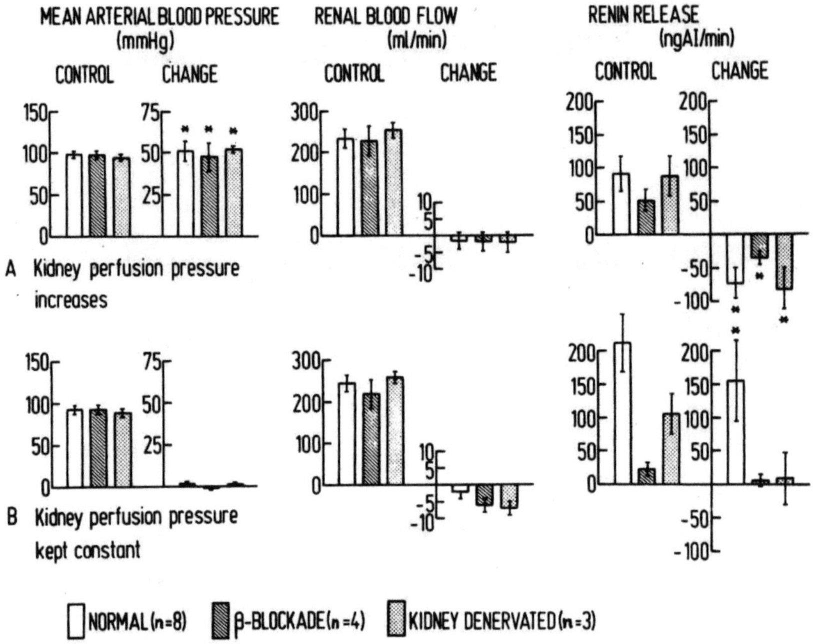

Fig. 6A, B. Effect of common carotid occlusion on blood pressure, renal blood flow, and renin release (renal-venous minus arterial plasma renin activity multiplied by renal plasma flow) with increasing (*A*) and constant (*B*) renal perfusion pressure in 15 conscious dogs (eight normal dogs, four dogs after β-blockade, and three dogs with a surgically denervated kidney). Depicted are the average control and the mean experimental change ± s.e.m. *AI*, angiotensin I. * denotes change is statistically significant from zero ($p < 0.05$)

We also studied the effect of carotid occlusion on renal function (implanted ureteral catheter). When renal perfusion pressure was allowed to rise (+49.6%), carotid occlusion caused an increase of urine output by 80.5% and of sodium excretion by 85.3% due to the fall in fractional sodium reabsorption (–0.9%); these changes occurred without a significant change in renal blood flow or glomerular filtration rate. Since neither an increase of diuresis or sodium excretion nor antinatriuresis was observed when renal perfusion pressure was controlled, it seems reasonable to conclude that the moderate sympathetic activation associated with carotid occlusion is too small to induce renal sympathetic vasoconstriction or antinatriuresis (Gross et al. 1981b). The diuresis and natriuresis with rising perfusion pressure is obviously a pressure-dependent phenomenon.

Since there were no changes in glomerular filtration rate and renal blood flow, the carotid occlusion experiment in the conscious dog seemed a suitable model for investigation of the influence of the sympathetic nerves independent of a number of other factors on renin release. Fig. 5 shows an experiment in which the animal blood was sampled from chronically implanted catheters in the renal vein and the abdominal aorta. Plasma renin activity was determined by radioimmunoassay, and renin release was calculated by multiplying renal plasma flow with renal-venous minus

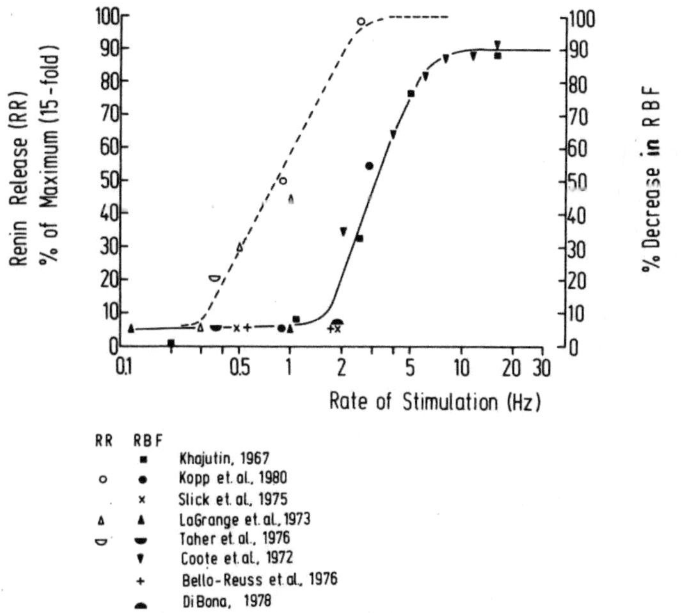

Fig. 7. Effect of electrical stimulation of the renal sympathetic nerves on renin release (*left ordinate*) and renal blood flow (*right ordinate*) in anesthetized dogs or cats. Data were collected from recent studies in which renin release (*RR*), renal blood flow (*RBF*), or both were measured. In the studies by Slick et al., Bello-Reuss et al., and DiBona renal nerve stimulation resulted in a decrease of sodium excretion independent of renal blood flow or glomerular filtration rate

arterial plasma renin activity. The left panel in Fig. 5A shows a common carotid occlusion experiment of 4 min duration without control of renal perfusion pressure. In the right panel (B), renal perfusion pressure was controlled at a level some 5 mmHg below the resting pressure (pneumatic cuff around abdominal aorta proximal to the renal arteries); subsequently, after this pressure level was stabilized, the carotid arteries were occluded for 4 min with renal perfusion pressure kept constant. The experiment in Fig. 5 demonstrates that with a rise in perfusion pressure there is a decrease in renin release from 104 to 32 ng angiotensin I/min. However, with control of renal perfusion pressure, carotid occlusion increases renin release from 190 to 310 ng angiotensin I/min. The average results of these experiments in three groups of dogs (normal, β-blockade, and surgical denervation of the kidney) are depicted in Fig. 6. The results show that the pressure-dependent mechanism reduces renin release in all three experimental groups. With control of renal perfusion pressure an increase in renin release is only observed in the intact dogs; this increase in renin release is abolished by β-adrenergic blockade or renal denervation. We conclude from these experiments that the reflex increase in sympathetic nerve activity augments renin release by a direct β-adrenergic stimulation of the juxtaglomerular cells without exerting vasomotor effects on the kidney, provided renal perfusion pressure is maintained at control level. The pressure-dependent vascular receptor mechanism can effectively counteract the stimulating influence of the renal sympathetic nerves when renal perfusion pressure is allowed to rise.

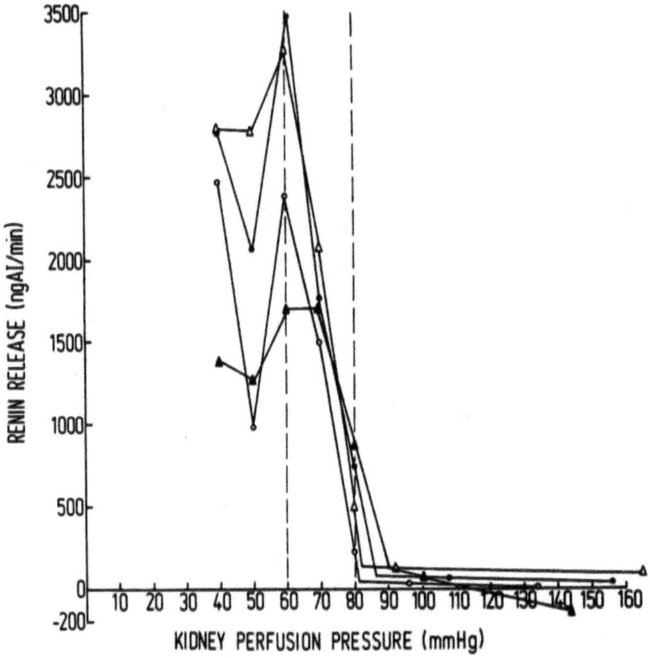

Fig. 8. Effect of kidney perfusion pressure (controlled independently from aortic blood pressure) on renin release in a resting conscious dog after β-adrenergic blockade (propranolol). *AI*: Angiotensin I. Depicted are four single experiments

In Fig. 7 we have collected data from experiments in which the renal sympathetic nerves were electrically stimulated and sodium excretion, renal blood flow, or renin release were recorded. A 15-fold increase in renin release was considered 100%. No significant change in renal blood flow was plotted at the 5% change level. The authors who measured sodium excretion invariably found a decrease in sodium excretion with a stimulation rate between 0.5 and 2 Hz without a change in renal blood flow (Slick et al. 1975; Bello-Reuss et al. 1976; DiBona 1978). Authors who studied both renal blood flow and renin release (Taher et al. 1976; Kopp et al. 1980; La Grange et al. 1973) found significant effects on renin release between 0.3 and 2 Hz without any effect on renal blood flow. Comparing the curves drawn in Fig. 7 with our experimental results, I should like to speculate that there are probably different thresholds for the sympathetic nervous influence on: (a) sodium reabsorption, (b) renin release, and (c) vasomotor effects.

In Fig. 8 four different experiments are shown, in which we have studied the pressure-dependent mechanism on renin release by decreasing renal perfusion pressure in a conscious dog under β-blockade. It demonstrates that renin release occurs within a very narrow pressure range (85–60 mmHg), although renal blood flow shows perfect autoregulation between 160 and 60 mmHg (not shown in Fig. 8). I should like to conclude from this experiment that renin release is in some way related to the pressure-dependent myogenic dilatation of the renal resistance vessels, since it fails to show a further rise once the vessels are maximally dilated (60 mmHg). However, it needs further experiments to present evidence for this suggestion and to explain why renin release increases once the blood pressure falls below a threshold pressure of some 88 mmHg.

Finally, to recapitulate very briefly, under physiological conditions in the conscious dog we have presented evidence that there is a resting sympathetic activity in the renal nerves. This is probably subthreshold to induce a significant resting vasomotor tone, but might have an important tonic influence on renin release and sodium reabsorption. A moderate activation of the sympathetic efferents by baroreceptors or "low-pressure system receptors" (blood loss) can influence renal function by increasing renin release and possibly sodium reabsorption without inducing major hemodynamic influences on the kidney. Both with regard to sodium excretion and renin release, blood pressure has an important direct influence, which in the case of renin release can counteract the sympathetic effects. The arterial baroreceptors and the receptors subserved by vagal afferents can thus in a first line of defense regulate systemic blood pressure by influencing renin release and sodium reabsorption with only minor effects on renal hemodynamics. This is not to deny that under extreme circumstances, such as severe blood loss associated with emotional stress or pain, an emergency situation develops with a vasomotor response in the kidney too.

References

Bello-Reuss E, Trevino DL, Gottschalk CW (1976) Effect of renal sympathetic nerve stimulation on proximal water and sodium reabsorption. J Clin Invest 57:1104–1107

Bernard C (1859) Leçons sur les propriètes physiologiques et les alterations pathologiques des liquides de l'organisme. Bailliere, Paris

Coote JH, Johns EJ, Macleod VH, Singer B (1972) Effect of renal nerve stimulation, renal blood flow and adrenergic blockade on plasma renin activity in the cat. J Physiol Lond 226:15–36

DiBona GF (1978) Neural control of renal tubular sodium reabsorption in the dog. Fed Proc 37:1214–1217

Freeman RH, Davis JO (1977) The control of renin secretion and metabolism. In: Genest J, Koiw E, Kuchel O (eds) Hypertension. McGraw-Hill, New York, pp 210–240

Gottschalk CW, Colindres RE, Moss NG, Rogenes P, Szalay L (1981) Neural control of renal function. In: Takacs L (ed) Advances of physiological sciences, vol 11, Kidney and body fluids. Pergamon Press – Akademiai Kiado, Budapest, pp 1–17

Gross R, Ruffmann K, Kirchheim H (1979) The separate and combined influences of common carotid occlusion and nonhypotensive hemorrhage on kidney blood flow. Pflügers Arch 379:81–88

Gross R, Kirchheim H (1980) Effects of bilateral carotid occlusion and auditory stimulation on renal blood flow and sympathetic nerve activity in the conscious dog. Pflügers Arch 383:233–239

Gross R, Hackenberg HM, Hackenthal E, Kirchheim H (1981a) Interaction between perfusion pressure and sympathetic nerves in renin release by carotid baroreflex in conscious dogs. J Physiol (Lond) 313:237–250

Gross R, Kirchheim H, Ruffmann K (1981b) Effect of carotid occlusion and of perfusion pressure on renal function in conscious dogs. Circ Res 48:777–784

Hardt D, Gross R, Kirchheim H (1971) Nachweis einer β-adrenergen Dilatation der Nierengefäße am wachen Hund. Pflügers Arch 327:152–166

Khajutin VM (1964) Physical properties of vessels and vasomotor regulation. In: Attinger EO (ed) Pulsatile blood flow. McGraw-Hill, New York Toronto London, p 331

Kirchheim H (1976) Systemic arterial baroreceptor reflexes. Physiol Rev 56:100–176

Kirchheim H, Gross R, Hackenberg M, Hackenthal E (1980) Effect of carotid sinus baroreceptors on kidney blood flow, renal sympathetic nerve discharge, and renin release in the conscious dog. In: Sleight P (ed) Arterial baroreceptors and hypertension. Oxford University Press, Oxford New York Toronto, pp 168–173

Kirchheim H, Ruffmann K, Gross R (1977) Mechanisms of baroreceptor reflex control of muscle blood flow in conscious dogs. Proc Int Union Physiol Sci 13:114

Kopp U, Aurell M, Nilsson IM, Åblad B (1980) The role of beta-1-adrenoceptors in the renin response to graded renal sympathetic nerve stimulation. Pflügers Arch 387:107–113

LaGrange RG, Sloop CH, Schmidt HE (1973) Selective stimulation of renal nerves in the anesthetized dog. Circ Res 33:704–712

Meckler RL, Macklem LJ, Weaver LC (1981) Excitation of cardiac sympathetic afferent nerves: effects on renal function. Am J Physiol 241:R267–R270

Melton LB, Frazier LW (1976) The effect of angiotensin II on electrolyte excretion by the renal tubules. Tex Rep Biol Med 34:289–295

Sadowski J, Kurkus J, Gellert R (1979) Denervated and intact kidney responses to saline load in awake and anesthetized dogs. Am J Physiol 237:F262–F267

Slick GL, Aguilera AJ, Zambraski GF, DiBona GF, Kaloyanides GJ (1975) Renal neuroadrenergic transmission. Am J Physiol 229:60–65

Smith HW (1951) The kidney: structure and function in health and disease. Oxford University Press, New York, pp 411–460

Taher MS, McLain LG, McDonald KM, Schrier RW (1976) Effect of beta-adrenergic blockade on renin response to renal nerve stimulation. J Clin Invest 57:459–465

Discussion

Brunner, Basel: According to older observations, renin release from the kidneys is mainly mediated through the mean arterial pressure, rather than through pulse pressure.

Kirchheim, Heidelberg: This very interesting problem has been studied in our laboratory. If, the abdominal aorta is constricted in such a way that blood pressure and pulse amplitude are reduced, the latter by only 5 mmHg, a remarkable release of renin can be seen. We believe that the decrease in pulse pressure is an important mechanism. Mellander and co-workers have shown that the myogenic response of the vascular wall is very sensitive to pulsatile changes.

Just, Freiburg: It has been suggested that the arterial hypertension in coarctation of the aorta may be due to the fact that the kidneys are situated in the poststenotic area and that in increased renin release is therefore a necessary consequence. Your studies would suggest that it is the decreased pulse pressure distal to the stenosis which may be the major and responsible mechanism.

Kirchheim, Heidelberg: It may well be that the pathogenetic mechanism of hypertension in coarctation is to be found here, but proof for this hypothesis is still lacking.

The Renin-Angiotensin-Aldosterone System in Congestive Heart Failure

G.A. Turini, H.R. Brunner

Patients with congestive heart failure often reach a state when the compensatory increase in vascular tone impairs pump performance. They fail to excrete sodium and lack the ability to maintain normal sodium balance. During recent years, it has been shown that renin, via angiotensin II, plays an important role in blood pressure homeostasis and in various forms of hypertensive diseases. It has also become apparent that it participates in the vasoconstriction of congestive heart failure, and may contribute to its impaired sodium-handling.

The Renin-Angiotensin-Aldosterone System

Renin is produced by the juxtaglomerular cells of the kidney. The primary signal inducing renin release involves a stretch-sensitive receptor in the afferent renal arterioles (1) and a receptor responding to changes in sodium transport located at the macula densa (2). A high sodium intake decreases renin release, whereas sodium deprivation increases renin levels (3). In normal volunteers, sodium intake is closely reflected by the 24-h urinary sodium excretion. The nomogram established for this renin-sodium relationship uses the 24-h urinary sodium excretion as an index of sodium balance (4). In addition to sodium deprivation, other factors, such as changes in sympathetic activity (5) and potassium balance (6), can also influence renin secretion by the kidney.

Renin is a proteolytic enzyme with a molecular weight of approximately 40,000 which, in plasma, interacts with renin substrate (angiotensinogen) synthetized by the liver to produce a decapeptide, angiotensin I. Angiotensin I has no clear vasoconstrictor effect of its own, but it is further transformed by a converting enzyme to the octapeptide angiotensin II. Angiotensin II has the following physiological actions: it constricts the arterioles and thereby increases blood pressure; it stimulates thirst by a direct effect on the brain; and it induces a prompt and sustained secretion of aldosterone by the adrenal cortex (7). In addition angiotensin II decreases renin release by a direct inhibitory feedback action on juxtaglomerular secretion. Angiotensin II is not the only stimulus for aldosterone secretion, which is also strongly influenced by the potassium ion (8) and, at least transiently, by adrenocorticotropic hormone (ACTH) (9). Aldosterone acts primarily on the distal renal tubules to increase the reabsorption of sodium and to enhance the elimination of potassium.

The renin-aldosterone system thus plays a dual role in maintaining blood pressure and regulating sodium and potassium homeostasis. Since renin substrate and converting enzyme are usually present in sufficient amounts, renin appears to be the major determinant of circulating angiotensin II levels.

Pharmacological agents inhibiting the renin-angiotensin system have opened a new and productive approach to the study of the role of this hormonal system both in normal subjects and in diverse pathological states (Fig. 1). Such inhibitors act either as competitive antagonists of the active hormone angiotensin II at the receptor level, or as inhibitors of the enzyme that converts angiotensin I into its active form, angiotensin II.

Specific Angiotensin II Inhibition by Saralasin in Congestive Heart Failure

In 1971, a new specific inhibitor of angiotensin II, saralasin (Eaton), became available for clinical and experimental use (10). This substance is a peptide analogue of angiotensin II with two aminoacid substitutions, which binds without activation to the angiotensin II receptors in blood vessels and adrenal cortex. There is no evidence that saralasin can lower the blood pressure by any mechanism other than by competitive antagonism of angiotensin II. However, this substance may have a pressor or agonistic action of its own (11), which becomes most apparent when plasma renin activity is markedly diminished.

Studies by Davis in dogs with constricted inferior vena cava have shown that the ensuing decrease in blood flow to the kidneys promotes renin secretion (12). This results in increased circulating angiotensin II and aldosterone. When given saralasin,

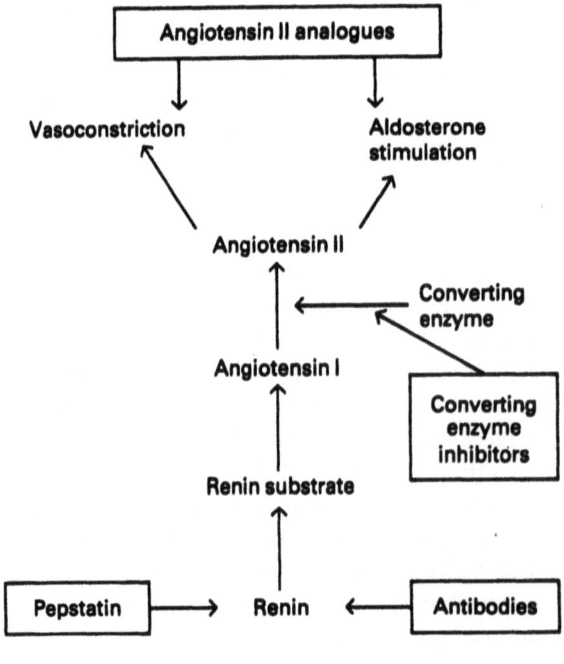

Fig. 1. Sites of blockade of renin-angiotensin inhibitors

these dogs showed a marked decrease in arterial pressure that was quickly reversible when the infusion of the antagonist was discontinued. An important vasoconstrictive action of angiotensin II in maintaining arterial pressure was thus demonstrated in this very particular experimental model of low-output heart failure.

We infused saralasin in 15 patients with congestive heart failure and normal arterial pressure and in five controls during cardiac catheterization (13). Fig. 2 shows that

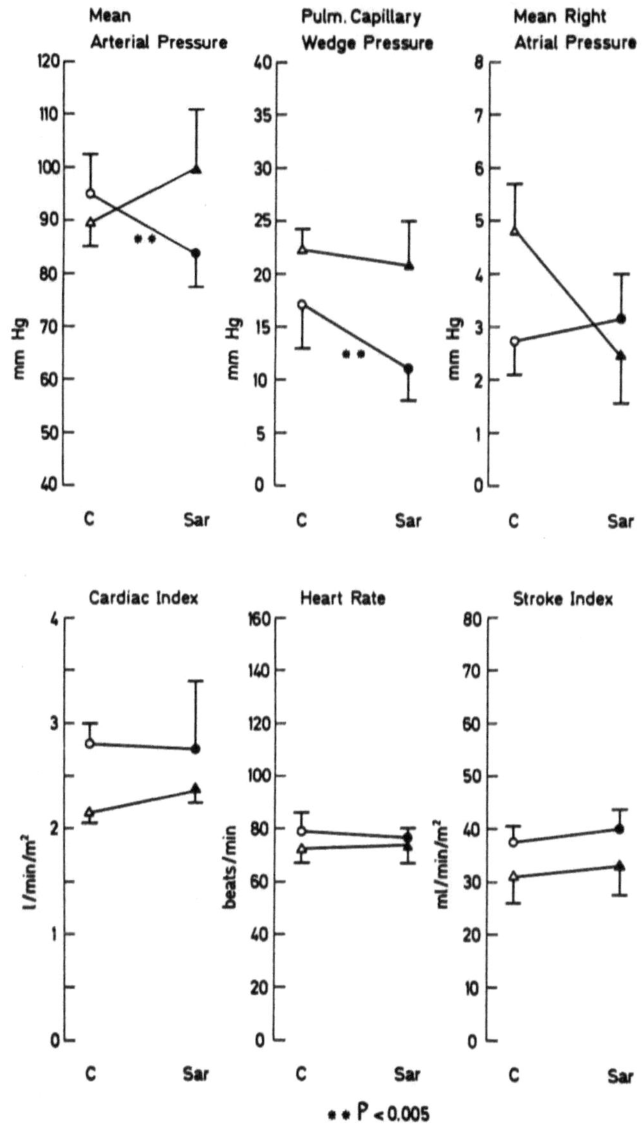

Fig. 2. Hemodynamic correlates before (*C, open symbols*) and after (*Sar, closed symbols*) saralasin infusion. This shows patients who had decreases in blood pressure (*circles*) in response to saralasin and patients who experienced no change or an increase in their blood pressure (*triangles*). Pulmonary capillary wedge pressure decreased significantly ($p < 0.005$) in the patients whose mean arterial pressure decreased. Turini et al. (13)

eight out of 14 patients (circles) decreased their blood pressure by a mean of 18 mmHg, the pulmonary capillary wedge pressure of each patient decreased simultaneously from a mean of 17 to 11 mmHg with no significant change in heart rate or stroke index. Such a response suggests that via angiotensin II renin has an important role in maintaining blood pressure in some patients with normotensive congestive heart failure. Of the other six patients (triangles), three did not change their arterial pressure, whereas three showed evidence of an agonistic effect with an increase in arterial pressure and no significant change in heart rate or stroke index. One patient, not included in the figure, exhibited a marked drop in arterial pressure with uneventful recovery after immediate discontinuation of saralasin infusion. The five controls did not present any hemodynamic change in response to saralasin.

Fig. 3 illustrates the initial plasma renin activity of all patients according to their arterial pressure response to saralasin. The six whose arterial pressure did not decrease had a mean plasma renin activity of 2 ng/ml/h – not significantly different from the values of five controls. The two patients who experienced the most striking agonistic response to saralasin had the lowest renin values of all. In contrast, the eight patients who responded to saralasin by a fall in arterial pressure had a significantly higher initial plasma renin activity of 8 ng/ml/h.

Converting Enzyme Inhibitors in Congestive Heart Failure

The nonapeptide teprotide (E.R. Squibb) (14), developed shortly after saralasin, inhibits the enzyme that converts the inactive angiotensin I to the active pressor hormone angiotensin II. In contrast to saralasin, it has no agonistic effect of its own.

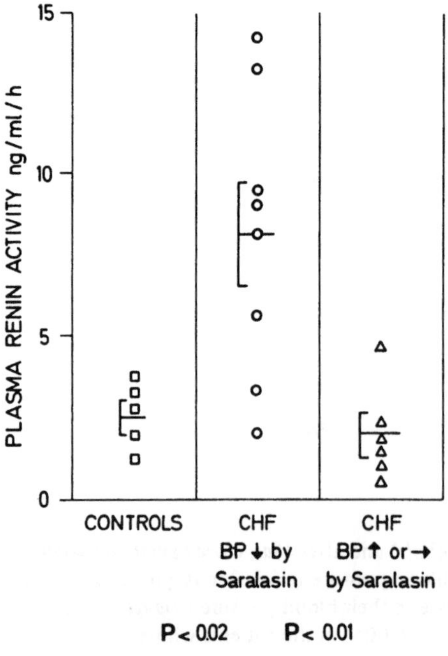

Fig. 3. Initial plasma renin activity of controls (*squares*), and patients with congestive heart failure who had a decrease (*circles*) or no change or an increase (*triangles*) in arterial pressure in response to saralasin infusion. The patients who had a decrease in blood pressure (*BP*) had significantly higher renin levels ($p < 0.01$) Turini et al. (13)

In dogs with constriction of the pulmonary artery or of the thoracic inferior vena cava, Watkins and co-workers observed a rapid reduction in arterial pressure, followed by a rise in plasma renin activity and aldosterone with progressive increase in blood volume and blood pressure (15). That maintenance of blood pressure was dependent on renin was shown by the hypotension induced when these animals received teprotide.

Curtis and co-workers evaluated the hemodynamic effects of teprotide in 14 patients with normotensive or hypotensive congestive heart failure (16). At peak effect, they observed a decrease in mean arterial pressure from 80 to 62 mmHg and in left ventricular filling pressure from 23 to 12 mmHg, whereas cardiac index rose by 20%, indicating improved cardiac performance. A similar response to teprotide was reported by Gavras et al. in four normotensive and four hypertensive patients (17). Being polypeptides, saralasin and teprotide both have to be administered parenterally, which makes chronic blockade of the system impractical in patients.

In 1976, captopril (E.R. Squibb) (Fig. 4), a new inhibitor of the angiotensin-converting enzyme, which is active when taken orally, became available (18–19).

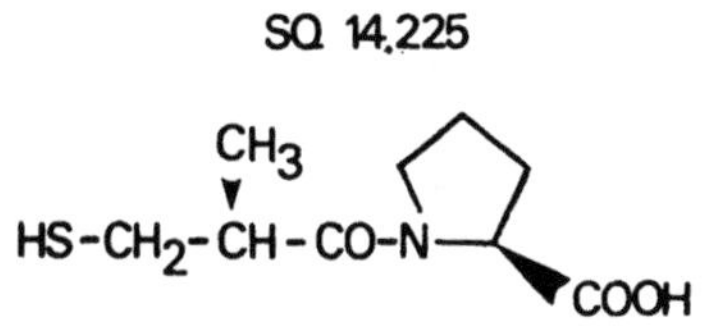

(D-2-methyl-3-mercaptopropanoyl - L - proline)

Fig. 4. Structure of captopril

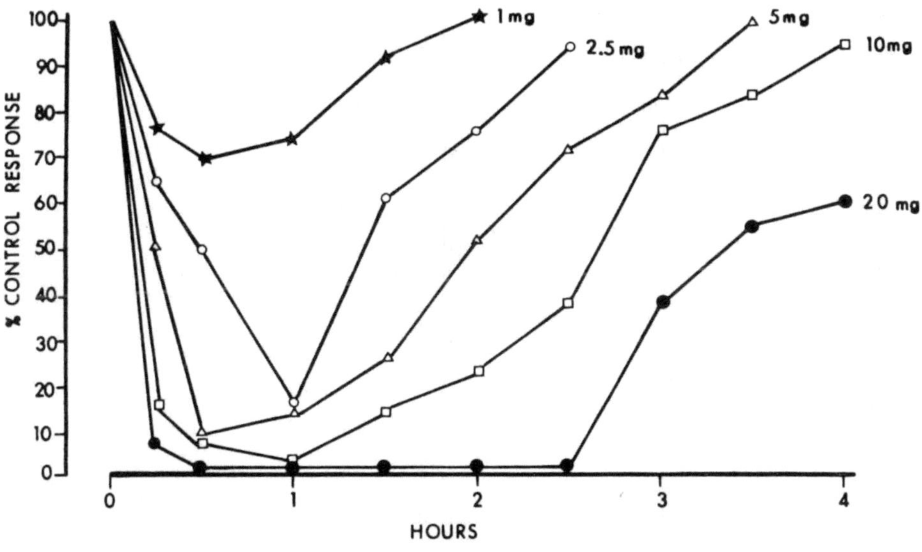

Fig. 5. Inhibition of systolic pressor responses to angiotensin I in 14 healthy men after incremental doses of captopril. Mean responses of three subjects are shown for each dose of captopril except for 20 mg where data are derived from two only. Ferguson et al. (19)

Captopril was first tested in 14 normal volunteers. A dose-response curve for angiotensin I was drawn up for each subject prior to the first oral ingestion of captopril. After administration of a single dose of the inhibitor, the maximum dose of angiotensin I that had been previously used was reinjected several times over the following hours. Increasing doses of captopril (from 1 to 20 mg) were tested in the same way. Fig. 5 shows the degree of inhibition of the systolic pressure response to angiotensin I obtained after ingestion of these various doses of captopril.

Both the degree and duration of converting enzyme inhibition appear to be dose-dependent. As little as 1 mg of captopril induces partial inhibition, whereas higher doses offered a marked blockade of the renin-angiotensin system. Thus the mechanism of action of this new inhibitor was confirmed in man, since 20 mg of captopril were sufficient to block completely the response to angiotensin I within 15 min and for over 2 h, and for doses of angiotensin I up to 160 or even 320 ng/kg

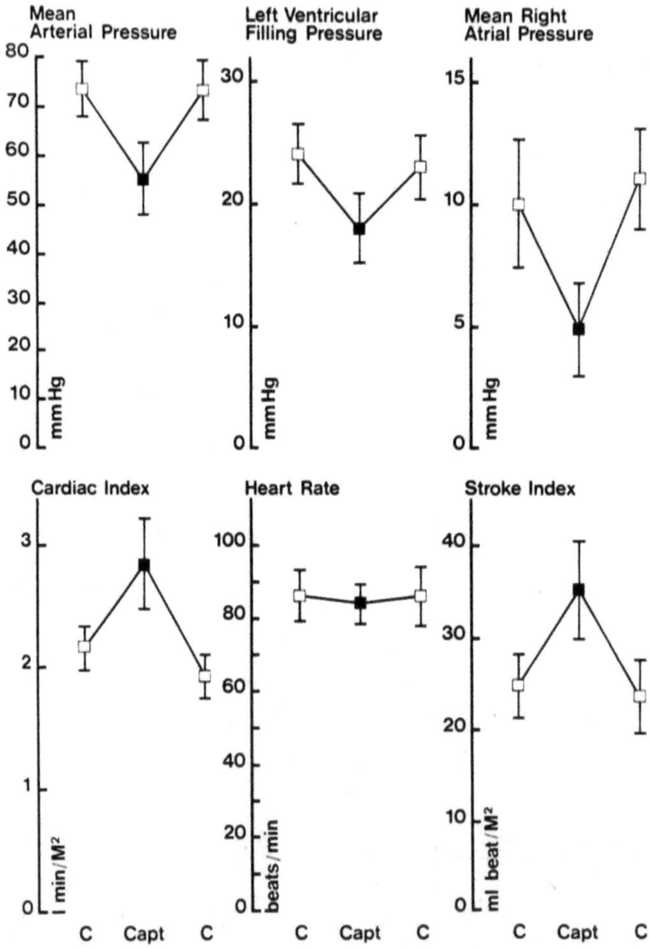

Fig. 6. Hemodynamic indices before captopril (*C, open symbols*), during maximum effect of captopril (*Capt, closed symbols*), and after return to baseline (*C, open symbols*). Turini et al. (20)

body wt. Consequently, it has to be kept in mind when treating patients that increasing the dose of captopril to over 20 mg will probably not bring about more blockade of the renin system, which is already complete for any possible level of circulating angiotensin II, but can only prolong the duration of the blockade.

The hemodynamic effect of 25 mg oral captopril in six patients with refractory heart failure is shown in Fig. 6. Within 1–3 h of captopril administration, mean arterial pressure fell from 74 to 55 mmHg, left ventricular filling pressure from 24 to 18 mmHg, and mean right arterial pressure from 10 to 5 mmHg. Simultaneously, the cardiac index increased from 2.1 to 2.9 l/min/m^2 due to a rise in stroke index from 25 to 35 ml/beat/m^2, while heart rate did not change. All these indices returned to baseline within 3–8 h after drug administration (20).

Other investigators have reported a similar hemodynamic response to the inhibitor, with increases in cardiac index of 15%–30% and decreases in left ventricular filling pressure ranging from 25% to 46% (21–24). All these studies show that captopril acts both as an arterial and as a venous vasodilator, shifting the left ventricular function curve upwards and to the left. This improvement in cardiac performance seems to persist on chronic administration (25–28).

Over the past year, different new converting enzyme inhibitors have been administered to normotensive volunteers (29). Fig. 7 shows the results obtained with one such new inhibitor, MK 421 (Merck, Sharp & Dohme). On the ordinate to the right, the mean blood pressure response to a given dose of angiotensin I is expressed in percent of the response observed prior to the administration of MK 421. To the left, plasma converting enzyme activity is given. Four different doses of the inhibitor were administered. Decreasing levels of plasma converting enzyme activity are associated with a decrease in blood pressure response to angiotensin I. Practically speaking, it appears that levels below 20 nmol/min of converting enzyme activity have to be obtained in order to get substantial decreases in the response to exogenous angiotensin I. This substance is now subject to clinical evaluation. It may become interesting because it clearly has a longer duration of action than captopril.

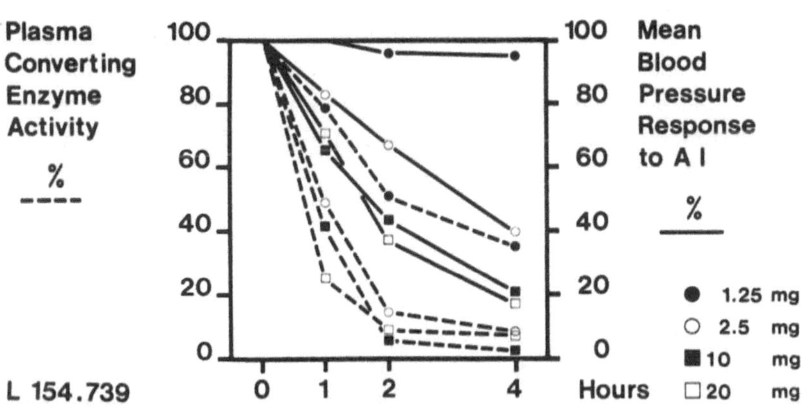

Fig. 7. Mean arterial pressure responses to angiotensin I (*AI*) (——) in normal volunteers after four doses of the converting enzyme inhibitor MK 421 (L 134, 739). Simultaneous changes in plasma converting enzyme activity are also depicted (– – –)

Potential Mechanisms of Renin Inhibition in Congestive Heart Failure

Studies of the renin-angiotensin system in congestive heart failure have raised many new questions. Angiotensin-converting enzyme is identical with kininase II, which is partly responsible for the inactivation of bradykinin (30). In addition to direct vasodilatation (31), bradykinin may also induce catecholamine secretion by the adrenal medulla and release prostaglandin from various organs (32). Thus the vascular effect of converting enzyme inhibition could be increased both by an accumulation of endogenous bradykinin due to kininase II inhibition, or/and by way of an enhanced action of prostaglandins. However, in a recent experimental study, no evidence in favor of an angiotensin-independent effect of captopril could be found (33).

Converting enzyme inhibition reduces aldosterone secretion (13, 20, 34, 36). At least theoretically, this should result in less sodium retention and could participate in the long-term improvement of cardiac performance. So far, there exists no clear evidence that aldosterone suppression indeed adds any benefit of its own.

As with many other vasodilating drugs, little is known about the effect of renin inhibition on the distribution of the cardiac output to different vascular beds. Preliminary work suggests that converting enzyme inhibition may increase coronary (37) and renal blood flow (38).

Finally, catecholamines are well known for their important role as compensatory inotropic and vasoconstrictor agents in heart failure (39). Catecholamines may stimulate renin secretion (40). On the other hand, angiotensin II has been shown to modulate the vascular effects of noradrenaline (41). Little is known so far about the interrelationships between these two systems in congestive heart failure.

Conclusion

In conclusion, both specific angiotensin II blockade and converting enzyme inhibition have clearly shown that in congestive heart failure the renin-angiotensin system participates in controlling vascular tone. Its inhibition improves cardiac performance by decreasing preload and afterload. Long-term studies suggest that these effects might be maintained with chronic administration of captopril, resulting in a marked increase n exercise capacity and persistent hemodynamic improvement. Double-blind studies now need to be performed to define the relative efficacy and side-effects of chronic converting enzyme inhibition.

References

1. Tobian L, Tomboullan A, Janecek J (1959) Effect of high perfusion pressures on the granulation of juxtaglomerular cells in an isolated kidney. J Clin Invest 38:605
2. Vander AJ, Miller R (1964) Control of renin secretion in the anesthetized dog. Am J Physiol 207:537
3. Davis JO (1973) The control of renin release. Am J Med 55:333–350
4. Brunner HR, Laragh JH, Baer L, et al. (1972) Essential hypertension. Renin and aldosterone, heart attack and stroke. N Engl J Med 286:441–449

5. Vander AJ (1965) Effect of catecholamines and the renal nerves on renin secretion in anesthetized dogs. Am J Physiol 209:659
6. Brunner HR, Baer L, Sealey JE, Ledingham JGG, Laragh JH (1970) Influence of potassium administration and of potassium deprivation on plasma renin in normal and hypertensive subjects. J Clin Invest 49:2128
7. Page IH, Bumpus FM (eds) (1974) Angiotensin. Springer, Berlin Heidelberg New York
8. Laragh JH, Stoerk HC (1957) A study of the mechanism of secretion of the sodium-retaining hormone (aldosterone). J Clin Invest 36:383–392
9. Ganong WF, Biglieri EG, Mulrow PJ (1966) Mechanisms regulating adrenocortical secretion of aldosterone and glucocorticoids. Recent Prog Horm Res 22:381
10. Pals DT, Masucci FD, Sipos F et al. (1971) A specific competitive antagonist of the vascular action of angiotensin II. Circ Res 29:664–672
11. Streeten DHP, Anderson GH, Freiburg JM et al. (1975) Use of an angiotensin II antagonist (saralasin) in the recognition of "angiotensinogenic" hypertension. N Engl J Med 292:657–662
12. Davis JO, Howell DS (1953) Mechanisms of fluid and electrolyte retention in experimental preparations in dogs. II. With thoracic inferior vena cava constriction. Circ Res 1:171
13. Turini GA, Brunner HR, Ferguson RK, Rivier JL, Gavras H (1978) Congestive heart failure in normotensive man. Haemodynamics, renin and angiotensin II blockade. Br Heart J 40:1134
14. Ondetti MA, Williams NJ, Sabo EF, Pluscec J, Weaver ER, Kocy O (1971) Angiotensin-converting enzyme inhibitors from the venom of Bothrops jararaca. Isolation, elucidation of structure and synthesis. Biochemistry 10:4033
15. Watkins L Jr, Burton JA, Haber E, Cant JR, Smith FW, Berger C (1976) The renin-angiotensin-aldosterone system in congestive failure in conscious dogs. J Clin Invest 57:1606–1617
16. Curtiss C, Cohn JN, Vrobel T, Franciosa JA (1978) Role of the renin-angiotensin system in the systemic vasoconstriction of chronic congestive heart failure. Circulation 58:763
17. Gavras H, Faxon DP, Berkoben J, Brunner HR, Ryan TJ (1978) Angiotensin converting enzyme inhibition in patients with congestive heart failure. Circulation 58:770
18. Ondetti MA, Rubin B, Cushman DW (1977) Design of specific inhibitors of angiotensin-converting enzyme: new class of orally active antihypertensive agents. Science 196:441–444
19. Ferguson RK, Turini GA, Brunner HR, Gavras H, McKinstry DN (1977) A specific orally active inhibitor of angiotensin-converting enzyme in man. Lancet I:775–778
20. Turini GA, Brunner HR, Grbic M, Waeber B, Gavras H (1979) Improvement of chronic congestive heart failure by oral captopril. Lancet 1:1213–1215
21. Levine TB, Carlyle PF, Gron KA, Franciosa JA, Cohn JN (1979) Hemodynamic and clinical response to captopril in congestive heart failure. Circulation 59, 60:II–39
22. Adler R, Chatterjee J, Ports T et al. (1979) Beneficial hemodynamic effects of an angiotensin converting enzyme inhibitor in chronic refractory heart failure. Am J Cardiol 43:404
23. Davis R, Ribner HS, Keung E, Sonnenblick EH, Lejemtel TH (1979) Effect of captopril in heart failure. N Engl J Med 301:117–121
24. Mason DT, Hermanovich J, Evensen M, Awan NA (1980) Oral captopril in ambulatory management of severe congestive heart failure; sustained beneficial effects of ventricular function with 6 months therapy shown by cardiac catheterization, nuclear scintigraphy, echocardiography, treadmill exercise and symptomatology. Am J Cardiol 45:411
25. Mason DT, Awan NA, Hermanovich J et al. (1979) Oral captopril in congestive heart failure: efficacy by cardiac catheterisation, forearm plethysmography, nuclear scintigraphy, echography, treadmill exercise and symptoms. Circulation 59, 60:II–230
26. Adler R, Chatterjee K, Ports T et al. (1979) Sustained beneficial haemodynamic effects of long term oral angiotensin converting enzyme inhibitor in chronic heart failure. Circulation 59, 60:II–130
27. Sharpe DN, Douglas JE, Coxon RJ, Long B (1980) Low-dose captopril in chronic heart failure: acute haemodynamic effects and long-term treatment. Lancet 2:1154–1157
28. Turini GA, Brunner HR (1980) Les inhibiteurs du système rénine-angiotensine dans l'insuffisance cardiaque chronique. Extrait des «Journées de Réanimation» de l'Hôpital Claude-Bernard. Librairie Arnette, Paris, pp 307–315

29. Biollaz J, Burnier M, Turini GA, Brunner DB, Porchet M et al. (1981) Three new long-acting converting-enzyme inhibitors: relationship between plasma converting-enzyme activity and response to angiotensin I. Clin Pharmacol Ther 29/5:665–670
30. Erdös EG (1976) Conversion of angiotensin I to angiotensin II. Am J Med 60:749–759
31. Murthy VS, Waldron TL, Goldberg ME (1978) The mechanism of bradykinin potentiation after inhibition of angiotensin-converting enzyme by SQ 14.225, in conscious rabbits. Circ Res [Suppl 1] 43:140–145
32. McGiff JC, Terragno NA, Malik KU, Lonigro AJ (1973) Release of a prostaglandin E-like substance from canine kidney by bradykinin. Circ Res 31:36–43
33. Textor SC, Brunner HR, Gavras H (1981) Converting enzyme inhibition during chronic angiotensin II infusion in rats. Evidence against a nonangiotensin mechanism. Hypertension 3/2: 269–276
34. Gavras H, Brunner HR, Turini GA et al. (1978) Antihypertensive effect of the oral angiotensin converting-enzyme inhibitor SQ 14,225 in man. N Engl J Med 298:991–995
35. Brunner HR, Gavras H, Turini GA et al. (1979) Oral angiotensin-converting enzyme inhibitor in long-term treatment of hypertensive patients. Ann Intern Med 90:19–23
36. Bravo EL, Tarazi RC (1979) Converting enzyme inhibition with an orally active compound in hypertensive man. Hypertension 1:39–46
37. Gavras H, Liang C, Brunner HR (1978) Redistribution of regional blood flow after inhibition of the angiotensin-converting enzyme. Circ Res [Suppl 1] 43:59–63
38. Mimran A, Brunner HR, Turini GA, Waeber B, Brunner DB (1979) Effect of captopril on renal vascular tone in patients with essential hypertension. Clin Sci 57:421s–423s
39. Braunwald E, Ross J, Sonnenblick EH (1968) Mechanisms of contraction of the normal and failing heart. Little, Brown, Boston, pp 157–163
40. Davis JO, Freeman RH (1976) Mechanisms regulating renin release. Physiol Rev 56:1–56
41. Spertini F, Brunner HR, Waeber B, Gavras H (1981) The opposing effects of chronic angiotensin-converting enzyme blockade by captopril on the responses to exogenous angiotensin II and vasopressin vs. norepinephrine in rats. Circ Res 48/5:612–618

Discussion

Starke, Freiburg: You mentioned the interrelationship of the renin-angiotensin-aldosterone system and the sympathoadrenal system.

Turini, Lausanne: Simultaneous activation of both systems might explain some of the hemodynamic effects of the inhibitors of the renin-angiotensin-aldosterone system, as well as the decreased effectiveness of receptivity toward noradrenaline.

Starke, Freiburg: The effect of noradrenaline on vascular smooth muscle cells is increased by angiotensin, as is the release of noradrenaline.

Turini, Lausanne: In six patients with congestive heart failure we have seen a slight, not statistically significant, decrease in plasma noradrenaline. This is well in keeping with observations of Cohn, who has also seen some decrease in plasma noradrenaline during treatment with vasodilators.

Cohn, Minneapolis: Selective blockade of the converting enzyme by captopril produces a pronounced venodilatation and a somewhat lesser arteriolar dilatation. In this situation in congestive heart failure the plasma catecholamines are seen to fall, thereby indicating an inhibition of sympathetic outflow. We are, however, so far unable to correlate these changes with alterations in venous tone, right atrial pressure, or other hemodynamic parameters. Probably the plasma catecholamines are really too insensitive in detecting changes in sympathetic interaction.

Jahrmärker, München: Could there be changes not only in plasma catecholamine levels, but also in vascular sensitivity towards endogenous catecholamines?

Cohn, Minneapolis: The facilitation of catecholamine response in the presence of angiotensin II would not really be refuted by changes in plasma levels.

Turini, Lausanne: According to clinical experience, patients receiving diuretics will become dependent on circulating blood volume if captopril is being given. Hypovolemia induced by diuretics will render the patients very sensitive to vasodilatation with captopril. Here profound hypotension with circulatory collapse may ensue.

On the other hand, we have observed that people with decreased renal function and markedly increased creatinine values show a very significant fall in arteriolar resistance and blood pressure without further deterioration of renal function. We conclude, therefore, that captopril also induces renovasodilatation with subsequent improvement of renal perfusion.

Endogenous Catecholamines Under Chronic Vasodilator Treatment

W. Klein

Today there is abundant evidence that sympathetic activity is enhanced in cardiac failure (7, 8, 11, 14, 34). The increased plasma concentration of the endogenous neurotransmitter noradrenaline augments heart rate and contractility of the failing heart. Since the cardiac muscle fibers are hypersensitive to endogenous or exogenous catecholamines in heart failure, these hormones seem to be essential in restoring left ventricular function. However, the elevated plasma level of noradrenaline, together with an activation of the renin-angiotensin-aldosterone system leads to an increased stiffness of the arteriolar wall due to salt and water retention. This results in an elevation of peripheral resistance to restore an adaequate perfusion pressure of the vital organs.

On the other hand, the noradrenaline content of the heart is reduced in cardiac failure (8, 34). The decreased amount of catecholamines in the heart muscle is, however, not the cause but the consequence of the impaired cardiac function. Although the depletion of the cardiac catecholamine stores in heart failure has been demonstrated by several workers in the experimental animal and in man, the exact mechanism remains obscure (34). It has been suggested that noradrenaline and dopamine β-hydroxylase are released by the cardiac β-receptors. The increased amount of catecholamine release in heart failure may the result in depletion of the cardiac stores (34).

Braunwald et al. (8) have postulated that the heart itself, like an endocrine organ, releases noradrenaline in critical situations to restore cardiac function. The amount of noradrenaline released, and therefore the plasma concentration, is according to Thomas and Marks (34) directly correlated with the degree of left ventricular dysfunction. A further increase in plasma catecholamines is noted under exercise. Chidsey et al. (7) have shown that the concentration of noradrenaline in the coronary sinus during physical work is higher in patients with heart failure than in normal controls. On the other hand, Francis et al. (14) have demonstrated that the overall plasma noradrenaline increases by 913% during work in normal man, but only by 171% in heart failure. Obviously, in cardiac failure the catecholamine stores are too depleted to permit a higher increase of the overall plasma concentration. The heart muscle in cardiac failure gets more and more dependant on the circulating catecholamines, since the cardiac stores are more and more depleted.

The stimulation of the autonomic system in heart failure is a reflex stimulation via the carotid sinus receptors: the activity of these receptors is determined not only by the systolic blood pressure, but also by the rate of pressure rise in the arteries and the arterial wall tonus. The carotid sinus receptors are therefore very sensitive in detecting cardiac failure. Like any other compensatory mechanism in heart failure, the activation of the sympathoadrenal system may overshoot. In this case the peripheral resistance increases too much, leading to an exaggerated increase in ventricular afterload and to further impairment of left ventricular function (9, 28). This vicious cycle can be interrupted by vasodilators. The peripheral vasodilatation reduces cardiac afterload, cardiac output increases, and the periphery is perfused sufficiently. The reflex activation of the sympathetic nervous system is thereby reduced or abolished and the circulating catecholamines should decrease.

The favorable effects of vasodilator treatment of acute and chronic heart failure are well established (5, 6, 13, 17, 18, 20, 21, 28, 35). For long-term oral treatment, prazosin, a postsynaptic α-blocking agent, has proved to be effective (1–4, 14–16, 29). However, peripheral vasodilators or α-blocking drugs themselves stimulate the baroreceptor system as a side-effect, resulting in an increase of the plasma noradrenaline and renin, tachycardia, and salt and water retention (6, 10, 12, 31). α-Blockers that act mainly presynaptically, like phentolamine, lead to a higher noradrenaline increase than predominantly postsynaptic α-blockers (4, 15, 16).

These humoral and hemodynamic changes attenuate the favorable effects of vasodilator treatment of heart failure, and may in fact result in tachyphylaxis (12). In the present study, therefore, the hemodynamics and the endogenous catecholamines representing the activity of the sympathetic nervous system in patients with chronic left heart failure and vasodilator treatment were evaluated. The findings reported in the literature are somewhat at variance: Curtis et al. (11) found a decrease in plasma noradrenaline under saralasin treatment, Maslovski et al. (27) no change after captopril, and Kersting et al. (19) an increase after long-term prazosin therapy. Manthey et al. (26) found different behavior in responders and nonresponders to vasodilator therapy with hydralazine and nitrates.

Methods

A total of 17 patients with chronic left heart failure due to coronary heart disease, hypertensive heart disease, or primary cardiomyopathy were included in the study. In the first part of the study, 11 patients received 0.8 mg nitroglycerin sublingually on the 1st day and 1–3 mg prazosin on the following day. Then patients were treated with 9 mg prazosin per day for a period of 4 weeks. Before and after therapy the following parameters were measured and calculated: blood pressure, cardiac output (thermodilution), pulmonary artery pressure, cardiac index, stroke index, peripheral resistance. After registration of the resting values, the measurements were repeated after exercise (25–50 W during 3–5 min in a sitting position on a bicycle ergometer). Moreover, the catecholamines adrenaline, noradrenaline, and dopamine were measured at rest and during maximal exercise according to the method of da Prada and Zurcher (33).

In the second part of the study in another six patients, a Swan-Ganz catheter was positioned in the pulmonary artery, a Lehman catheter in the left ventricle, and a Gensini catheter in the coronary sinus. Exercise was now performed in a supine position up to exhaustion. Hemodynamic parameters and catecholamines from the aorta, right atrium, and coronary sinus were determined at rest and after exercise, before and after 1.6 mg nitroglycerin.

Results

The results of the first study are shown in Tables 1 and 2. After 0.8 mg nitroglycerin there was a slight decrease in systemic blood pressure at rest ($p < 0.02$) and after exer-

Table 1. Hemodynamic effects of 0.8 mg nitroglycerin in 11 patients with congestive heart failure. *HR*, heart rate; *CAP*, central aortic pressure; *DPAP*, diastolic pulmonary artery pressure; *CO*, cardiac output; *HI*, heart index; *SI*, stroke index; *TPR*, total peripheral resistance

Parameter	Control		Nitroglycerin			
	Rest	Work	Rest	*P*	Work	*P*
HR (beats/min)	86 ± 6	115 ± 8	95 ± 5	N.S.	115 ± 7	N.S.
CAP (mmHg)	95 ± 2	103 ± 2	89 ± 1	0.02	98 ± 3	0.06
DPAP (mmHg)	24 ± 2	31 ± 4	20 ± 1	0.02	28 ± 3	0.05
CO (l/min)	4.1 ± 0.2	9.3 ± 0.7	4.1 ± 0.3	N.S.	9.7 ± 0.9	N.S.
HI (l/min/m²)	2.1 ± 0.1	4.8 ± 0.3	2.1 ± 0.1	N.S.	5 ± 0.4	N.S.
SI (ml/m²)	25 ± 2	45 ± 6	22 ± 1	N.S.	45 ± 5	N.S.
TPR (dyn · s · cm^{-5})	1983 ± 115	991 ± 115	1827 ± 176	N.S.	958 ± 160	N.S.

Table 2. Effect of 0.8 mg nitroglycerin on venous catecholamine values in 11 patients with congestive heart failure

	Control		Nitro			
	Rest	Work	Rest	*P*	Work	*P*
Dopamine (pg/ml)	30 ± 15	88 ± 50	29 ± 7	N.S.	26 ± 2	N.S.
Noradrenaline (pg/ml)	257 ± 26	372 ± 40	291 ± 39	N.S.	383 ± 55	N.S.
Adrenaline (pg/ml)	112 ± 16	121 ± 27	77 ± 8	N.S.	91 ± 16	N.S.

cise ($p < 0.06$) and a significant decrease in diastolic pulmonary artery pressure at rest ($p < 0.02$) and after exercise ($p < 0.05$). Heart rate, cardiac index, and stroke index remained unchanged; peripheral resistance was only slightly diminished. Plasma noradrenaline concentration was augmented during physical work during the control period and after nitroglycerin; however, there was no significant difference between control values and those after nitroglycerin (Table 2).

Prazosin given in a dose of 1–3 mg on the following day also led to a decrease in mean blood pressure at rest and after exercise ($p < 0.03$) and in disatolic pulmonary pressure at rest and after work ($p < 0.03$); moreover, after prazosin peripheral resistance was diminished significantly ($p < 0.006$). Heart rate, cardiac index, and stroke index remained constant. After chronic prazosin therapy the peripheral resistance showed a further decline ($p < 0.001$) and cardiac output as well as stroke index increased significantly ($p < 0.001$) (Table 3). Catecholamines were not altered significantly during acute or chronic prazosin therapy; only adrenaline was augmented after long-term treatment ($p < 0.01$) (Table 4).

In the second part of the study the catecholamines were measured in the arterial, venous, and coronary venous blood separately. Adrenaline and dopamine showed no significant changes. However, noradrenaline values were significantly elevated in the arterial (762 pg/ml) and coronary venous blood (1028 pg/ml) in those patients with congestive heart failure, and there was a further dramatic increase after exercise up to three- or fourfold the control value (Table 5). Nitroglycerin at rest caused no significant change of the noradrenaline concentrations in the arterial or coronary sinus blood, although the values were moderately increased (Fig. 1). After exercise, however, the arterial and coronary sinus concentrations were significantly lower under nitroglycerin (Figs. 2 and 3).

Discussion

Nitroglycerin and prazosin, like other vasodilators, lead to a reflex stimulation of the sympathetic system with increase of catecholamines, provided that there is no congestive heart failure (23, 24, 30, 31). De Leeuw et al. (24) found an increase in noradrenaline from 120 to 350 pg/ml after prazosin in hypertensive patients. Similar results were obtained by Muiesan et al. (30).

Kersting et al. (19) observed an increase of catecholamines under prazosin therapy also in patients with heart failure; however, the control values of noradrenaline in these patients were low, within the normal range, and the authors point out in their study that the patients did not have severe heart failure. On the other hand, Curtis et al. (11) reported a significant decrease in markedly elevated noradrenaline levels in their heart failure patients after saralasin.

Manthey et al. (26) found a decrease in noradrenaline after hydralazine and isosorbide dinitrate in severe heart failure and an increase in patients without overt cardiac failure. Overall, they saw no significant changes. In our study the population was different, and in general there were no marked changes in catecholamine levels. One might assume that the noradrenaline concentration would be elevated in severe heart failure, and after vasodilator treatment one would expect a further increase due

Table 3. Behavior of circulatory parameters under prazosin therapy. *HR*, heart rate; AO_m, mean aortic pressure; *DPAP*, diastolic pulmonary artery pressure; *CO*, cardiac output; *HI*, heart index; *SI* stroke index; *TPR*, total peripheral resistance

Parameter	Control		Short-term therapy				Chronic therapy			
	Rest	Effort	Rest	*P*	Effort	*P*	Rest	*P*	Effort	*P*
HR	86 ± 6	115 ± 8	94 ± 4	N.S.	127 ± 7	N.S.	84 ± 7	N.S.	123 ± 8	N.S.
AO_m	95 ± 2	103 ± 2	83 ± 3	0.006	88 ± 5	0.01	96 ± 3	N.S.	105 ± 7	N.S.
DPAP	24 ± 2	31 ± 4	20 ± 1	0.03	25 ± 3	0.05	28 ± 3	0.18	38 ± 4	N.S.
CO	4.1 ± 0.2	9.3 ± 0.7	4.3 ± 0.2	N.S.	9.9 ± 0.8	N.S.	5.7 ± 0.4	0.001	13 ± 0.9	0.0001
HI	2.1 ± 0.1	4.7 ± 0.3	2.3 ± 0.1	N.S.	5 ± 0.4	N.S.	3 ± 0.2	0.001	6.9 ± 0.4	0.0001
SI	25 ± 2	45 ± 6	24 ± 1	N.S.	41 ± 4	N.S.	37 ± 3	0.001	58 ± 5	0.001
TPR	1983 ± 115	1009 ± 111	1530 ± 114	0.007	805 ± 118	0.006	1432 ± 105	0.001	676 ± 63	0.001

Table 4. Behavior of the catecholamines under prazosin therapy

	Control		Short-term therapy				Chronic therapy			
	Rest	Effort	Rest	*P*	Effort	*P*	Rest	*P*	Effort	*P*
Dopamine	29 ± 15	34 ± 8	20 ± 4	N.S.	33 ± 7	N.S.	35 ± 9	N.S.	33 ± 7	N.S.
Noradrenaline	317 ± 66	340 ± 43	332 ± 36	N.S.	347 ± 40	N.S.	326 ± 40	N.S.	347 ± 40	N.S.
Adrenaline	113 ± 16	93 ± 17	101 ± 9	N.S.	99 ± 16	N.S.	169 ± 15	0.011	99 ± 16	N.S.

Table 5. Effect of 1.6 mg nitroglycerin on the arterial, venous, and coronary venous catecholamine concentrations

Catecholamine (pg/ml)	Control Rest	Control Work	*P*	Nitro Rest	*P*	Nitro Work	*P*
Noradrenaline							
Art.	762 ± 451	2199 ± 1424	N.S.	1099 ± 569	N.S.	1342 ± 404	N.S.
Ven.	696 ± 436	2524 ± 1876	0.05	881 ± 445	N.S.	1684 ± 685	N.S.
Cor. v.	1028 ± 480	4229 ± 4309	N.S.	1097 ± 405	N.S.	3091 ± 3426	0.01
Adrenaline							
Art.	363 ± 273	705 ± 659	N.S.	293 ± 210	N.S.	484 ± 439	N.S.
Ven.	312 ± 208	1810 ± 3096	N.S.	395 ± 497	N.S.	767 ± 996	N.S.
Cor. v.	926 ± 1077	644 ± 647	N.S.	488 ± 657	N.S.	600 ± 674	N.S.
Dopamine							
Art.	1901 ± 3490	619 ± 1040	N.S.	551 ± 480	N.S.	515 ± 240	N.S.
Ven.	401 ± 361	1498 ± 689	N.S.	592 ± 312	N.S.	875 ± 714	N.S.
Cor. v.	565 ± 795	804 ± 602	N.S.	816 ± 610	N.S.	405 ± 401	N.S.

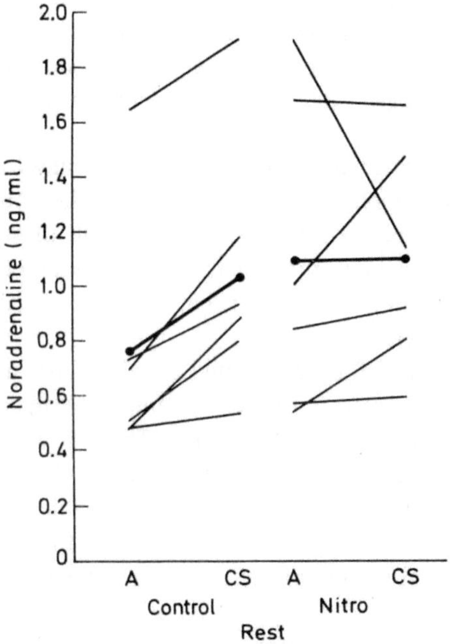

Fig. 1. Arterial (*A*) and coronary venous (*CS*) concentration of noradrenaline before (*control*) and after nitroglycerin (*nitro*) at rest

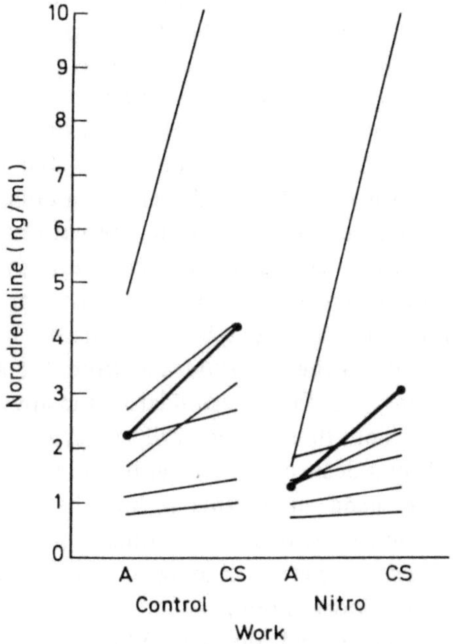

Fig. 2. Arterial (*A*) and coronary venous (*CS*) concentration of noradrenaline before (*control*) and after nitroglycerin (*nitro*) during exercise

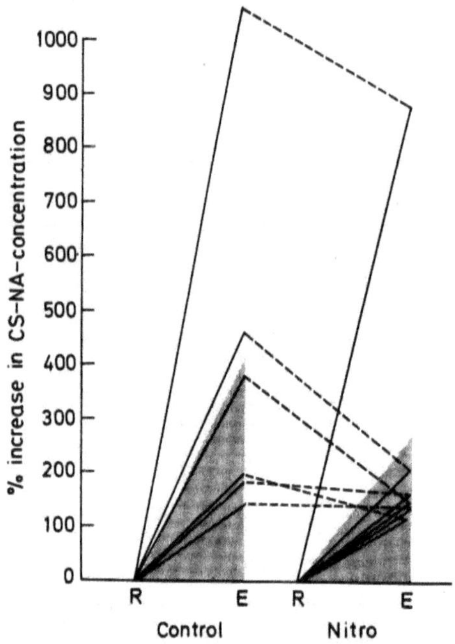

Fig. 3. Percentile increase of coronary venous noradrenaline concentration (CS-NA-concentration) under exercise (*E*) before (*control*) and after nitroglycerin (*nitro*). Resting (*R*) values are 100%

to reflex stimulation of the sympathetic system; however, this was not observed in our study or in others dealing with severe heart failure patients.

The reason may be that in severe heart failure stroke volume is increased after vasodilator treatment, and blood pressure may even increase paradoxically, due to afterload reduction. The periphery is perfused again and there is no longer a reflex stimulation from the baroreceptor site.

It is obvious from our study that the myocardial noradrenaline release may be stopped or at least reduced by vasodilator treatment (22). The decrease in the noradrenaline concentration in the coronary sinus blood could also be the result of an increase in coronary flow under nitroglycerin; we did not measure the coronary blood flow in our experiments, but it is clear from other studies (25) that coronary blood flow is not increased, but reduced after nitroglycerin.

From the literature and from our study, therefore, one would conclude that in severe heart failure vasodilator therapy augments cardiac output and improves organ perfusion, thus paradoxically reducing baroreceptor reflex stimulation. This is followed by a decrease in myocardial noradrenaline release, especially after exercise. On the other hand, in patients without heart failure or with only slightly disturbed hemodynamics, vasodilator treatment may aggravate the sympathetic drive by further reflex stimulation, leading to further myocardial noradrenaline release and depletion of the cardiac stores.

Packer and Meller (32) have pointed out that vasodilators are contraindicated if the cardiac index is over 2.5 l/min/m^2, or if the left ventricular end-diastolic pressure is less than 15 mmHg. In these patients tachycardia, hypotension, and a decrease in cardiac output may result. From our study it should be added that the endogenous noradrenaline may increase under treatment and that the myocardial stores may be depleted.

Summary

Congestive heart failure is usually accompanied by a reflex stimulation of the sympathetic nervous system as a compensatory mechanism. Vasodilator therapy may further enhance sympathetic activity resulting in an increase of plasma noradrenaline and renin, tachycardia, and fluid retention. These changes attenuate the favorable effects of vasodilator treatment of congestive heart failure and may lead to tachyphylaxis. In the present study, therefore, the hemodynamic data and plasma catecholamines were measured in 17 patients with chronic left heart failure, before and after vasodilator treatment.

Administration of prazosin (9 mg/day) over a period of 4 weeks led to a reduction of peripheral resistance ($p < 0.001$) and elevation of cardiac output and stroke index ($p < 0.001$). Noradrenaline concentrations were moderately elevated (up to 500 pg/ml) in the venous blood, but were not significantly altered during prazosin therapy. Only adrenaline was augmented slightly after therapy ($p < 0.01$). When cardiac noradrenaline release was measured directly by coronary sinus catheterization during rest and supine exercise, noradrenaline was markedly elevated (to 1028 pg/ml) in the coronary sinus blood, and there was a further dramatic increase after exercise, suggesting that left ventricular function in heart failure is mainly maintained by release of noradrenaline from the cardiac nerve endings. After nitroglycerin (1.6 mg) noradrenaline values were not altered at rest, but were significantly lower during exercise ($p < 0.01$), indicating a diminished sympathetic drive of the heart by pre- and afterload reduction.

It is concluded from this study that use of vasodilator treatment should be confined to patients with severe heart failure who respond to vasodilator treatment with a decrease in plasma catecholamines. In patients with mild or moderate degree of heart failure, vasodilators may enhance myocardial noradrenaline release, and cardiac catecholamine stores may be depleted completely.

References

1. Aronow WS, Donahy DT (1978) Efficacy of trimazosin and prazosin therapy on cardiac and exercise performance in outpatients with chronic congestive heart failure. Am J Med 65:155
2. Aronow WS, Lurie M, Turbow M, Whittakter K, Van Camp S, Hughes D (1979) Effect of prazosin vs placebo on chronic left ventricular heart failure. Circulation 59:344
3. Awan NA, Miller RR, Miller MP, Specht K, Vera Z, Mason DT (1978) Clinical pharmacology and therapeutic application of prazosin in acute and chronic refractory congestive heart failure. Am J Med 65:146
4. Cavero I, Fenard S, Gomeni R et al. (1978) Studies on the mechanism of the vasodilator effects of prazosin in dogs and rabbits. Eur J Pharmacol 49:259
5. Chatterjee K, Parmley WW, Massie B, Greenberg B, Werner J, Klausner S, Norman A (1976) Oral hydralazin therapy for chronic refractory heart failure. Circulation 54:879
6. Chatterjee K, Parmley WW (1977) The role of vasodilator therapy in heart failure. Prog Cardiovasc Dis 19:301

7. Chidsey CA, Harrison DC, Braunwald E (1962) Augmentation of the plasma norepinephrine response to exercise in patients with congestive heart failure. N Engl J Med 267:650
8. Chidsey CA, Braunwald E, Morov AG et al. (1963) Myocardial norepinephrine concentration in man: effects of reserpine and of congestive heart failure. N Engl J Med 269:653
9. Cohn JN (1973) Vasodilator therapy for heart failure. Circulation 48:5
10. Cohn JN, Taylor N, Vrobel T, Moskowitz R (1978) Contrasting effect of vasodilators on heart rate and plasma catecholamines in patients with hypertension and heart failure. Clin Res 26:547 A
11. Curtis C, Cohn JN, Vrobel T, Franciosa JA (1978) Role of the renin-angiotensin system in the systemic vasoconstriction of chronic heart failure. Circulation 58:763
12. Fahmy NR, Sunder N, Moss J et al. (1979) Tachyphylaxis to nitroprusside. Role of the renin-angiotensin system and catecholamines in its development. Anesthesiology [Suppl] 51:72
13. Franciosa JA, Blank RC, Cohn JN (1978) Nitrate effects on cardiac output and left ventricular outflow resistance in chronic congestive heart failure. Am J Med 64:207
14. Francis GS, Goldsmith SR, Cohn JN (1981) The sympathetic response to exercise in patients with congestive heart failure. Am J Cardiol 47:395
15. Graham RM, Pettinger WA (1979) Prazosin. N Engl J Med 300:232
16. Graham RM, Kennedy P, Stephenson W, Pettinger WA (1978) In vivo evidence for the presynaptic α-adrenergic receptor controlling norepinephrine release. Fed Proc 37:308
17. Himmler FC, Wirtzfeld A, Klein G, Volger E, Schmidt G (1980) Hämodynamische Wirkung von Prazosin bei Patienten mit schwerer Herzinsuffizienz. Herz/Kreisl 12:317
18. Imhof P (1977) Behandlung der Herzinsuffizienz mit Phentolamin. Schweiz Med Wochenschr 107:1041
19. Kersting F, Samosny G, Kasper W, Meinertz T, Gilfrich H-J, Just H (1980) Clinical pharmacology of prazosin and phentolamine in patients with heart failure. J Cardiovasc Pharmacol 2:373
20. Klein W, Brandt D, Harpf H, Goebel R (1979) Fortschritte in der Therapie der akuten und chronischen Herzinsuffizienz durch den Einsatz systemischer Vasodilatatoren. Wien Med Wochenschr 129:625
21. Klein W, Goebel R (1980) Hämodynamik und Katecholaminstoffwechsel unter einer Vasodilatatorentherapie der chronischen Linksinsuffizienz. Z Kardiol 69:120
22. Klein W, Goebel R, Brandt D, Maurer E (1981) Effect of nitroglycerin on hemodynamics and catecholamines in patients with left heart failure. In: Engel/Lichtlen (eds) Nitrates IV. Springer, Berlin Heidelberg New York, p 244
23. Koch G (1979) Combined beta-receptor and calcium blockade in ischemic heart disease: effect on hemodynamics and plasma catecholamines (Abstr). 4th International Adalat-Symposium, Paris, 1979
24. De Leeuw PW, Wester A, Willemse PJ, Birkenhäger WH (1980) Effects of prazosin on plasma noradrenaline and plasma renin concentrations in hypertensive subjects. J Cardiovasc Pharmacol 2:361
25. Lichtlen P (1975) Die Wirkung von Nitriten und Nitraten auf die linksventrikuläre und koronare Dynamik in Ruhe und während dynamischer Belastung. In: Rudolph W et al. (ed) Nitrate-Wirkung auf Herz und Kreislauf. Urban & Schwarzenberg, Baltimore Munich, p 80
26. Manthey J (1981) Effect of vasodilator therapy on sympathetic activity in patients with impaired left ventricular function (Abstr). International Symposium "Catecholamines and the Heart", München, 1981
27. Maskowski AH, Nicholls MG, Ikram H, Espiner EA (1981) Haemodynamic hormonal and electrolyte responses to captopril in resistant heart failure. Lancet 71
28. Mason DT (1978) Ventricular afterload reduction in management of congestive heart failure: a rational new concept that has rapidly come of age by vasodilator drugs. Clin Cardiol 1:55
29. Miller RR, Awan NA, Maxwell KS, Mason DT (1977) Sustained reduction of cardiac impedance and preload in congestive heart failure with the antihypertensive vasodilator prazosin. N Engl J Med 297:303
30. Muiesan E, Agabiti-Rosei E, Alicandri C, Beschi M, Castellano M, Pasini C, Platto L (1980) Influence of verapamil on catecholamines, renin and aldosterone in hypertensive patients

(Abstr). International Symposium Calcium Antagonism in Cardiovascular Therapy, Florenz, 1980
31. Muiesan E (1980) Diskussionsbemerkung zu 24. J Cardiovasc Pharmacol [Suppl 3] 2:368
32. Packer M, Meller J (1978) Oral vasodilator therapy for chronic heart failure: a plea for caution. Am J Cardiol 42:686
33. Da Prada M, Zurcher G (1976) Simultaneous radioenzymatic determination of plasma and tissue adrenaline, noradrenaline and dopamine within the femtomole range. Life Sci 19:1161
34. Thomas JA, Marks BH (1978) Plasma norepinephrine in congestive heart failure. Am J Cardiol 41:233
35. Williams DD, Bommer WJ, Miller RR, Amsterdam EA, Mason DT (1977) Hemodynamic assessment of oral peripheral vasodilator therapy in chronic congestive heart failure: prolonged effectiveness of isosorbide dinitrate. Am J Cardiol 39:84

Discussion

Schröder, Berlin: Plasma catecholamine concentrations are frequently elevated in congestive heart failure. With vasodilator treatment a fall in plasma catecholamines can be observed according to your studies. How is this reflex mediated?

Klein, Graz: I cannot answer that question; I assume, however, that improved perfusion of the cerebral tissue accounts for this effect.

Just, Freiburg: Our current knowledge does allow an interpretation of this effect: improved stroke volume and improved organ perfusion with increased pulse pressure activates the carotid sinus reflex and thereby reduces sympathomimetic activity. The baroreflex mechanism would suffice as an explanation.

Kersting, Koblenz: Tolerance to vasodilator treatment with prazosine is according to some authors, often related to an increase in peripheral renin activity. Here an activation of the renin-angiotensin-aldosterone system can be assumed. Activation of the sympathoadrenal system in congestive heart failure is the rule. You have, however, found relatively low values for the plasma noradrenaline concentrations in your heart failure patients.

Klein, Graz: We have indeed seen a number of patients with low values of plasma catecholamines. This, however, was only a minority within the group of patients studied. Six showed strongly elevated values. In these patients, a lowering of the catecholamine values was observed.

Bussmann, Frankfurt: In your studies you assume that in congestive heart failure under treatment with nitroglycerin and prazosin the coronary perfusion remains unchanged. Otherwise you would not be able to draw conclusions as to the liberation of catecholamines in the heart.

Klein, Graz: This is a very important point. We cannot, however, from our studies evaluate the behavior of the coronary flow. Therefore, we have no clear evidence as to possible liberation of catecholamines from the heart in this situation.

Bassenge, Freiburg: We have measured coronary flow in dogs with and without nitroglycerin. There a very brief increase of flow was observed and then the values returned towards the normal range. Prazosin shows a clear-cut increase in coronary flow which can also be found where there is an increment of coronary venous oxygen. The increase in coronary flow is quite different from animal to animal. In some it may increase by 100% or more, and here the coronary venous oxygen saturation may come up to 70%. In others a much smaller increment of coronary flow is observed without obvious reason.

Auditorium: With regard to increased coronary flow after prazosin, the site of action of the drug needs to be clarified.

Bassenge, Freiburg: Prazosin dilates the larger epicardial coronary arteries as well as the coronary arterioles.

Effects of Hydralazine and Isosorbide Dinitrate on Vasoconstrictor Mechanisms in Patients with Heart Failure*

J. Manthey, R. Dietz, E. Hackenthal, N.Y. Ke, H. Leinberger, A. Schömig, F. Schwarz, W. Kübler

Introduction

Compensatory stimulation of vasoconstrictor systems may contribute to the rise in systemic vascular resistance observed in patients with heart failure (1). Because the elevation of peripheral resistance may further depress left ventricular function, vasodilators have been used to counteract this vasoconstriction and to improve cardiac performance (2–4). This study examines the effects of isosorbide dinitrate and/or hydralazine on the activity of the sympathetic nervous system, on the renin – angiotensin system, and on vasopressin release in patients with congestive heart failure. In particular, we have assessed whether increased activity of these vasoconstrictor systems may limit the drug-mediated effects in these patients.

Patients and Methods

Thirty-one patients (27 men and four women) with congestive heart failure and with an average age of 47 years (range 21–58 years) were studied. The etiology of cardiac failure included congestive cardiomyopathy (28 patients) and coronary heart disease (three patients). Ejection fractions were measured by contrast angiography; the average was 27% with a range of 8%–46%. A triple-lumen thermistor-equipped Swan-Ganz catheter (Edwards Laboratories) was inserted percutaneously via an antecubital vein and advanced into the pulmonary artery. Hemodynamic parameters [heart rate, arterial blood pressure (cuff technique), pulmonary artery pressure, cardiac output (thermodilution)] and humoral parameters (plasma norepinephrine and plasma vasopressin concentrations, plasma renin activity) were measured at supine rest before and 60 min after vasodilator administration. After control measurements the patients were given either hydralazine (1 mg/kg orally) in combination with isosorbide dinitrate (40 mg orally) (15 patients, group I) or hydralazine (0.4 mg/kg intravenously) without nitrate (16 patients, group II).

---

* Supported by the Deutsche Forschungsgemeinschaft, SFB 90 (Kardiovaskuläres System) Heidelberg, Federal Republic of Germany

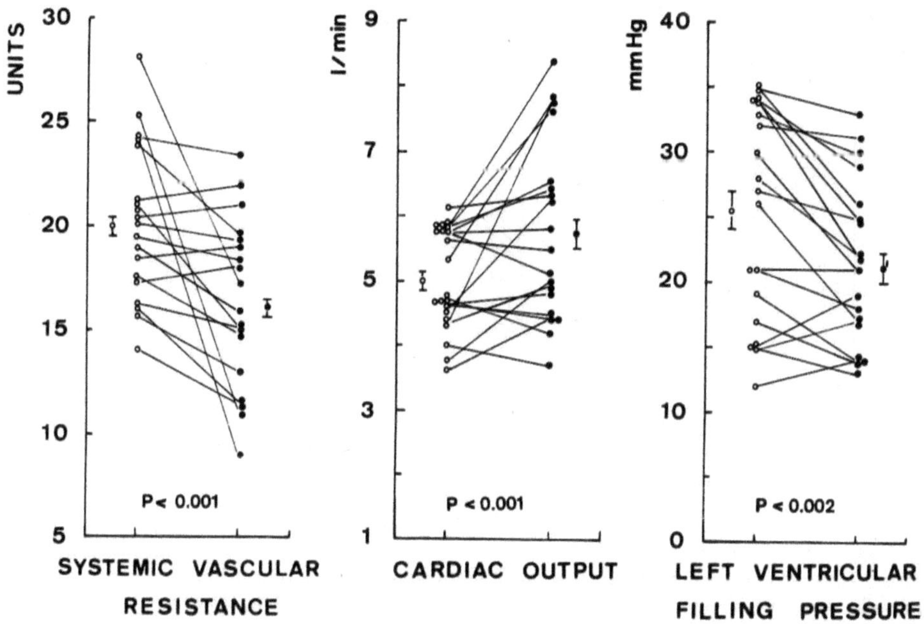

Fig. 1. Effects of combination of hydralazine and isosorbide dinitrate on hemodynamics in patients of group I

Results and Discussion

The effects of the hydralazine and nitrate combination on hemodynamics in patients of group I are shown in Fig. 1. With therapy, mean systemic vascular resistance decreased by 20%, mean cardiac index rose by 14%, and left ventricular filling pressure fell by 25% of pretreatment control values. Comparison of individual values, however, demonstrated that several patients did not show a marked hemodynamic improvement after the hydralazine-nitrate combination. Therefore, the patients were classified into responders and nonresponders according to whether the vasodilator regimen produced hemodynamic improvement or not. Responders were defined by a decrease in systemic vascular resistance of 10% or more. Nonresponders had a decrease in systemic vascular resistance of less than 10% of pretreatment values.

Fig. 2 shows the effects of the hydralazine and nitrate combination on the activity of the vasoconstrictor systems investigated in this study. Plasma vasopressin levels and plasma renin activity increased markedly under the treatment, but no significant differences between responders and nonresponders were observed. In contrast, plasma norepinephrine levels did not show a unidirectional response to the hydralazine-nitrate combination in all patients. In responders mean plasma norepinephrine concentrations remained unchanged after therapy. In nonresponders, however, norepinephrine levels increased significantly ($p < 0.01$) after drug administration. This increase in circulating norepinephrine in nonresponders may explain in part why vasodilator therapy in

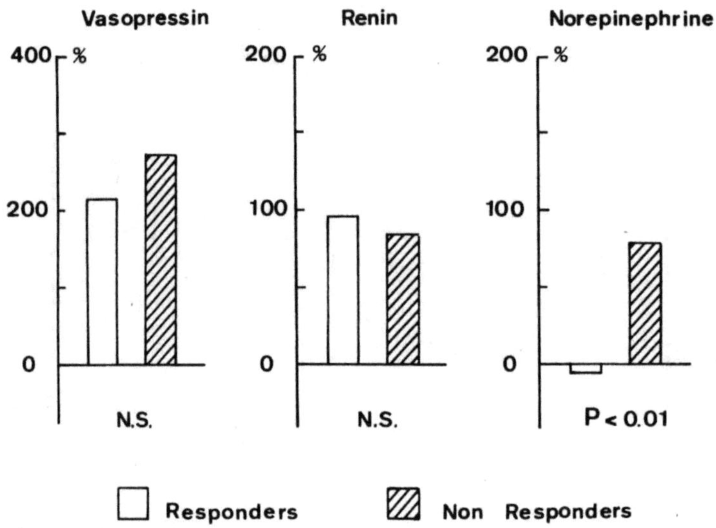

Fig. 2. Changes in plasma vasopressin concentration, plasma renin activity, and plasma norepinephrine concentration in patients of group I before and after combination of hydralazine and isosorbide dinitrate in responders and nonresponders. Pretreatment values = 100%

these patients was less effective than in the subgroup of patients who responded adequately.

To determine to what extent the minimal hemodynamic response to an oral regimen of vasodilators in some patients might be improved by administration of higher quantities of hydralazine, in a further series of investigations hydralazine was given intravenously in high dosage (patients of group II). After intravenous hydralazine administration, nonresponders were not observed. All patients showed a marked improvement in hemodynamics, with a 50% reduction in systemic vascular resistance and a rise in cardiac index to 160% that of controls. Left ventricular filling pressure fell by 25% of pretreatment values after therapy. Both vasopressin levels and plasma renin activity increased after intravenous hydralazine. In general, these results were quite comparable to those obtained after administration of the hydralazine-nitrate combination, except that the increase in plasma vasopressin levels was less pronounced after intravenous hydralazine than after the vasodilator combination ($p < 0.1$).

The effect of intravenous hydralazine administration on plasma norepinephrine depended on pretreatment hemodynamics. In patients with a stroke volume index >30 ml/m^2 and a diastolic pulmonary artery pressure <20 mmHg, an increase in plasma norepinephrine levels after hydralazine administration was observed. In contrast, in patients with a stroke volume index ≤ 30 ml/m^2 and a diastolic pulmonary artery pressure ≥ 20 mmHg there was a decrease in plasma norepinephrine levels. A significant correlation was noted between the change in plasma norepinephrine due to hydralazine administration and the corresponding change in systemic vascular resistance with an r value of 0.63 ($p < 0.01$).

Summary and Conclusions

In patients with congestive heart failure isosorbide dinitrate and/or hydralazine administration is accompanied by an increase in renin and vasopressin release. The stimulation of these vasoconstrictor mechanisms does not appear acutely to limit the drug-mediated vasodilating effects. In contrast, plasma norepinephrine levels do not show a unidirectional response to vasodilator therapy. In patients with severely depressed stroke volume, hydralazine administration produces a significant fall in plasma norepinephrine concentrations. This decrease in sympathetic activity after therapy further increases the vasodilator effect of hydralazine by additional reduction of catecholamine-induced elevation of systemic vascular resistance, and hence is of beneficial effect in these patients. In patients with mildly depressed stroke volume, hydralazine administration produces a significant rise in plasma norepinephrine levels. This drug-induced rise in sympathetic activity may limit the beneficial effects of hydralazine in these patients.

References

1. Cohn JN, Franciosa JA (1977) Vasodilator therapy of cardiac failure. N Engl J Med 297:27
2. Franciosa JA, Cohn JN (1979) Immediate effects of hydralazine isosorbide dinitrate combination on exercise capacity and exercise hemodynamics in patients with left ventricular failure. Circulation 59:1085
3. Manthey J, Dietz R, Leinberger H, Schmidt Gayk H, Schömig A, Schwarz F, Kübler W (1980) Vasodilator therapy in heart failure: limited by activation of vasoconstrictor mechanisms? (Abstr) Circulation [Suppl III] 62:259
4. Packer M, Meller J, Medina N, Gorlin R, Herman MV (1980) Importance of left ventricular chamber size in determining the response to hydralazine in severe chronic heart failure. N Engl J Med 303:250

Discussion

Bussmann, Frankfurt: From your studies you conclude that only patients with severe heart failure in class IV benefit from treatment with hydralazine.

Manthey, Heidelberg: It is indeed my impression that patients in NYHA classes III and IV benefit most from hydralazine treatment. Here stroke volume index is markedly reduced at rest and plasma catecholamine concentrations are elevated. We studied these patients at rest and during exercise. The elevation of cardiac output was maintained under both conditions. It remains to be seen, however, whether this is really to the benefit of the organism as a whole, or whether we are dealing with a luxury perfusion which does not constitute a real benefit to the patient.

Kreye, Heidelberg: You have expressed scepticism about the effectiveness of treatment with hydralazine alone. The vasodilatation induced by hydralazine will evoke a number of regulatory processes. It would seem conceivable to block some of these counterregulations by the administration of β-blockers.

Manthey, Heidelberg: This has been attempted with patients in heart failure due to congestive cardiomyopathy. Swedish authors have reported good results. We have attempted this mode of treatment in a few patients, but we have not had good experience. Therefore we have not followed this concept. There are reasons to assume that patients with high plasma noradrenaline concentrations will benefit most from hydralazine treatment alone. A possible additional treatment with β-blockers will probably only be useful in patients with mildly reduced stroke volume index and slightly elevated filling pressure. Here such a concept would seem conceivable, but we have not studied this aspect.

Auditorium: Does the stimulation of the renin-angiotensin-aldosterone system during long-term treatment with vasodilators produce significant side-effects?

Manthey, Heidelberg: With chronic treatment with hydralazine fluid retention is always observed, and this is probably mediated through the activation of the renin-angiotensin-aldosterone system. This secondary fluid retention tends to attenuate the effect of the vasodilator. However, a pronounced fall in peripheral vascular resistance and subsequent increase in stroke volume can be observed after each single dose of hydralazine even after an 8-week treatment.

Bödigheimer, München: More recent studies of Packer have shown that in patients with severe congestive heart failure and very high filling pressure the response to hydralazine may be insufficient.

Manthey, Heidelberg: We cannot substantiate this view. In our observations all patients have improve stroke volume independently of the level of filling pressure. It may be, however, that with very high right atrial filling pressures the reabsorption of hydralazine in the gastrointestinal tract may be insufficient, and here may lie the explanation of certain nonresponders to hydralazine therapy in congestive heart failure.

Klein, Graz: Packer and co-workers have shown that the end-diastolic cross-sectional diameter of the left ventricle may be an important determinant of the reaction of the heart to hydralazine. Could this be a factor separating responders form nonresponders?

Just, Freiburg: In general it can be said that the more severe the damage to the myocardium, the less response will be evoked by the vasodilating drug. The same, by the way, holds true for the effectiveness of inotropic drugs.

Manthey, Heidelberg: It is my impression that patients with very low ejection fractions show the least long-term response. We have not studied this aspect, however, so the final answer cannot be given.

Hertting, Freiburg: Heusler and co-workers have shown that hydralazine exerts its vasodilating effect mainly through activation of the prostaglandin system. In dogs this seems to be the major route of action, whereas in the human situation approximately 50% of the effect is mediated directly, and 50% through the prostacyclin mechanism. If this is so, than drug interference could easily account for nonresponse in your patients. Do you know whether the nonresponders had been taking headache tablets or other drugs interfering with prostacycline and prostaglandin metabolism?

Manthey, Heidelberg: Our patients took only digitalis and no other drugs during the study.

Drexler, Freiburg: The differentiation of responders and nonresponders to hydralazine therapy seems to be very difficult. If the influence of the drug on peripheral resistance is taken as the main mode of action, than only the behavior of this parameter should be used for differentiation.

II. Pharmacology

Cellular Mechanism of Action of Directly Acting Vasodilator Drugs*

Volker A.W. Kreye

Introduction

Directly acting vasodilator drugs relax vascular smooth muscle by interference with the cellular process of excitation-contraction coupling. Although pharmacological antagonists, mostly of the competitive type, such as α-adrenoceptor blockers, conform with this definition, they are usually not included in the group of vasodilators.

Any progress in the elucidation of the mode of action of vasodilators is strictly related to progress in our understanding of the excitation-contraction coupling process. The general outlines of this process are known, but a deeper knowledge of several details is lacking. Hence it is not surprising that the final elucidation of the mode of action has not been achieved so far for the vast majority of these substances.

Excitation-Contraction Coupling in Vascular Smooth Muscle (Fig. 1)

The excitation of the cellular membrane, e.g., by noradrenaline, results in an increased concentration of intracellular free calcium ions. This activation of calcium is probably in part the result of an increased membrane permeability for extracellular calcium ions, and may be linked either to a drug-induced depolarization of the cell membrane (electromechanical coupling) leading to an opening of "voltage-dependent calcium channels" (Bolton 1979), or to an opening of "drug-dependent calcium channels" without obvious changes of the electric membrane potential. This latter process, together with a drug-induced release of calcium ions from intracellular storage sites, e.g., from the sarcoplasmic reticulum, which adds to the calcium activation, constitutes the basis of "pharmacomechanical coupling." Once the intracellular calcium concentration is raised above a threshold level of about $2 \cdot 10^{-7}$ M, it interferes with the inhibitory action of certain regulatory proteins which, in the absence of calcium, prevent the contractile proteins actin and myosin from reacting with each other. In contrast to that of skeletal muscle, this regulatory process is not well understood in vascular smooth muscle at present. In the former type of muscle, calcium binds to

* Supported by the Deutsche Forschungsgemeinschaft, SFB 90 (Kardiovasculäres System) Heidelberg, Federal Republic of Germany

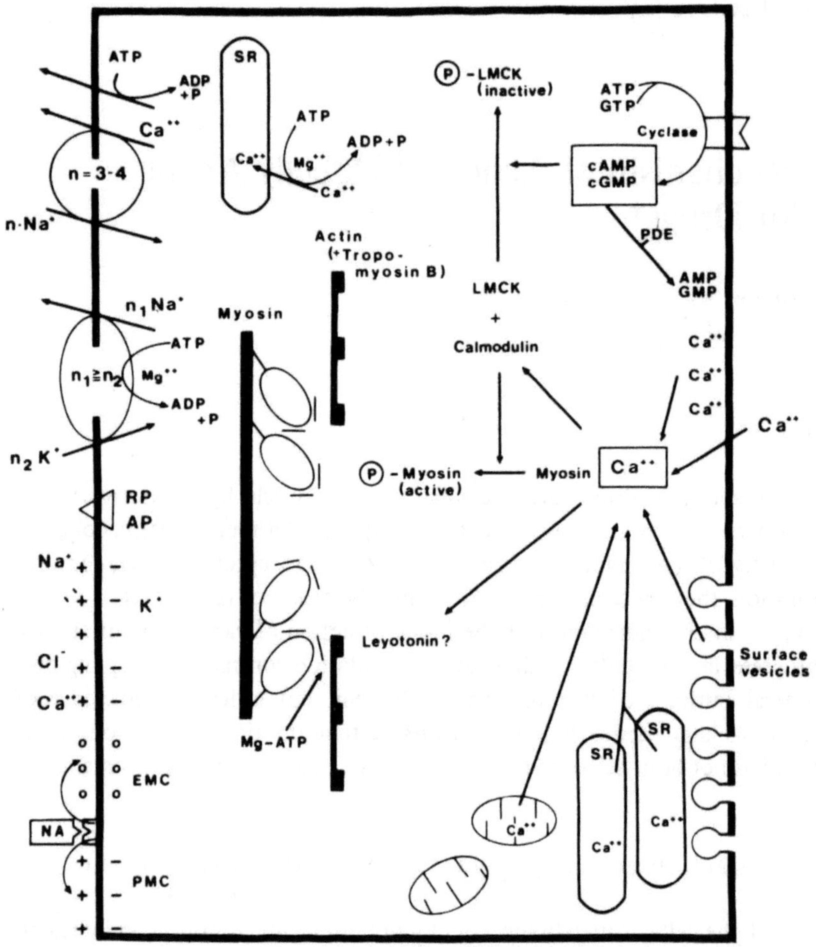

Fig. 1. Functional relationships in excitation-contraction coupling of vascular smooth muscle. *AP*, action potentials; *EMC*, electromechanical coupling; *LMCK*, light chain myosin kinase; *NA*, noradrenaline; *PDE*, phosphodiesterase; *PMC*, pharmacomechanical coupling; *RP*, resting membrane potential; *SR*, sarcoplasmic reticulum

troponin, which then influences another actin-bound protein, tropomyosin, in such a way that the heads of the myosin molecules can form "cross bridges" with actin resulting in the "sliding filament mechanism" that underlies both the isometric and isotonic mechanical responses of all types of muscle cells. From vascular smooth muscle, however, no troponin could be isolated, and it seems that two other intracellular proteins assume its regulatory role in the excitation-contraction coupling, i.e., probably the actin-linked "leyotonin" and the myosin-linked "light chains" (for review see Perry and Grand 1979). In addition to the role that the cellular calcium level plays as a mediator in excitation-contraction coupling in vascular smooth muscle, the degree of the cyclic-nucleotide-induced phosphorylation of these

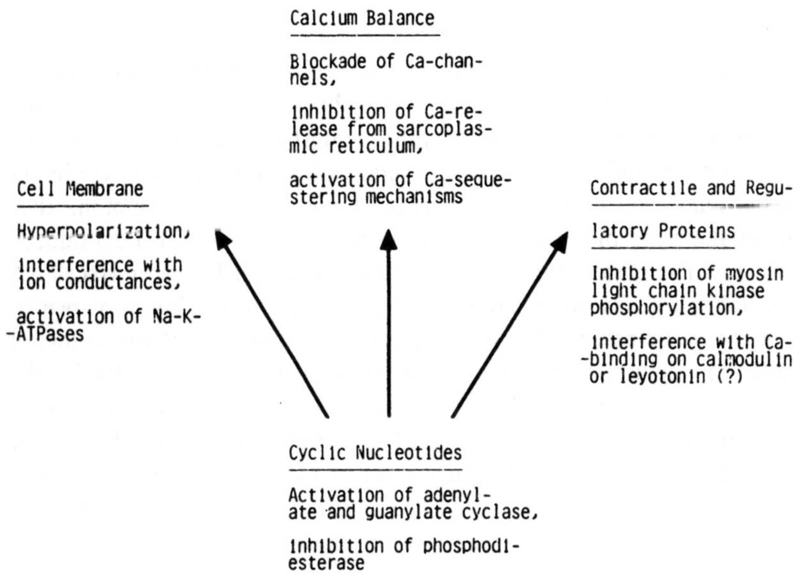

Fig. 2. Hypothetical major points of attack of vasodilator drugs in the vascular smooth muscle cell and presumed modes of action

regulatory proteins seems to be of similar, but as yet undefined, importance in the sequence of events leading to smooth muscular contraction.

General Principles of Vascular Smooth Muscle Relaxation (Fig. 2)

Principally, any interference at any site in the chain of events of the excitation-contraction coupling process should produce relaxation of the vascular smooth muscle cells, and thereby vasodilatation, but four major points of attack have evolved from the work with vasodilator drugs during the last decade.

The Cell Membrane

Any drug-induced hyperpolarization of the cell membrane should interfere with the opening of voltage-dependent calcium channels and thereby produce vasodilatation. If one looks at the cell membrane as a parallel connection of four electrical batteries, three of them constituting the equilibrium potentials of K^+, Na^+, and Cl^-, and one of them being the electrogenic component of the Na-K-ATPase, and if one takes into account the known intracellular and extracellular ion concentrations, membrane hyperpolarization may result from an increase in K^+ conductivity, from decreases in Cl^- and Na^+ conductivity, and from activation of the Na-K-ATPase (for details see Kreye 1980, 1981).

The Calcium Balance

A decrease in the concentration of the intracellular free calcium reduces the degree of coupling between excitation and contraction, and thereby leads to vasodilatation. In the past, this point of attack has been given the greatest attention by the scientists working in the field of vasodilator drugs, and several authors still adhere to the hypothesis that vascular relaxation is inevitably preceded by intracellular calcium inactivation. However, on the basis of recent insight it seems quite likely that drug-induced interference with vascular muscle tone may be possible at an unchanged intracellular calcium concentration, e.g., by modulating the degree of phosphorylation of regulatory proteins via the cyclic nucleotide system. The best example of a pharmacological interference with the calcium balance in vascular smooth muscle is provided by the "calcium antagonists," such as D 600 and verapamil, which block transmembrane calcium channels and thereby induce vasodilatation. Drugs which either directly or via the cyclic nucleotide system could induce enhanced calcium uptake into intracellular storage sites, e.g., by stimulating the calcium pump located in the membranes of the sarcoplasmic reticulum, would likewise produce vasodilatation.

Contractile and Regulatory Proteins

With the growing knowledge that the degree of calcium-dependent phosphorylation of myosin-linked light chains is one determinant of the contractile state of vascular smooth muscle, and that this process is inhibited in the presence of cAMP, it may be anticipated that those vasodilators which increase cellular cAMP levels (see below) have their ultimate site of action at this point of action (Rüegg et al. 1981). Whether vasodilator drugs can directly interfere with the regulatory proteins of vascular smooth muscle is an open question; it has been shown recently that vasodilator drugs of the Ca^{++} antagnostic type, including fendiline, prenylamine, verapamil, D 600, and diltiazem, can bind to the hydrophobic chains of calmodulin thereby inhibiting its interaction and Ca^{++}-dependent activation of myosin light chain kinase (Johnson et al. 1982).

Cyclic Nucleotide System

Several vasodilator drugs increase either cellular cAMP (e.g., β-adrenergic agonists, theophylline, papaverine), or cGMP (sodium nitroprusside, organic nitrates). For the former "second messenger," effects on the regulatory proteins as well as on the calcium-sequestering system in the vascular smooth muscle cell have been found by various authors that interfere with excitation-contraction coupling and produce relaxation. For cGMP, the intracellular actions are not known as yet, but mechanisms similar to those of cAMP may be anticipated.

Mode of Action of Two Selected Vasodilator Drugs

What follows is a short survey of the present knowledge of the mode of action of two vasodilator drugs, sodium nitroprusside and hydralazine. Predominantly those findings will be considered that have evolved from the work in our own laboratory during recent years.

Sodium Nitroprusside

In vascular smooth muscle, nitroprusside is active as a relaxant regardless of the agonist used (e.g., norepinephrine, angiotensin II, vasopressin, barium chloride, potassium chloride). Any inhibitory or stimulating effects of nitroprusside on α- or β-adrenergic receptors, respectively, have been excluded. Upon electrical stimulation, the efflux of norepinephrine from canine saphenous veins has been found to be unchanged in the presence of nitroprusside, suggesting that the drug does not interfere with transmitter release from sympathetic nerve endings. Nitroprusside relaxes isolated smooth muscle preparations of various origins, but considerable differences in sensitivity to the drug have been found. For instance, vascular and tracheal preparations, which belong to the group of tonic smooth muscle, are highly responsive. In contrast, phasic smooth muscle from the splanchnic region is less sensitive and may develop tachyphylaxis (duodenum and portal vein of the rat), or is even resistant to the action of nitroprusside (uterus and vas deferens of the rat). As regards the differential effects on tonic and phasic smooth muscle, nitroprusside resembles nitroglycerin and other nitrates, and there exist several other similarities between these drugs, including the speed of the relaxant response of isolated preparations, as well as their common effects on ion fluxes and on cyclic nucleotide levels in vascular smooth muscle (see below). Thus from a cellular point of view, and despite their well-known hemodynamic differences, it seems justified to put nitroprusside and the nitrates into the same group of vasodilator drugs.

As outlined above, the intracellular level of free calcium ions regulates the contractile tone of vascular smooth muscle. In organ bath studies on isolated arteries, an apparent competitive antagonism between nitroprusside and extracellular calcium ions has been found, suggesting an interference of the drug with calcium influx. However, controversial results were obtained from ^{45}Ca-flux studies with nitroprusside. In the absence of calcium ions and the presence of calcium-antagonistic compounds (e.g., lanthanum, verapamil), contractions elicited by norepinephrine or potassium chloride, which depend on the activation of intracellular calcium ions, were still depressed by nitroprusside (Kreye et al. 1975), and the speed of relaxation of rat aortic strips was doubled by 10^{-4} *M* nitroprusside, irrespective of whether calcium was present or absent in the extracellular medium. Thus nitroprusside relaxes vascular smooth muscle independently of extracellular calcium, and a possible interference with transmembrane calcium fluxes by the drug is not a necessary prerequisite for its vasodilator action. Neither has the assumption that nitroprusside may reduce vascular smooth muscle tone by stimulation of the intracellular calcium sequestration into the sarcoplasmic reticulum, nor the stimulation of a sarcolemmal outward pump for calcium ions (Kreye and Schlicker 1980) been substantiated on vascular smooth muscle microsomes. Nitroprusside causes a concentration-dependent hyperpolarization in the rabbit main pulmonary artery, and this finding might underlie its relaxant action on vascular smooth muscle. In a study on ion fluxes, nitroprusside proved to be a powerful inhibitor of the norepinephrine-induced increase in the ^{36}Cl-efflux from rabbit aorta, indicating a possible interference of the drug with the chloride permeability of vascular smooth muscle cells (Kreye et al. 1977). Since in this type of tissue the resting membrane potential is positioned between the potassium and the chloride equilibrium potentials, such an effect would explain the documented hyperpolarization by nitroprusside, but the recent finding of an increased membrane conductivity in the pulmonary artery in the presence of sodium nitroprusside (Ito et al. 1978) is not consistent with this hypothesis. In conclusion, nitroprusside probably acts as a vasodilator by affecting the cell membrane of the vascular smooth muscle cells, but the final mechanism of this action remains to be established.

Nitroprusside has no effect on cellular cAMP levels. In contrast, it increases cGMP levels in various smooth muscles including canine femoral artery, rabbit carotid and main pulmonary artery, and hog spleen artery. The increases in cGMP levels are large, e.g., 50-fold higher than controls in hog spleen artery with 10^{-4} *M* nitroprusside. 8-bromo-cGMP, a lipophilic derivative of cGMP, relaxes rat aortic strips (Schultz et al. 1979), suggesting a role of cGMP in nitroprusside-induced vasodilatation. On the other hand, several nonrelaxant drugs such as carbachol, acetylcholine, or histamine also increase cGMP levels in smooth muscle. Also, nitroprusside does not relax rat vas deferens, although it increases cGMP considerably in this tissue. Although this discrepancy remains to be solved, recent evidence gives strong support for the causal role of cGMP in nitroprusside-induced vasodilatation; it was shown that inactivators of the NO group released intracellularly from the nitroprusside molecule inhibited both the relaxant effect of nitroprusside and the nitroprusside-induced activation of the guanylate cyclase in coronary arterial muscle. A predominant inhibitor of cGMP-phosphodiesterase, M&B 22948, increased the sensitivity of coronary smooth muscle to the relaxant potency of nitroprusside in parallel with a more pronounced increase of the cellular cGMP level. In addition, the same compound augmented the relaxant response of the tissue to 8-bromo-cGMP (Kukovetz et al. 1979).

In conclusion, in vascular smooth muscle nitroprusside hyperpolarizes the cell membrane resistance. Also, it increases cellular cGMP levels, the response to it can be modified by compounds that interfere with the cellular cGMP metabolism, and its relaxant effects can be mimicked by exogenous 8-bromo-cGMP. So far, the link between these actions is missing, and the ultimate discovery of the mode of action of sodium nitroprusside remains to be made.

Hydralazine

It is a major drawback of the study of the vasodilator action of hydralazine that most of the arterial preparations commonly used in organ bath studies are little or inconsistently affected by the drug. In our own studies, this was found to be true for rat and rabbit aortic strips. The intact tail artery of the rat was sensitive to hydralazine in perfusion experiments (Worcel 1978), but required 10^{-4} *M* or even higher concentrations when we used it as helical strip preparations. In another study, $3 \cdot 10^{-4}$ *M* hydralazine was needed as a threshold dose in rabbit aortic strips contracted by norepinephrine, and with 10^{-3} *M* only a minor degree of relaxation was achieved. The in vitro effects of hydralazine at such doses are probably little or not at all related to its pharmacodynamic action in man, in whom the serum levels are in the range of 10^{-7} to 10^{-6} *M* at an average therapeutic dose of 50 mg of hydralazine given orally.

In contrast to other large arteries commonly used in organ bath studies, the rabbit renal artery proved to be highly sensitive to the vasodilator action of hydralazine. Helical strips contracted by 10^{-7} *M* norepinephrine started to relax at a threshold concentration of $3.9 \cdot 10^{-8}$ *M* hydralazine (IC_{10}). The IC_{60} was $1.4 \cdot 10^{-7}$ *M* and at $3 \cdot 10^{-6}$ *M* hydralazine the preparations had relaxed almost completely. The development of relaxation was slow, lasting up to 1 h. Dose-response curves to norepinephrine in the presence of hydralazine were shifted to the right in parallel fashion at an only slight reduction of the maximum response. Contractions of rabbit renal artery strips induced by 45 m*M* KCl or by 10^{-7} *M* norepinephrine superimposed on the KCl-induced tone were almost unresponsive to hydralazine. Similarly, prolonged incubation of the arterial strips with 10^{-5} *M* ouabain abolished the response to hydralazine, probably as a result of a ouabain-induced depolarization. Neither indomethacin, an inhibitor of tissue prostaglandin synthesis, nor sulpiride, an antagonist to dopamine, interfered significantly with the vasodilator action of hydralazine. This suggests that the relaxant effect of hydralazine is not mediated by locally formed prostaglandins, and that it is not dopaminergic in nature. The present findings indicate that hydralazine is a potent direct vasodilator with a predominant action on pharmacomechanical coupling but little effect on electromechanical coupling.

Recently, indirect mechanisms of action for hydralazine have also been discussed, such as the induction of a release of the vasodilators prostaglandin E_2 and I_2 (Haeusler and Gerold 1979), or an interference with sympathetic transmitter release at the level of the vascular smooth muscle cell (Chevillard et al. 1980). Thus it seems conceivable that the in vivo effects of hydralazine are not exclusively the result of its direct vasodilator effect.

References

Bolton TB (1979) Mechanisms of action of transmitters and other substances on smooth muscle. Physiol Rev 59:606–718

Chevillard C, Mathieu MN, Saiag B, Worcel M (1980) Hydralazine: effect on the outflow of noradrenaline and mechanical responses evoked by sympathetic nerve stimulation of the rat tail artery. Br J Pharmacol 69:415–420

Haeusler G, Gerold M (1979) Increased levels of prostaglandin-like material in the canine blood during arterial hypotension produced by hydralazine, dihydralazine, and minoxidil. Naunyn-Schmiedebergs Arch Pharmacol 310:155–167

Ito Y, Suzuki J, Kuriyama H (1978) Effects of sodium nitroprusside on smooth muscle cells of rabbit pulmonary artery and portal vein. J Pharmacol Exp Ther 207:1022–1031

Johnson JD, Vaghy PL, Crouch TH, Potter JD, Schwartz A (1982) A hypothesis for the mechanism of action of some of the Ca^{++} antagonist drugs: calmodulin as a receptor. Advances in Pharmacology and Therapeutics II. Vol 3. In: Yoshida H, Hagihara Y, Ebashi S (eds) Cardiorenal and cell pharmacology, Pergamon Press, Oxford New York Toronto Sydney Paris Frankfurt, pp 121–138

Kreye VAW (1980) Sodium nitroprusside: approaches towards the elucidation of its mode of action. Trends Pharmacol Sci 1:384–388

Kreye VAW (1981) Role of the membrane potential in the function of vascular smooth muscle. In: Vanhoutte PM, Leusen I (eds) Vasodilatation. Raven Press, New York, pp 299–305

Kreye VAW, Baron GD, Lüth JB, Schmidt-Gayk H (1975) Mode of action of sodium nitroprusside on vascular smooth muscle. Naunyn-Schmiedebergs Arch Pharmakol 288:381–402

Kreye VAW, Kern R, Schleich I (1977) ^{36}Chloride efflux from noradrenaline-stimulated rabbit aorta inhibited by sodium nitroprusside and nitroglycerine. In: Casteels R, Godfriend T, Rüegg JC (eds) Excitation-contraction coupling in smooth muscle. Elsevier, North-Holland Biomedical Press, Amsterdam New York, pp 145–150

Kreye VAW, Schlicker E (1980) Effects of vasodilator drugs alkaline phosphatase, and cyclic AMP-dependent protein-kinase on the ^{45}calcium uptake of sarcolemmal microsomes from human umbilical arteries. Br J Pharmacol 70:537–544

Kukovetz WR, Holzmann S, Wurm A, Pöck G (1979) Evidence for cyclic GMP-mediated relaxant effects of nitro-compounds in coronary smooth muscle. Naunyn-Schmiedebergs Arch Pharmacol 310:129–138

McLean (1978) J Pharmacol Exp Ther 205:418–425

Perry SV, Grand RJA (1979) Mechanisms of contraction and the specialized protein components of smooth muscle. Br Med Bull 35:219–226

Rüegg JC, Sparrow MP, Mrwa U (1981) Cyclic-AMP mediated relaxation of chemically skinned fibres of smooth muscle. Pflügers Arch 390:198–201

Schultz K-D, Böhme E, Kreye VAW, Schultz GD (1979) Relaxation of hormonally stimulated smooth muscular tissues by the 8-bromo derivative of cyclic GMP. Naunyn-Schmiedebergs Arch Pharmacol 306:1–9

Worcel M (1978) Relationship between the direct inhibitory effects of hydralazine and propyldazine on arterial smooth muscle contractility and sympathetic innervation. J Pharmacol Exp Ther 207:320–330

Discussion

Auditorium: In rat experiments different concentrations of hydralazine exert widely varying effects on the cell membrane. It may be that investigators working with higher concentrations will obtain quite different results from those operating with low concentrations of hydralazine.

Kreye, Heidelberg: Three years ago, McLean and co-workers (1978) studied hydralazine in the rabbit aorta. They used very high concentrations of the drug, which had a predominant effect on potassium-induced contractions, contrasting with our own findings. Furthermore, they studied the effect of hydralazine at varied concentrations of extracellular calcium and concluded that hydralazine may have calcium-antagonistic properties. However, it must be made clear that experiments with doses as high as those employed in that study, do not correlate well with the clinical situation, but rather represent the toxic range, which is governed by quite different principles.

Auditorium: Hydralazine may stimulate the prostaglandin system. Is this stimulation a nonspecific effect, which can also be demonstrated with other vasodilators, for example, with nitroglycerin?

Kreye, Heidelberg: We have studied the effect of indomethacin on the hydralazine-induced relaxation of rabbit renal artery preparations. Since we found no influence of indomethacin it may be assumed that at the level of the arterial wall, prostaglandins are not the mediators of the hydralazine action. Personally, I would suspect that any vasodilator drug which has such a potent effect on the renal vascular bed as has been documented for hydralazine liberates prostaglandin-like material from the renal tissue.

Hertting, Freiburg: Chemically speaking, the three vasodilators nitroprusside, nitroglycerin, and hydralazine are entirely different. They do not have any common feature. On the other hand, they all have active nitrogen groups. You have mentioned that active nitrogen compounds or metabolites may appear after hydralazine administration. Do you have evidence that certain nitrogen compounds could represent a common active principle of the three vasodilators mentioned?

Secondly, I should like to mention that hydralazine accumulates in the vascular wall. This may explain that a discrepancy is found between the duration of action and the persistence of the plasma concentration of hydralazine.

Kreye, Heidelberg: A basic molecular configuration common to all nitrogen-containing vasodilator compounds has not yet been defined. It is of interest, however, that at least for the nitrates including sodium nitroprusside, the intracellular formation of S-nitrosothiols has been suggested recently which may constitute the active vasodilator principle.

The selective accumulation of hydralazine in the vascular wall has been known for long, but this observation has not led to a better understanding of the vasodilator action of the drug.

Greef, Düsseldorf: Hydralazine is a potent antihypertensive agent. Lowering of blood pressure will not be accomplished by relaxation of the aortic wall alone. Has it been discovered whether, and if so how, hydralazine acts upon the peripheral vascular bed, including the capillaries?

Kreye, Heidelberg: The larger arteries, especially the aorta, are particularly resistent to the relaxant action of hydralazine. This has been one of the major obstacles to the elucidation of the mechanism of action of hydralazine. The effectiveness of the drug is particularly pronounced at the arteriolar level, which has been studied in different pharmacological models.

Schröder, Berlin: Nitroprusside and nitroglycerin clearly show different points of action insofar as the effectiveness of nitroprusside is more pronounced on the arteriolar side and that of nitroglycerin action on the venous side. Are there any pharmacological observations pertinent to this apparent preferential action?

Kreye, Heidelberg: The predominant action of the nitrates on the venous system and of nitroprusside on the arteriolar system remains unexplained and surprising to the pharmacologist. This problem has not been solved as yet, but an experimental study devoted to this particular question is in progress in our laboratory at present.

Pharmacokinetics of Nitrates

Thomas Meinertz

This paper does not aim to give a complete overview of the pharmacokinetics of isosorbide dinitrate (ISDN) or other organic nitrates. I would rather concentrate on discussing some preliminary results that we obtained with ISDN in patients with chronic heart failure.

The pharmacokinetics of ISDN have been previously studied in healthy subjects and in patients with coronary artery disease. Recently, ISDN was introduced in the therapy of chronic heart failure. Since it is well known that drug pharmacokinetics can be markedly influenced by the state of disease, we studied the pharmacokinetics and pharmacodynamics of ISDN in patients with coronary artery disease with and without chronic heart failure.

The data that I will present were obtained in collaboration with Dr. H.-L. Fung, Department of Pharmaceutics, School of Pharmacy, State University of New York at Buffalo, Amherst; Dr. U Wiegand and Dr. E. Jähnchen from the Department of Pharmacology, University of Mainz; and our own group at the Second Department of Internal Medicine in Mainz. Plasma concentrations of ISDN and of the mononitrate metabolites (5-ISMN and 2-ISMN) were determined by a specific and sensitive gas-liquid chromatographic method.

We studied the disposition of ISDN and of the mononitrate metabolites in patients with coronary artery disease with and without chronic heart failure after intravenous and topical administration of the drug. After intravenous administration of ISDN (2 mg over 15 min) to patients without chronic heart failure, peak plasma concentrations of the parent drug declined with a distribution half-life of 5–16 min. The total clearance was calculated to be in the range of 3.4–8.5 l/min. Interestingly, and somewhat unexpectedly high plasma concentrations of 5-ISMN were found during and after the infusion of ISDN. While the peak plasma concentration of 5-ISMN was similar to or even higher than that of the parent compound, only traces of the 2-ISMN metabolite were found in the plasma under these conditions.

We have then looked to the fate of ISDN in patients with chronic heart failure. There were no pronounced changes in the plasma concentrations, the distribution half-life, or the systemic clearance of ISDN between patients with and without chronic heart failure. Furthermore, the disposition of the ISMNs was not different in the two categories of patients. It is, however, not clear at the moment whether minor differences in the pharmacokinetics of ISDN exist between the two groups.

We also studied the pharmacokinetics of ISDN and the mononitrate metabolites in the two groups of patients following cutaneous application of the drug. Isosorbide dinitrate oinment (100 mg) was applied to 40 cm^2 of the skin on the upper left quadrant or the chest. Maximum plasma concentrations of ISDN ranged from 1 to 10 ng/ml and were achieved 5–12 h after application of the ointment. Thereafter, plasma concentrations declined slowly to 0.4–4 μg/ml at 30 h. Plasma concentrations of 5-ISMN were higher and those of 2-ISMN were lower than those of the parent drug. Marked differences in the systemic availability of ISDN were found between patients. In patients with chronic heart failure similar results were obtained. There were no pronounced differences in the plasma concentrations of ISDN and of the mononitrate metabolites between the two groups of patients.

One should, however, be aware of the fact that the clinical effects of ISDN are determined not only by the pharmacokinetics but also by its pharmacodynamic properties. The response of the vascular system may be basically different in the two groups of patients. Our preliminary data indicate that differences in the response to ISDN between patients with and without chronic heart failure cannot be explained by differences in the pharmacokinetics of ISDN or its mononitrate metabolites.

Isosorbide Dinitrate: Pharmacokinetics After Intravenous Administration*

R.A. Morrison, D. Höhmann, T. Meinertz, E. Jähnchen, Ho-Leung Fung

Isosorbide dinitrate (ISDN) is an important organic nitrate found therapeutically useful in its sublingual and oral forms in various cardiovascular diseases such as angina pectoris (1) and congestive heart failure (2). Recently Distante et al. (3) showed that an intravenous infusion of this drug, at 0.021–0.083 mg/min is also effective in managing unstable angina. The availability of an intravenous dosage form of ISDN not only affords the opportunity to characterize the pharmacokinetics of this drug after this particular mode of therapy, but also gives the possibility of assessing the bioavailability of this drug after other (e.g., oral) routes of administration in patients. This latter point has been a subject of major controversy ever since Needleman et al. (4) made the assertion that oral nitrate therapy is irrational because of its complete first-pass metabolism.

A preliminary study has appeared which provided some initial information on this important issue. Taylor et al. (5) showed that in two normal young subjects who received intravenous ISDN the plasma clearance was found to be 0.32 and 0.16 l/min. Based on these values, these authors concluded that sublingual and oral administration of ISDN are only bioavailable to the extent of 6% and 3% respectively. If these data can be confirmed in a larger group, they could represent an unusual example of very poor sublingual bioavailability of a small neutral compound given at a relatively low dose. Further, if shown applicable to all patients, these data could also have wide-ranging implications regarding the clinical use and the regulatory policy of oral and sublingual dosage forms of ISDN. It follows that there may be a necessity (and perhaps urgency) for further research to identify the reasons for, and ways of circumventing, the poor sublingual bioavailability of this drug.

We have expanded on the study by Taylor et al. (5) through an examination of the pharmacokinetics of ISDN in 11 cardiac patients after intravenous infusion at two different rates. These data were obtained as part of a comprehensive study which examined the pharmacokinetics of ISDN after intravenous, sublingual, oral, and percutaneous administration. Because the intravenous data are crucial in the strategic planning of further pharmaceutical research on this drug, we present here preliminary results on this aspect for the interest of workers in this area.

---

* Supported in part by NIH grants HL 22273 and GM 20852 and by Pharma-Schwarz GmbH, Federal Republic of German

Eleven angina patients were infused intravenously with an ISDN solution (Isoket Ampullen, Pharma-Schwarz GmbH, 4019 Monheim, West Germany) diluted such that a total dose of 2 mg in 10 ml was delivered over 15 min (infusion rate 0.133 mg/min). Two of these patients received ISDN again on another day at a slower infusion rate (0.083 mg/min), one for 1 h and the other for 2 h. This latter infusion rate was identical to that used by Taylor et al. (5). Via an HPLC analysis developed in our laboratory (H.-L. Fung and S.L. Chan, unpublished work), the strength of the dosing solution was determined to be a mean of 95% of the theoretical value. The concentration of ISDN in the infusion solution emerging from the polyethylene infusion tubing was essentially identical before and after the experiment, indicating that adsorptive loss to the infusion system was absent. Serial plasma samples were drawn, anticoagulated with sodium citrate, and frozen (–20 °C) until assayed for ISDN by gas chromatography-electron capture detection (6).

The terminal disappearance phase rate constant (β) was obtained by least-square regression of the last four concentration time points. The area under the plasma concentration vs time curve from time zero to the last plasma sample collected (C_t) was determined by Lagrange integration (7). Total area under the curve (AUC) was estimated by addition of the residual area (ca. 15%) which was calculated by dividing C_t by β. The systemic clearance (Cl_{iv}) was calculated by dividing the administered dose by AUC. The apparent volume of distribution (Vd_{β}) was then determined by dividing the systemic clearance by β.

Fig. 1 shows the plasma concentration of ISDN after intravenous infusion in the patient population studied. Results from the short infusion (2 mg over 15 min) showed that steady-state concentrations were not achieved after this infusion regimen and that decline of plasma drug was apparently biexponential. Unfortunately, the initial

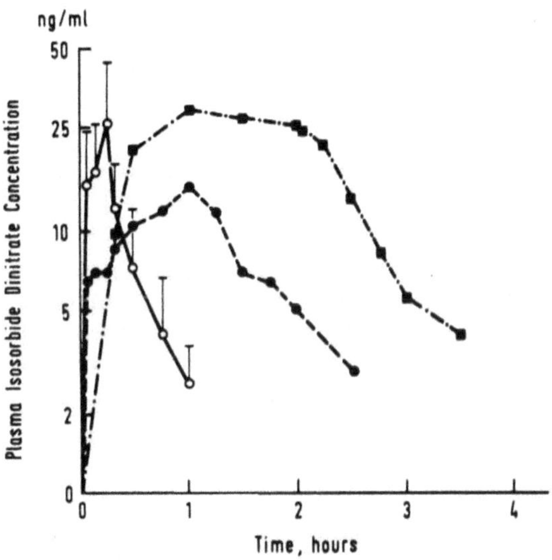

Fig. 1. Plasma ISDN concentrations observed after different rates and duration of intravenous infusion (o——o) 0.133 mg/min for 15 min, mean ± SD, n = 11; •---•, 0.083 mg/min for 1 h; ▪–·–▪, 0.083 mg/min for 2 h)

rapid distributive phase observed after cessation of the infusion could not be quantitated due to insufficient sampling. The mean half-life of the terminal disappearance phase, however, was found to be 18 ± 7 min (mean ± SD, n = 11), with a range of 10–30 min. Cl_{iv} was estimated at 3.4 ± 1.4 l/min (n = 11) with a range of 1.8–6.3 l/min. Mean Vd_{β} was estimated to be 101 ± 67 l (range 46–241 l). Infusion rate and duration did not appear to affect the pharmacokinetics of ISDN: in one patient, Cl_{iv} was estimated at 4.0 and 4.7 l/min at infusion rates of 0.083 mg/min for 1 h and 0.133 mg/min for 15 min respectively; in a second patient, Cl_{iv} was 2.9 and 2.7 l/min at infusion rates of 0.083 mg/min for 2 h and 0.133 mg/min for 15 min respectively.

The clearance values observed in our patient population were, in general, at least tenfold larger than those obtained by Taylor et al. (5) in two normal subjects. The reasons for this discordance are not known at present. However, it appears necessary to reassess the validity of the low bioavailability estimates of sublingually and orally administered ISDN thus far reported, particularly regarding its applicability in cardiac patients. A study addressing this issue is approaching completion and will be reported in detail in the near future.

References

1. Sweatman T, Strauss G, Selzer A, Cohn KE (1972) Am J Cardiol 29:475
2. Franciosa JA, Mikulic E, Cohn JN, Jose E, Fabie A (1974) Circulation 50:1020
3. Distante A, Maseri A, Severi S, Biagini A, Chierchia S (1979) Am J Cardiol 44:533
4. Needleman P, Lang S, Johnson EM (1972) J Pharmacol Exp Ther 181:489
5. Taylor T, Chasseaud LF, Doyle E (1980) Biopharm Drug Dispos 1:149
6. Morrison RA, Fung H-L (1980) Abstract of papers presented before the APhA Academy of Pharmaceutical Sciences. 10:126
7. Yeh KC, Kwan KC (1978) J Pharmacokinet Biopharm 6:79
8. Gibaldi M, Perrier D (1975) Pharmacokinetics. Marcel Dekker, New York

Discussion (zu den Seiten 82–87)

Just, Freiburg: With regard to the pharmacological activity of the metabolites of ISDN: Goodman an Gilman report that the mononitrates are less potent than ISDN proper by a factor of 500. Your studies suggest a much higher biological activity of the ISDN metabolites.

Kreye, Heidelberg: I have studied the relaxing action of ISDN, 2-ISMN, and 5-ISMN in isolated rat aortic strips. It is beyond doubt true that ISDN is by far the most active substance. This, however, is of little use, since in vivo very little of the intact molecule is left. The second most active substance is 2-ISMN. The least activity is shown by 5-ISMN, this substance is now being introduced.

Meinertz, Mainz: The notion proposed in Goodman and Gilman cannot be supported by recent evidence. The differences between the classic pharmacological evaluation of ISDN and its metabolites certainly have to be reviewed in the light of observations in the intact human organism. Here a much higher activity of the metabolites is to be found.

Bussmann, Frankfurt: 5-ISMN is a highly active substance. It shows identical actions to those of ISDN. It is biologically active in comparable doses, that is, 20 mg ISDN have a similar vascular effect in the clinical situation to 20 mg 5-ISMN. In comparison, 5-ISMN shows a slightly delayed onset of action. The duration of action is approximately comparable to that of ISDN that is 4–6 h. The pharmacodynamic effects correlate well with the plasma levels.

Meinertz, Mainz: A major disadvantage of 5-ISMN is to be found in the fact that this drug is not absorbed sublingually, which is an important feature in antianginal drugs.

Scholz, Hannover: It is true that ISDN is not much more active in vivo than its metabolite 5-ISMN, the reason being that in vivo ISDN is metabolized quite rapidly and the breakdown product 5-ISMN therefore exerts its full action. In vitro the activity of ISDN is approximately a hundred-fold that of ISMN. The difference between the in vitro and the in vivo behavior is not at all surprising and not unexplained.

Auditorium: How is the metabolism of the nitrates altered, if the liver is in a congested state? To what extent does a cirrhosis of the liver influence metabolism and activity of the nitrates?

Meinertz, Mainz: It is surprising to see that the pharmacokinetic behavior of the nitrates, especially ISDN, is not altered under the impaired hepatic perfusion in congestive heart failure. However, our knowledge in this field is only limited and a final statement cannot be given as yet.

Kersting, Koblenz: I should like to point out that the extraction rate is not entirely independent of the hepatic perfusion, which may be impaired in congestive heart failure. It may well be that the total clearance for ISDN is reduced in patients with severe failure. Are there other observations on the relative concentrations of the metabolites of ISDN in the systemic venous blood as opposed to the hepatic venous blood?

Meinertz, Mainz: This very interesting question cannot be answered as yet. As far as the clinical situation is concerned, we can confidently state today that the bioavailability of ISDN is largely independent of the state of congestion of the liver.

Prazosin and α-Blocking Agents

F. Kersting

Interest in applying vasodilator treatment to congestive heart failure by use of α-adrenergic receptor blocking agents has been stimulated mainly for two reasons:

1. Sympathetic nervous activity is increased in congestive heart failure. Therefore, interruption of vasoconstriction by α-adrenergic blocking agents represents a therapeutic rationale.
2. Development of novel α-adrenergic receptor blocking drugs has reduced the side-effects of this class of agents to an acceptable degree.

Of the various compounds being introduced as α-adrenergic receptor blocking drugs in vasodilator therapy, prazosin is considered an important advancement. Some of its pharmacological properties and side-effects will be discussed in the following. In addition, similar aspects of phentolamine and phenoxybenzamine will briefly be presented.

In an attempt to discover a directly acting vasodilator agent for antihypertensive treatment Hess (5) and Constantine (4) evaluated prazosin. The mechanism of a direct action on vascular smooth muscle could be derived from its chemical structure, in which it is possible to recognize the dimethoxybenzo-moiety of papaverine (Fig. 1), a classical myorelaxant, and the aminopyrimidine portion of cyclic GMP (cGMP) and cyclic AMP (cAMP). However, during recent years it has been realized that prazosin possesses a high selectivity for postsynaptic α-1-adrenergic receptors (2). And in fact, because of its high degree of selectivity and high affinitiy for α-1-adrenergic receptors, tritiated prazosin has become a valuable ligand for the identification and study of α-1-adrenergic receptors.

Evidence of prazosin's selectivity for postjunctional α-1-receptors is mainly based on the concept of noradrenaline release regulated through a negative feedback mechanism, i.e., noradrenaline released from the postganglionic nerve terminal can inhibit its own further release by activation of prejunctional α-2-receptors. This concept, basically worked out by Rand et al. (11), Langer et al. (9) and Starke (12), prompted

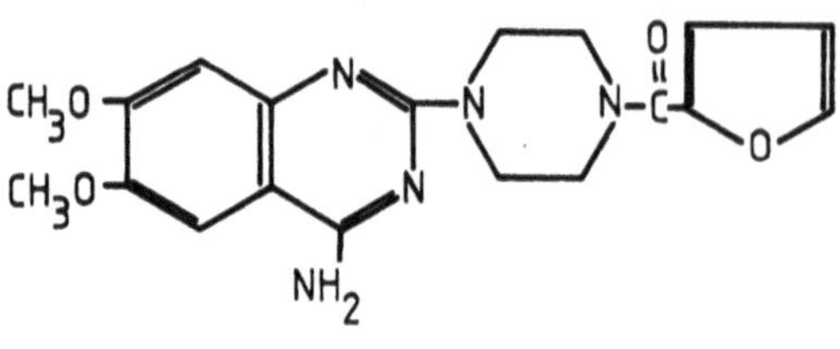

Fig. 1. Structure of prazosin

Cambridge et al. (2) to compare the affinity of prazosin for pre- and postjunctional α-receptors, especially as prazosin did not cause the tachycardia and the increase in plasma renin activity that would result from blockade of cardiac and juxtaglomerular prejunctional α-2-receptors.

The experimental preparation used initially was the transmurally stimulated superfused rabbit pulmonary artery preparation prelabeled with tritiated noradrenaline. To study the prejunctional effects of drugs it was necessary to label the neuronal noradrenaline stores with tritiated noradrenaline by prior incubation, and measure the stimulation-induced overflow of tracer noradrenaline. Postjunctional effects were assessed by measuring the tension changes in the tissue caused by the noradrenaline liberated in response to nerve stimulation. In an experiment Cambridge et al. (2) demonstrated that increasing concentrations of phenoxybenzamine – an α-receptor blocker with partial affinity to prejunctional α-receptors – caused the expected dose-related reduction of the stimulus-evoked tension, while at the same time there was an increase in the overflow of tritiated noradrenaline during stimulation. In contrast, the same experiment performed with prazosin showed no increase of the output of tritiated noradrenaline during stimulation.

In addition to these antagonist data, which confirmed that prazosin lacks affinity for prejunctional α-2-adrenergic receptors, there are agonist data available that are consistent with the idea of prazosin's preferential postjunctional affinity. Phenylephrine is an α-receptor agonist that can be used to demonstrate preferential postjunctional action. The competitive action of prazosin as antagonist and phenylephrine as agonist at postjunctional α-receptor sites can even be shown in a clinical study, an example of which was given by Pitterman et al. (10), who infused phenylephrine at increasing rates. Comparing the pressor responses before and 3 h after 5 mg prazosin, a nearly parallel shift of the dose-response curve indicates a competitive antagonism of prazosin, i.e., the effect of the antagonist prazosin can be overcome by the agonist phenylephrine. Pharmacokinetically, prazosin is characterized by an elimination half-life of about 2.9 h, a plasma protein binding of 94%, and a significant first pass metabolism. It is readily absorbed from the gastrointestinal tract, with a peak level occurring 1–3 h after administration. This may be the reason why, e.g., Bateman et al. (1) were able to find a reasonable correlation between the plasma drug level and the hypotensive effect after infusion of prazosin. However, after the same dose of prazosin given orally, no correlation at all could be seen between plasma drug level and effect. This has recently been confirmed by Pitterman et al. (10). When they measured prazosin plasma concentration after a 5-mg oral dose of prazosin they found a continuous decline with a half-life of approximately 3 h. However, the α-adrenergic blocking effect, as assessed by the drug sensitivity ratio, remained essentially the same over a period of 7 hours. Thus there is no correlation between α-adrenergic receptor blockade and prazosin plasma concentration after oral treatment. This is mainly due to the extensive metabolism to active metabolites, e.g., 6-0-demethylprazosin, 7-0-demethylprazosin, and 2,4 diamino-6,7 dimethoxyquinazoline. In a study by Jaillon et al. (6) it was demonstrated that in patients with congestive heart failure the amount of prazosin absorbed from the gastrointestinal tract was less than that found in normal volunteers. Furthermore, the peak plasma level was lower. The elimination half-life was prolonged up to 6–8 h in these patients.

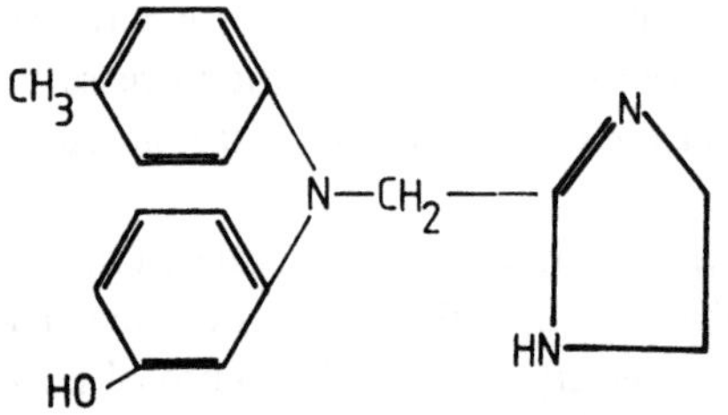

Fig. 2. Structure of phentolamine

The major side-effect of prazosin is the so-called first dose phenomenon, which comprises an acute reaction characterized by dizziness, palpitations, and sometimes loss of consciousness for periods of time ranging from a few minutes to over an hour. This syndrome is not as yet completely understood. The hemodynamic mechanism involved in this side-effect may be a massive pooling of blood in the venous system as a consequence of the abrupt α-blockade in critical vascular regions. It has been shown by Jauernig et al. (7) that prazosin has differential effects on the various vascular beds, the visceral being particularly sensitive to the sympatholytic action of prazosin.

Generally, the side-effects most frequently experienced with prazosin seem mild and tolerable. In a series of 934 patients, the figures for reported side-effects were dizziness in 10.3%, headache in 7.8% followed by drowsiness in 7.6%, and lack of energy in 6.9% (3).

The most common feature of prazosin (and of phentolamine, which is structurally an imidazoline derivative) is in a wider sense their competitive α-receptor antagonism (Fig. 2). Thus the pressor response curve of noradrenaline can characteristically be shifted rightward in a parallel manner. However, it appears that at least part of its vasodilator activity in man is due to direct action on blood vessels rather than to specific blockade of α-adrenergic receptors. This and the fact that it has partial affinity to pre- and postjunctional α-receptors makes phentolamine in a strict pharmacological sense a less distinct α-receptor blocking agent than prazosin. Phentolamine is poorly absorbed and has been administered in patients with congestive heart failure in the dosage range of 150–450 mg/day in a slow-release form (8). It appears that the more pronounced the congestion the more likely side-effects are to occur. Given parenterally as an infusion in a dosage of 10–40 μg/kg/min, phentolamine has proved to be a potent vasodilator. The maximum effects start to occur 15 min after starting infusion and approximately 1 h after ingestion. Mean half-life is about 6 h. The main side-effects are tachycardia and gastrointestinal disturbances.

Phenoxybenzamine is a haloalkylamine with potent α-adrenergic receptor antagonism and virtually no direct relaxant action on vascular smooth muscle. It has a slow onset, and a prolonged duration of action. Once α-blockade has fully developed it cannot be overcome by any antagonist treatment. Therefore, it is classified as a noncompetitive α-adrenergic receptor antagonist. Orally, it is administered in a dosage range of 10–240 mg/day and intravenously at 1–2 mg/kg/h. Duration of action is 3–4 days. Major side-effects are postural hypotension, miosis, and nasal stuffiness.

In conclusion, the use of α-adrenergic receptor blocking agents in vasodilatory treatment of congestive heart failure has a considerable rationale. Of the α-receptor blocking agents at present available for chronic treatment, prazosin – and perhaps its relative trimazosin – appear the most useful: (a) because of their relatively distinct pharmacological properties; and (b) because they can reasonably be handled in clinical practice.

Finally, however, one must use caution in extrapolating experimental data on prazosin to man.

References

1. Bateman DN, Hobbs DC, Twoney TM et al. (1979) Prazosin, pharmacokinetics and concentration effect. Eur J Clin Pharmacol 16:177–181
2. Cambridge D, Davey MJ, Massingham R (1977) The pharmacology of antihypertensive drugs with special reference to vasodilators, alpha-adrenergic blocking agents and Prazosin. Med J Aust [Suppl] 2:2–6
3. Cavero J, Roach AG (1980) Minireview: the pharmacology of Prazosin, a novel antihypertensive agent. Life Sci 27:1525–1549
4. Constantine JW (1974) Analysis of the hypotensive action of Prazosin. In: Cotton DWK (ed) Prazosin-evaluation of a new antihypertensive agent. Excerpta Medica, Geneva, pp 16–36
5. Hess HJ (1974) Biochemistry and structure-activity studies with Prazosin. In: Cotton DWK (ed) Prazosin-evaluation of a new antihypertensive agent. Excerpta Medica, Geneva, pp 3–15
6. Jaillon P, Rubin P, Yee YG et al. (1979) Influence of congestive heart failure on Prazosin kinetics. Clin Pharmacol Ther 25:790–794
7. Jauernig RA, Moulds RFW, Shaw J (1978) The action of Prazosin in human vascular preparations. Arch Int Pharmacodyn Ther 231:81–89
8. Kersting F, Samosny G, Kasper W, Meinertz T, Gilhich HJ, Just H (1980) Clinical pharmacology of Prazosin and Phentolamine in patients with heart failure. J Cardiovasc Pharmacol [Suppl 3] 2:373–383
9. Langer SZ, Adler E, Enero MA, Stefano FJE (1971) The role of the alpha-receptor in regulating noradrenaline overflow by nerve stimulation. Proc 25th Int Congr Physiol Sciences, p 325. Referred to in Davey MJ, J Cardiovasc Pharmacol 2:298
10. Pitterman AB, Rollins DE, Shen DD et al. (1981) Alpha adrenoceptor blockade with oral Prazosin. Clin Pharmacol Ther 29:143–148
11. Rand MJ, Story DF, Allen GS et al. (1973) In: Usdin E, Snyder SH (eds) Frontiers in catecholamine research. Pergamon, London, pp 579–581
12. Starke K (1972) Alpha sympathomimetic inhibition of adrenergic and cholinergic transmission in the rabbit heart. Naunyn Schmiedebergs Arch Pharmacol 274:18–45
13. Taylor SH (1975) Physiology and clinical pharmacology of phentolamine. In: Taylor SH, Gould LA (eds) Phentolamine in heart failure and other cardiac disorders. Huber, Berne Stuttgart Vienna, pp 2–10

Discussion

Just, Freiburg: Would you mind to comment on the drug sensitivity ratio of prazosin?

Kersting, Koblenz: The drug sensitivity ratio is defined as the dose of phenylephrine required to increase systolic blood pressure by 25 mmHg. The higher the drug sensitivity ratio, the higher the blockade will be.

Bassenge, Freiburg: What is the evidence for the fact that phentolamine acts directly on vascular smooth muscle, apart from its action on α-1-receptors?

Kersting, Koblenz: Quite old pharmacological experiments have shown that phentolamine has a direct vasodilatory effect, which is independent of the α-adrenergic blocking action. Taylor (13) was the first to describe this twofold effect in the clinical situation.

Bussmann, Frankfurt: What is the current opinion on the clinical usefulness of α-blocking agents in chronic congestive heart failure?

Kersting, Koblenz: The application of phentolamine to patients with congestive heart failure in classes III and IV gave quite satisfying results in doses up to 300 mg/day. However, gastrointestinal side-effects were quite common. In a comparative study of phentolamine and prazosin, no significant differences in the hemodynamic effects and the incidence of side-effects were observed. However, I believe that patients with advanced heart failure will quite frequently develop severe gastrointestinal disturbances which may necessitate interruption of the treatment programme. I do believe that there is still a clinical indication for the use of phentolamine, because it can be administered intravenously. Phentolamine is certainly a very potent vasodilator with dominant arteriolar action.

Scholz, Hannover: With α-blocking agents a rather pronounced increase in heart rate is observed. If you compare the increase in heart rate induced by phentolamine with that seen after administration of β-2-stimulators such as pirbuterol, is there a difference?

Kersting, Koblenz: It is very difficult to answer this question, because there are no studies comparing pirbuterol and phentolamine. Recently a paper was published in the British Medical Journal describing the effects of pirbuterol in more than 15 patients. Here, a reasonable reduction in heart rate was seen. With phentolamine in severe congestive heart failure a fall in heart rate is also seen. A true comparison of the two drugs in the situation of heart failure has, however, not been published as yet.

Meinertz, Mainz: As far as the gastrointestinal absorption and bioavailability of prazosin is concerned, I should like to draw your attention to two recently published papers reporting quite opposite findings. What is the current view on reabsorption and bioavailability of prazosin?

Kersting, Koblenz: There is no ready explanation of these discrepancies. One of the studies was done in normal volunteers, the others in patients with heart failure. The answer may lie there. It may well be that in congestive heart failure a lower peak plasma level of prazosin is found.

Auditorium: It needs to be mentioned that the area under the curve may increase in the presence of lower maximum plasma concentrations of a drug, if the terminal slope is delayed.

Pippig, Gütersloh: Phenoxybenzamine has a prolonged duration of action of up to 3 days. Can this substance be used with particular advantage in heart failure because of this prolonged availability.

Kersting, Koblenz: Phenoxybenzamine has an established place in the preoperative treatment of patients with pheochromocytoma. There are a few studies describing the use of phenoxybenzamine in congestive heart failure. However, these have not yet been conclusive and the usefulness of phenoxybenzamine for the treatment of congestive heart failure cannot be determined today. I do believe, however, that a drug with a long-lasting action of over 3–4 days may be difficult to control. I would therefore be cautions in the use of a drug of this kind in fluctuating situation such as heart failure.

III. Clinical Experience

Promises and Disappointments of Vasodilator Treatment of Chronic Heart Failure

S.H. Taylor

A physiological approach to treatment is always academically satisfying but the results rarely live up to the early expectations. This has gradually become apparent with the increasing application of vasodilators in the treatment of patients with chronic heart failure. The exquisite implicity of the concept that reduction of the pressure against which the left ventricle has to pump would rejuvenate its performance is immensely attractive but sadly superficial. The immediate benefit to the failing heart of acute reduction in stretch of the myocardial fibres by reduction in after-load has been amply demonstrated in the isolated preparation, in the experimental animal and even in acute heart failure in man. Unfortunately, such reasoning ignores the crucial fact that in chronic heart failure vasodilator treatment is undertaken in the face of an innervated heart and largely intact neuro-humoral reflexes, the counter-effects of which may only become apparent after long-term treatment. The initial psychological resistance by physicians to such unusual treatment was followed by a surge of immense and uncritical enthusiasm for the therapeutic exhibition of these drugs in patients with chronic heart failure. However, the inadequate design and over-extended interpretation of many of the early studies in this field gradually eclipsed this rising wave of enthusiasm. Equally unfortunately, the waning popularity of this new approach has been accentuated by unwarranted disappointment, fuelled by the now fashionable cult of therapeutic nihilism (Fig. 1).

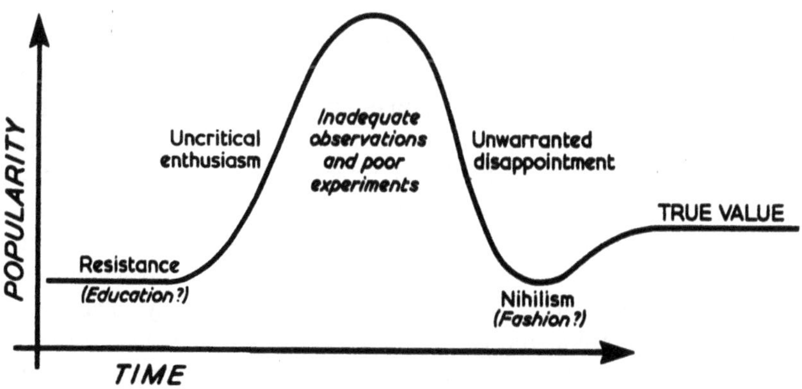

Fig. 1. Idealized scheme of potential therapeutic popularity of a drug, including the vasodilators

New methods of treatment for heart failure are urgently required. It is a syndrome so common as to be seen in either acute or chronic form by virtually every practising physician, and the problem is multiplied in an ageing community. It is a condition that gives rise to immense suffering; distressing symptoms prohibit a normal dynamic life. Moreover, it is a condition of progressive deterioration that is rarely spontaneously reversible; its early recognition and treatment is therefore vital. Diuretics and drugs with positive inotropic activity have long formed the main corner-stones of treatment. An attraction of the vasodilators was that they offered a new and totally different physiological approach to the treatment of the failing heart in man. But in such a chronic and progressive disease it is worthwhile first to address the question whether it is possible to alter the natural history of the syndrome with any drug, and if so by how much.

Is it Possible to Alter the Natural History of Chronic Heart Failure?

Chronic heart failure is characterized by factors which tend to ensure its progression:

1. The contractile activity of the individual myocardial cells in the afflicted ventricle is permanently and progressively reduced.
2. The ensuing dilatation of the heart results in further "slippage" of myocardial contractile reserve and ventricular pumping efficiency.
3. The extreme work-load on the myocardial cells in the dilated heart results in their progressive functional, and subsequently histological, demise.
4. These lost cells are not replaced and the impeded hypertrophy of the remainder is unable to replace the functional potential of those lost.

The hemodynamic progression of heart failure is thus one of crescendo deterioration; the aims of pharmacological treatment must be considered in this light. Ideally, drugs with positive inotropic activity are more physiologically justifiable than those that do nothing more than modify pre-load or after-load, but this theoretical approach is overridden by practical considerations. At best treatments can only be expected to improve cardiac performance temporarily, away from the low perfusion state incompatible with life and towards a relief of symptoms (Fig. 2). However, whether different treatments alter the subsequent rate of progression of the disease differently is quite unknown.

From the hemodynamic point of view, the conventional consideration of left ventricular performance by relating its output to its filling pressure also becomes progressively obscured by the continued deterioration of myocardial contractile performance. The initial response to treatment is frequently dramatic, but sustained treatment or the institution of new drugs tends to produce a progressively less marked effect as the disease progresses (Fig. 3). This illustrates the increasing resistance to treatment inherent in the progressive deterioration in cardiac pumping function that characterizes the disease. It also hides a change in meaning of the conventional hemodynamic relationships that apply in the normal heart. As the heart progressively dilates, the relationship between filling pressure and volume of the ventricle becomes increasingly distorted, and "forward" flow forms a decreasing component of total pumped output. Thus the conventional interpretation of "standard" hemodynamic measurements

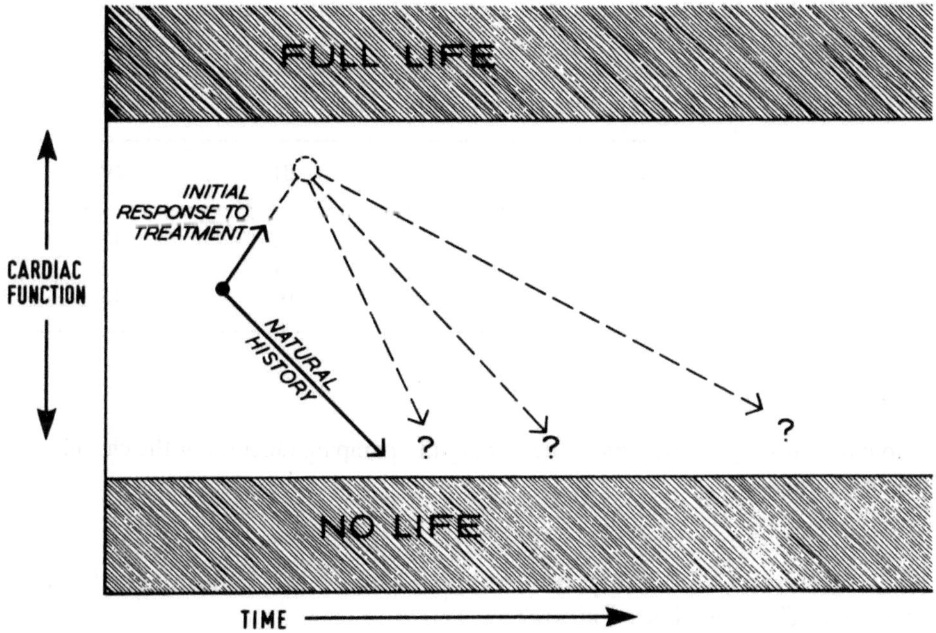

Fig. 2. Possible influences of treatment on the natural history of heart failure

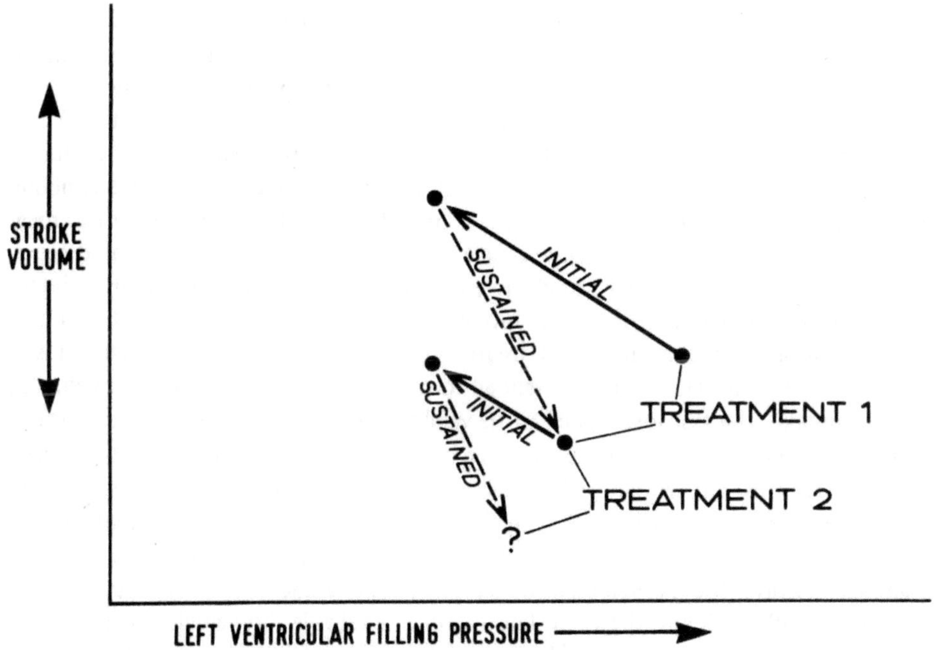

Fig. 3. Possible hemodynamic influences of treatment when first given and then sustained in chronic heart failure

Table 1. Variability in historical account of symptoms given to three cardiologists by 30 patients with chronic heart failure

Cardiologist	New York Heart Association Grade			
	I	II	III	IV
1	0	1	12	17
2	0	0	8	22
3	0	1	14	15

becomes increasingly unreliable in describing the pumping function of the chronically failing ventricle.

What Methods Can Validly be Used to Evaluate the Usefulness of Vasodilators in Chronic Heart Failure?

In the clinic the low specificity and low sensitivity of symptoms and physical signs in patients with chronic heart failure underlines thair failure to furnish a reliable quantitative measure of the influence of treatment on the degree of pumping failure of the heart. Patient communication to the physician of the measure of their disability is open to wide errors of expression and interpretation (Table 1). The major symptoms of patients with chronic heart failure are those of breathlessness and fatigue during exercise, but rarely does the physician observe his patient under these conditions. This also explains the common disparity that exists between the patient's historical account of his symptoms, which are usually related to those he develops during dynamic exercise, and the degree of heart failure detected by the physician examining him at rest. Even the standard chest X-ray is of little use in this regard, as for technical reasons this invariably affords only a measure of heart size at rest. It would seem, therefore, that an appropriate upright exercise test, such as treadmill-walking, should offer a reasonable method of clinical monitoring of the effects of treatment. In terms of observation of the influence of treatment on breathlessness and fatigue, this is almost certainly so. However, if extended to other measurements, such as maximum oxygen uptake, the test loses much of its validity for two reasons. Total body oxygen uptake must take into occount the oxygen consumed during the post-exercise oxygen debt period; this is often ignored. Moreover, reliability of total body oxygen uptake as a measure of work-load is valid only in the aerobic range of muscle metabolism. As the majority of patients in heart failure enter the anaerobic phase during even minimal exertion, the measurement of oxygen uptake alone does not reflect the total metabolic work-load, and therefore fails to afford a reliable test vehicle for the evaluation of the effects of treatment. Measurements of respiratory rate and minute ventilation do not suffer from these limitations, but despite their technical ease and relevance they do not appear to have been fully exploited in this field. The measurement of

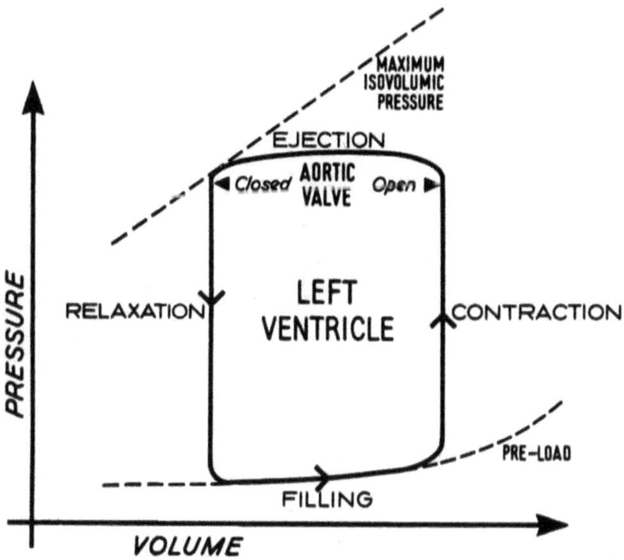

Fig. 4. Idealized left ventricular pressure-volume relationships in the normal subject

fatigue, the other major symptom in chronic heart failure, offers an impossibly vague end-point.

In the laboratory none of the measurements at present available furnishes a physiologically reliable measurement of left ventricular performance in chronic heart failure. Many techniques of hemodynamic evaluation have been suggested for the diagnosis and monitoring of treatment, but few meet the necessarily stringent criteria. Ideally, description of the left ventricular pressure-volume relationships are the most edifying (Fig. 4). However, although such a system is ideal for describing the pumping function of the normal ventricle. It suffers from a major disadvantage in describing that in the failing heart, due to the changed systolic and diastolic relationships that often prevail.

The segmental dyskinesia so often met in the ischaemic heart, imposes severe limitations on the interpretation of the systolic component of the pressure-volume characteristics of the left ventricle. Thus the idealized changes induced by increased contractility and reduced aortic impedance, which are so apt in the normal heart (Fig. 5), are often inappropriate as a measure of systolic performance when the heart is grossly dilated. Diastolic function, too, may be grossly distorted by disease. The reduced compliance of the ventricular wall induced by fibrosis is aggravated by the intense sympathetic drive that often accompanies heart failure, particularly during exercise. This reduced relaxation of the ventricle impedes its filling. Thus impairment of systolic and diastolic functions in the failing heart distorts the interpretation of many hemodynamic measurements.

What measures then can practically be applied to the measurement of left ventricular function in chronic heart failure with a view to monitoring the effects of drugs? Despite its limitations, the common method of relating left ventricular filling

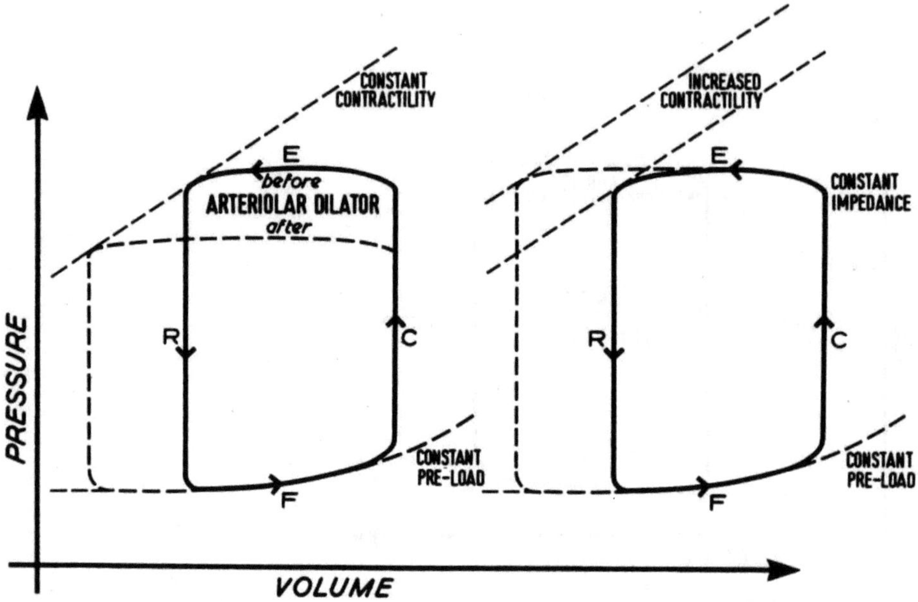

Fig. 5. Effect of changes in contractility and aortic impedance on left ventricular pressure-volume relationships in the normal subject. *F*, filling; *C*, isovolumic contraction; *E* ejection; *R*, relaxation

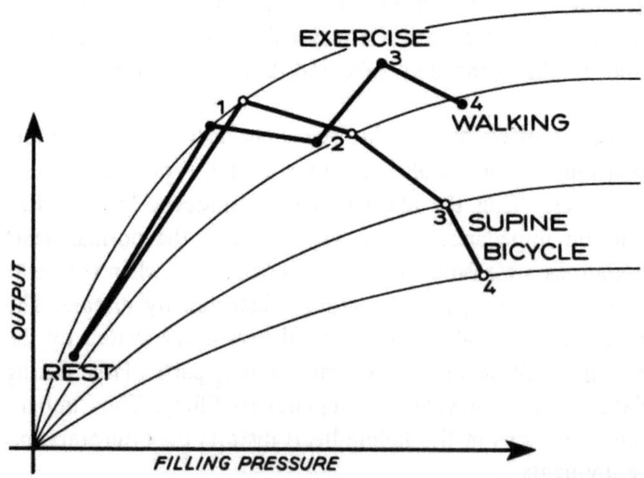

Fig. 6. Illustration of the possible changes induced by posture and time on the relationship between left ventricular filling pressure and output in heart failure. Numerals *1–4* relate to minutes of exercise at constant speed and work-load

pressure to its output is still the method most widely practised. The validity of this test is immensely improved by the addition of increased ventricular work-load, the most common being that imposed by dynamic exercise. However, the effects of time on the result are often ignored, and the importance of posture rarely considered (Fig. 6). This clearly illustrates the frailty of the left ventricle when any additional load is imposed and maintained. The greatest pre-load is experienced during supine bicycle exertion with the exercising muscles above heart level; this probably accounts for the progressive deterioration in left ventricular pumping function if such exercise is maintained, even at constant load.

Hemodynamic Basis for Use of Vasodilators in Heart Failure

The primary abnormality in heart failure of whatever cause is reduction of myocardial contractile activity, resulting in failure of the left ventricle to expel the blood presented and consequent dilatation. This has two immediate consequences. Dilatation results in increased stretch of the individual myocardial fibres, so that their contractile ability is increased (Frank-Starling mechanism). The second important compensatory mechanism involves the reflexes from high-pressure baroreceptors in the aortic arch and carotid sinus, and from low-pressure stretch receptors in the left atrium and pulmonary veins. Centrifugal impulses from these receptors are integrated in the central vasomotor centres, with the effector mechanism executed through the sympatho-adrenal pathways to the heart and peripheral vasculature. These reflexes, however, vital to the continued perfusion of the cerebral and coronary circulations, impose a paradoxically increased load on the already failing heart. Sympathetically mediated constriction of the arteriolar resistance vessels augments aortic impedance and increases left ventricular pressure work. Constriction of the small veins reduces venous capacity and augments pre-load and ultimately left ventricular dilatation. The resulting increase in left ventricular after-load and wall tension rapidly exaggerates the myocardial oxygen requirements of the already failing left ventricle. Reduction in left ventricular pressure work by lowering aortic blood pressure through arteriolar dilatation, and reduction in left ventricular volume by reducing venous return through dilatation of the veins, are obvious ways in which the function of the failing heart can be improved. This is the prime rationale for the use of vasodilators in heart failure.

Such a concept is attractive, but takes no account of the fact that the patient with chronic heart failure is often ambulant and with largely intact circulatory reflexes. Many complex neuro-humoral mechanisms are involved. Thus it is to be expected that any treatment of chronic heart failure, other than that directly applied to correction of the deficit in myocardial contractile ability, is likely to result in the initiation of reflexes that tend to offset the intended benefits of such treatment. This is particularly noticeable when the treatment produces such striking circulatory changes as many of the vasodilators accomplish. The early promises of vasodilators in patients with heart failure were based on a simplistic physiological argument that paid scant attention to these considerations. This is the source of many of the disappointments that have arisen from the practical results of their use in clinical practice.

What Factors Limit the Hemodynamic Value of Vasodilators?

The prime aim of the vasodilators is to reduce the aortic impedance against which the left ventricle has to pump, if possible without diverting blood from those tissues whose demands need to be met despite the reduced cardiac output. Unfortunately, vasodilator drugs are non-selective in terms of the regional distribution of the cardiac output; they do not divert blood to the territories demanding increased flow, and may even impede such benefit. This can be deduced from studies of vasodilators with predominantly arteriolar activity (e.g. hydralazine); cardiac output is increased but without an associated increase in total oxygen uptake, particularly during exercise. This indicates that the increased cardiac output is not selectively diverted to the areas with maximum demand: it is distributed to all areas. In heart failure, vasoconstriction is most pronounced in the percutaneous renal and splanchnic circulations, and however, beneficial such vasodilatation in these regions may be, it does nothing to enhance the performance of the exercising limb muscles. This is not to deny that such increased flow to non-muscular territories is not important (e.g. diuretic treatment is more effective after hydralazine). But the relief of one of the major symptoms of heart failure, namely fatigue of exercising muscles, is unlikely to be relieved by obligatory sharing of the very small increase in cardiac output with other regional circulations bought at the price of a reduction in the major determinant of their flow, namely the perfusion pressure.

Pharmacodynamic Conflicts in the Use of Vasodilators

Many of the pharmacodynamic conflicts originate in the omnibus expression "vasodilators". The vasodilators at present available are non-specific for any macro- or micro-circulation, and understanding of their mode of action is dependent upon a knowledge of the physiology of vascular smooth muscle. Vascular dependent upon a knowledge of the physiology of vascular smooth muscle. Vascular smooth muscle is far from uniform in its functional behaviour. Smooth muscle in veins differs markedly in its functional activity from that in arteries and arterioles. Moreover, the behavior of vessels of similar anatomical type is often specific to the vascular region in which they reside. Their functional activity also differs markedly. Arterioles have a well-marked degree of intrinisc tone and show no spontaneous phasic activity. Stimulation results solely in the developement of sustained contraction. Although vasodilator drugs are often grouped for therapeutic convenience according to their predominant site of action and resulting effects on the cardiovascular system, such considerations furnish only guide-lines of activity; their hemodynamic effects differ markedly between patients with apparently similar degrees of chronic heart failure. This is a reflection not only of the non-specificity of pharmacodynamic action of these drugs, but also of the complex neuro-humoral influences on the systemic vessels in the different stages of heart failure. In general, however, the circulatory mode of action of these drugs determines their overall effects in heart failure in the majority of patients, allowing the physician the choice between drugs that predominantly reduce pre-load (e.g. nitrates, prazosin) and those which predominantly result in arteriolar dilatation (e.g. hydralazine)

Table 2. Regional and central hemodynamic effects of vasodilator drugs

Drug	Vascular bed		Cardiac output		Pressures			
	Arterial	Venous	Stroke	Minute	Pulmonary venous		Systemic arterial	
					Rest	Exercise	Supine	Erect
Glyceryl trinitrate	±	++	(↑↓)	(↑↓)	↓↓	↓↓	↓	↓↓
Isosorbide dinitrate	±	++	(↓↑)	(↓↑)	↓↓	↓	↓	↓↓
Prazosin	+	++	↑	↑	↓	↓↓	↓	↓↓
Nitroprusside	+(+)	++	↑↑	↑↑	↓	↓	↓	–
Phentolamine	++	++	↑↑	↑↑	↓	↓	↓	–
Hydralazine	++	±	↑↑	↑↑	(↓)	↓	(↓)	(↓)

(Table 2). The very existence of these different sites of action in the systemic circulation brings in its train different postural effects. Drugs with predominant venodilator activity are more likely to result in postural falls in blood pressure than those with arteriolar dilator activity. However, once more the picture is complicated by the state of the patient, particularly the height of the venous and arterial pressures when these drugs are first given.

Finally, the concept of receptor-blocking drugs must be kept constantly under review, particularly during their sustained exhibition in patients with gross circulatory changes. Even drugs with specific receptor-blocking activity in the circulation may fail to exert a similar effect during their long continued administration, due to changes in the receptor population (aptly if inelegantly described as "up-regulation" or "down-regulation"). This applies not only to receptors in the heart but also to those in arteries and veins. Changes in receptor population are also related to the functional state of the circulation and almost certainly subject to wide individual variation. These are some of the problems underlying many of the pharmacodynamic conflicts that exist at present.

In Which Patients are the Vasodilators Most Effective?

The multitude of structural cardiac lesions, each giving rise to a different functional upset, immensely complicates any attempt to determine in which patients vasodilators are most effective. The picture becomes even more confused when it is considered that none of the vasodilator drugs is entirely specific for any given macro- or micro-region of the circulation, and their exhibition is prone to alter the responsiveness of the vessels they initially influence. Moreover, pharmacodynamic manipulation of one regional circulation is usually followed by neuro-circulatory reflexes which tend to offset the circulatory changes induced, so that the status quo of the circulation overall is maintained. It would seem reasonable to suggest, therefore, that if the hemodynamic profile in a patient was known, then the different vasodilator drugs available could more rationally be exhibited. Not only is this largely impractical, but more importantly, such expensively gained knowledge may benefit the patient but little. The mating of vasodilator treatment of hemodynamic profiling is impractical until the immediate and long-term pharmacodynamic effects of specific vasodilators have been clearly defined in the many functional subgroups of patients with chronic heart failure. The problem is further complicated by the lack of knowledge concerning the quantitative neuro-humoral responses throughout normal daily activity in the individual patient and the operation of these reflexes under the influence of vasodilators. The final complication for the majority of physicians without access to extensive investigative techniques is the all-important relationship of the clinical manifestations of the syndrome of heart failure to the magnitude of the underlying hemodynamic disorder.

As yet we have little or no knowledge that would allow us to predict the long-term benefits of any of the vasodilators in any of the infinite variety of functional subgroups of left ventricular failure. That crucial element of clinical practice, namely the matching of the individual to the most efficacious treatment, will continue to elude us until infinitely more is known of exactly how these drugs work in the patient with chronic cardiac disability.

Are the Immediate Benefits of Vasodilator Drugs Maintained During Long-Term Treatment?

To the surprise of many physicians, the immediate and often striking benefit of the vasodilator drugs in patients with acute heart failure is not reflected in similar benefit during sustained treatment of those with more chronic disease. There is a paucity of reliable information on the benefits or hazards of long-term treatment with these drugs (Table 3). Due to the inhomogeneity of the groups of patients studied by various investigators and the small numbers in each report, it would be invalid to pool the results for analysis. However, many studies indicate that the hemodynamic effectiveness and clinical efficacy of the vasodilator drugs in common use are maximum when these are first prescribed (Fig. 7). With continued treatment, their benefits appear to decline, but this statement must be heavily qualified. It must not be assumed that all vasodilators follow the same path of progressive therapeutic impotence. Whether attenuation of benefit occurs with all drugs or whether it is confined to certain pharmaco-

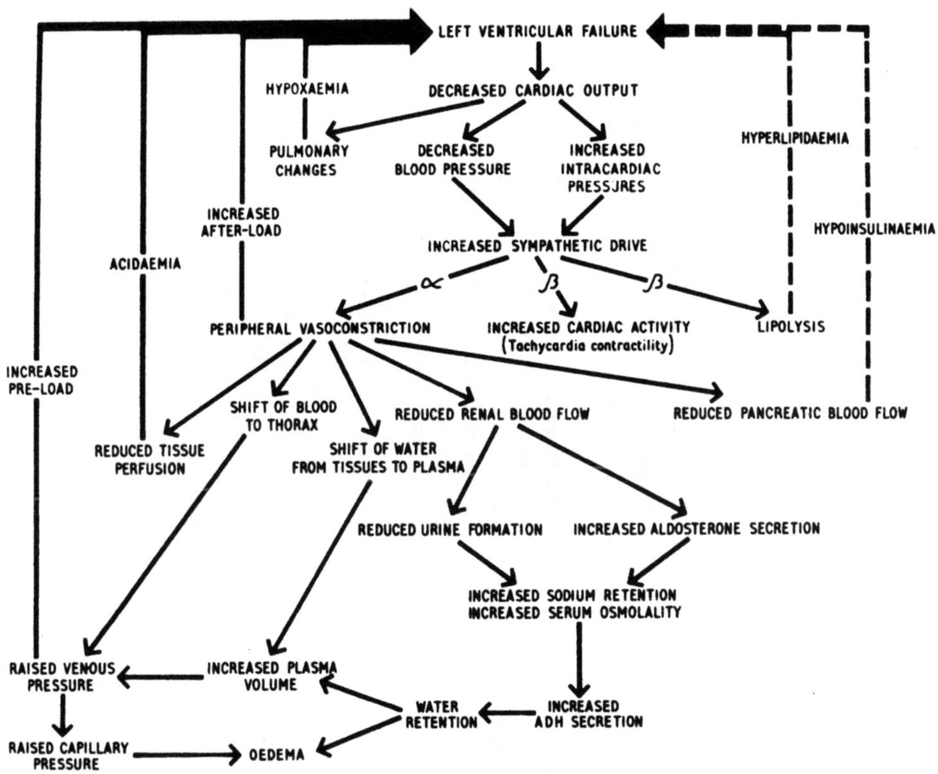

Fig. 7. Attenuation of the hemodynamic effects during sustained treatment with prazosin in patients with chronic ambulant heart failure (Silke et al., to be published)

Table 3. Hemodynamic and clinical results of sustained treatment in chronic heart failure. (*A*), abstract; *ISDN*, isosorbide dinitrate; *DB*, double-blind; *R*, randomized; *CO*, cross-over

Drug	Author(s)	No. of patients	Design of study	Hemodynamic effectiveness	Clinical efficacy
Phentolamine	Taylor 1975	4	Open		Maintained
Hydralazine	Chatterjee et al. 1976	9	Open		Maintained (7)
	Chatterjee et al. 1979 (A)	8	Open	Maintained	
ISDN + phenoxybenzamine	Kovick et al. 1976	9	Open	Maintained	Maintained
Nitrate	Franciosa and Cohn 1978	14	DB R CO		Maintained (6)
	Franciosa and Cohn 1978 (A)	11	Open	Maintained	
Prazosin	Packer et al. 1980	8	Open	Reduced	Reduced
	Arnold et al. 1979 (A)	13	Open	Reduced	
	Elkayam et al. 1979 (A)	8	Open	Reduced	
	Aronow et al. 1979	24	DB R	Maintained	Maintained
	Joye et al. 1979 (A)	6	Open		Maintained
	Foster et al. 1979 (A)	8	Open		Maintained
	Colucci et al. 1980	22	DB R		Reduced
	Silke et al. to be published	12	Open	Reduced	Reduced

dynamic subtypes is quite unknown. The relationship of the loss of benefit to the magnitude of the hemodynamic upset is equally obscure.

Few studies have been directly aimed at this problem. The majority of scattered reports available do little more than highlight the lack of knowledge. Many more studies of the effectiveness of the many different drugs available for long-term treatment of chronic heart failure are required before the physician can be certain of their continued clinical benefit.

What Extracardiac Effects Result from Vasodilator Drugs in Chronic Left Ventricular Failure?

Failure of the heart to pump sufficient blood to meet the metabolic demands of tissues results in a complex series of pathopysiological events (Fig. 8). The results of the superimposition of the effects of hemodynamically active vasodilator drugs on these integrated reflexes is largely unknown.

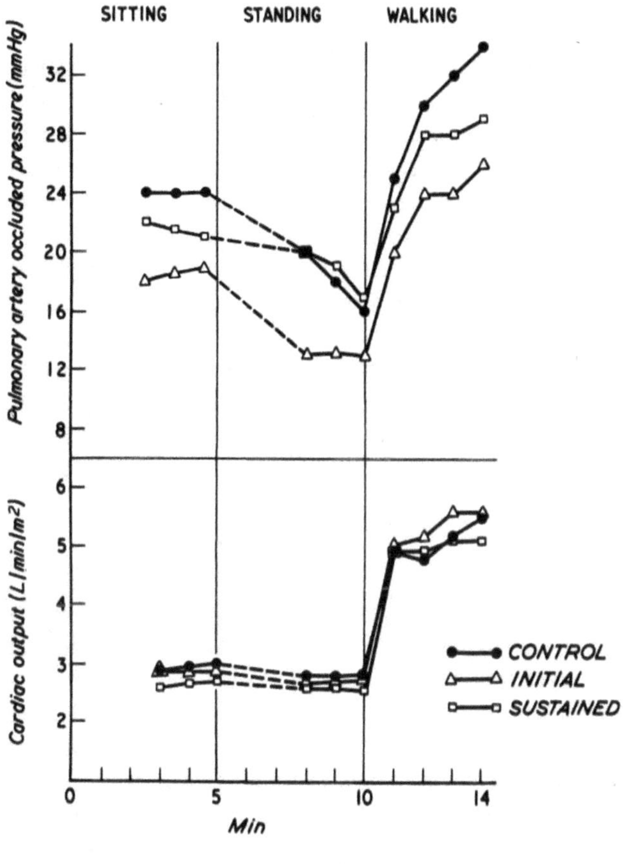

Fig. 8. Some of the pathophysiological consequences of left ventricular failure. *Solid lines* indicate events for which there is firm evidence; *broken lines* imply a more speculative relationship. [Taylor 1973 (6)]

Regional Blood Flow

The primary site of action of the vasodilator drugs is the peripheral blood vessels, yet surprisingly few reports have concerned their effects either immediately or during sustained treatment on regional blood flow in patients with chronic heart failure. This may be particularly important in a number of circumstances in ambulant patients. Drugs with predominant venodilator activity, such as the nitrates and prazosin, have marked postural effects which presumably must be reflected in substantial changes in distribution of the cardiac output on assumption of the erect posture. Drugs with predominant dilator effects on arteriolar resistance vessels almost certainly alter the distribution of the cardiac output to the various regional vascular beds during physical exertion. These therapeutically induced changes in regional blood flow may also be associated with marked metabolic and humoral effects. It is therefore important to know the particular regional vascular activity of each type of vasodilator during sustained treatment in chronic severe heart failure. Up to the present time no composite analysis of such effects has been reported.

Metabolic Effects

Left ventricular failure is characterized by reflex vasoconstriction in the majority of the regional vascular territories; this vasoconstriction is greatly enhanced during physical exertion. These changes induce important metabolic effects in many tissues, with the result that metabolic acidosis is an invariable consequence of severe heart failure, particularly during exercise. Excessive increase in arterial pH depresses myocardial contractile activity; relief of excessive vasoconstriction in heart failure by the vasodilator drugs may have important metabolic benefits, but systematic examination of this important aspect does not appear to have been undertaken. A further consequence of the persistent reduction in peripheral blood flow to the majority of regional vascular territories probably accounts for the excessive catabolism of many actively metabolizing cells, particularly those of striated muscle. The effect of the vasodilator drugs on this excessive catabolism is another avenue of investigation that as yet appears to be unexplored, despite the cachexia so common in chronic heart failure.

Humoral Effects

Heart failure is characterized by a number of important humoral effects. The endocrine glands are not immune to the generalized reduction in blood flow experienced by other territories; one of the most important effects of severe heart failure is reduction in insulin secretion, even in response to physiological stimulation. This is presumably a consequence of sympathetic stimulation and reduction in splanchnic blood flow. Insulin is of crucial importance in myocardial metabolism in the failing heart. The vasodilator phentolamine relieves the insulin suppression associated with sympathetic vasoconstriction, but whether all vasodilator drugs will achieve this during sustained treatment in chronic heart failure is quite unknown.

Another possibly important humoral disorder in heart failure is the enhanced secretion of renin, due presumably to a combination of reduced renal perfusion pressure and increased renal sympathetic stimulation. In subjects with normal cardiovascular function, vasodilator drugs enhance renin stimulation. It is important to know, therefore, whether such effects summate with those intrinsic to those of the heart failure syndrome; activation of the renin-angiotensin-aldosterone system could be expected to enhance sodium retention and offset the direct hemodynamic benefit of these drugs. Should such a finding be confirmed, then long-term treatment with renin antagonists would appear to be indicated. Increased thyroid activity and growth hormone secretion have also been reported in heart failure; their significance in the pathogenesis of heart failure is obscure and the effect of vasodilator drugs on these functions is unknown.

There is an immense hiatus of knowledge about the activity of these drugs on many important physiological functions intrinsic to the syndrome of chronic heart failure.

Are There Clinical Disadvantages with Vasodilator Drugs in Chronic Heart Failure?

The vasodilator drugs have undoubtedly added a new dimension to the treatment of chronic heart failure, but as with all drugs their efficacy is not absolute. Drugs with predominant venodilator activity afford relief of breathlessness by reduction of venous return and hence reduction in pulmonary venous pressure. These benefits are particularly marked in patients suffering from orthopnoea or acute exacerbation of heart failure, and during dynamic exercise. However, the price of these benefits is the increased muscle fatigue during exercise that results from failure of the normal increase in systemic blood pressure. Moreover, drugs with predominant venodilator activity are prone to result in marked postural changes in the blood pressure which may give rise in some instances to serious hypotension (Fig. 9). Drugs with predominant arteriolar dilator activity are less prone to result in increased fatigue or postural hypotension. However, in the majority of patients they tend to afford less reduction in breathlessness, despite their ability to benefit pumping function of the heart.

Both groups of compounds can be expected to result in increased renin secretion and activation of the angiotensin-aldosterone system, with effects on blood pressure and sodium excretion respectively that could offset the direct hemodynamic benefits of pre-load and after-load reduction. But whether such compounds do chronically increase renin secretion in patients, and what would be the clinical outcome of such an event, is quite unknown.

Disadvantagous pharmacodynamic interactions may also result from different drug combinations used in heart failure. The most common is enhancement of the postural effects of venodilator drugs by diuretics, particularly those with rapid onset of activity, e.g. frusemide. Combination of both arteriolar and venodilator drugs with inotropic agents possesses both theoretical and practical advantages in terms of enhancement of left ventricular pumping activity. However, their combination with drugs that possess inotropic activity, particularly those that exert their activity through stimulation of β-1-adrenoceptors, raises the theoretical problem of enhanced tendency to precipitation of malignant ventricular dysrhythmias.

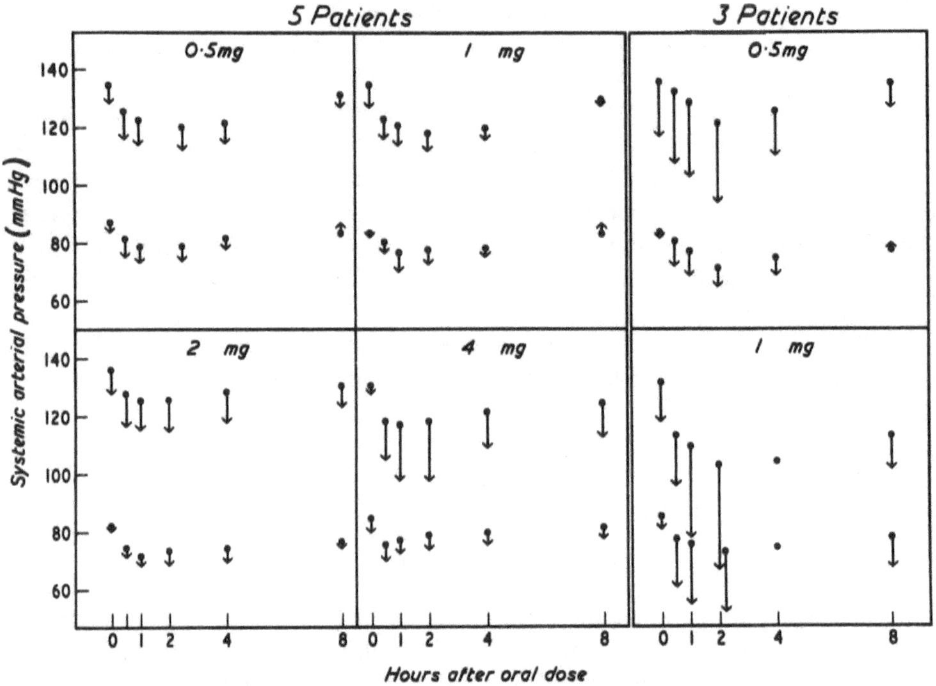

Fig. 9. Blood pressure changes recumbent (*dot*) and standing (*arrow*) after four single oral doses of prazosin in eight patients with similar hemodynamic profile; three were unable to tolerate more than 1 mg due to severe postural hypotension (Silke et al. 1980)

Finally, withdrawal of patients from treatment with vasodilator drugs cannot be lightly undertaken. Although the mechanisms are unclear, there is some evidence of serious hazard from their abrupt discontinuation.

Do the Vasodilator Drugs Prolong Life?

Despite the extensive clinical use of these drugs during the past decade and the high mortality risk associated with chronic heart failure, there is still relatively little dependable information on their influence on length of patient survival. This is perhaps not too surprising; there is even less information available on the effects of the far longer used and more widely exhibited conventional heart failure drugs digitalis and diuretics on survival of these patients. A solitary report highlights many of the problems encountered in attempting to evaluate this aspect of the long-term treatment of heart failure patients with vasodilator drugs (Franciosa et al. 1978). In this short-term randomized double-blind cross-over study, the results of treatment with nitrates were compared with placebo results on morbid events and length of survival in 16 patients with severe heart failure. However, only six patients were able to complete the study and one death occurred in each treatment group. Although reduction in morbid events was claimed for the patients during nitrate treatment, the evaluation was largely dependent upon patient description. The potential hazards of depending upon patients' accounts for the evaluation of treatment has already been mentioned. This study emphasizes that point, as the apparent benefits claimed by patients were not accompanied by significant hemodynamic differences and a simple objective exercise test showed little difference between the patients on the placebo and those on nitrates. This is not to say that vasodilator drugs may not reduce morbid cardio-vascular events and prolong life in these terminal patients; but it must be emphasized that as yet there are no reports that would lead to such prognostic promise. A number of major studies devoted to this aspect are now under way, but consideration of the biology of the problem does not encourage optimism.

Conclusions

There is no reasonable doubt that the introduction of the vasodilator drug was a major advance in the medical management of patients with heart failure. In acute heart failure and when first given to patients with more chronic disease the results are often strikingly impressive. Moreover, the physiological rationale on which their use is based is attractive, and their clinical deployment is facilitated by the relative ease which they may be handled in the individual patient. It is unfortunate for physican and patient alike that this promise also encompasses disappointments.

What are the factors underlying these disappointments? The simplistic concept which stimulated their introduction has long been surpassed by the immensely complex and sophisticated pathophysiological reflex mechanisms that their use has uncovered. It is of little comfort that such is the rapid state of development of knowledge in this field that many of the problems now being encountered could not possibly have been foreseen with the knowledge available when vasodilators were frist introduced a decade ago. A major disappointment in the diagnostic field has been the inability to predict the individual patient's response to these drugs by his circulatory profile. The limitations of the hemodynamic measurements available for the study of the circulation in these patients are only now becoming widely appreciated. Similarly, the failure to appreciate that increase in cardiac output is not synonymous with improvement in exercise performance or relief of symptoms is only now achieving clinical credence. It may be that much of our disappointment resides in confining our treatment to patients too ill to benefit. It is possible that less severely ill patients would benefit more from these drugs with lesser degrees of ventricular dilatation than those at the end stages of the disease.

Finally, have we enthusiastically but uncritically expected too much from these drugs? The natural history of heart failure is one of progressive deterioration in cardiac function, and unless the initiating cause is rapidly corrected, all treatments can only be expected to be palliative. But whether the primary aim of treatment in such a condition should be to increase life expectancy as against achieving some symptomatic improvement is debatable. There is no doubt that these drugs help to relieve symptoms in many of the patients to whom they are administered; to expect that they will cure patients or even extend their lives takes no account of the remorselessly progressive nature of chronic heart failure. In the future many new vasodilator drugs with different pharmacokinetic and pharmacodynamic profiles will become available. It is unlikely that they will do more than to form a useful adjuvant to other cardiac supportive remedies; their physiological role precludes them from a more prominent role. Many expected these drugs to become the mainstay of treatment, relieve all symptoms and even extend life in terminally ill patients. Too much was asked on the basis of the limited promises made, and thus disappointments arose.

References

I. General Reviews

1. Awan NA, Amsterdam EA, Mason DT (1979) Impedance reduction in severe left ventricular failure. J Cardiovasc Med 4:933
2. Braunwald E (1977) Vasodilator therapy – a physiologic approach to the treatment of heart failure. N Engl J Med 297:331
3. Chatterjee K, Parmley WW (1977) The role of vasodilator therapy in heart failure. Prog Cardiovasc Dis 19:301
4. Mason DT (1978) Afterload reduction and cardiac performance. Am J Med 65:106
5. Taylor SH, Majid PA (1971) Insulin and the heart. J Mol Cell Cardiol 2:302
6. Taylor SH (1973) Circulatory consequences of heart failure and shock. In: Hamer J (ed) Recent advances in cardiology, 6th edn. Churchill Livingstone, Edinburgh London
7. Taylor SH (1980) Vasodilator drugs in the treatment of heart failure. J Cardiovasc Pharmacol [Suppl 3] 2:407

II. Relevant Original Reports

Arnold SB, Williams RL, Ports TA, Baughman RA, Benet LZ, Parmley WW, Chatterjee K (1979) Attenuation of prazosin effect on cardiac output in chronic heart failure. Ann Intern Med 91:345

Aronow SW, Lurie M, Turbow M, Whittaker K, Van Camp S, Hughes D (1979) Effect of prazosin – vs – placebo on chronic left ventricular heart failure. Circulation 59:344

Chatterjee K, Parmley WW, Massie B, Greenberg B, Werner J, Klausner S, Norman A (1976) Oral hydralazine therapy for chronic refractory heart failure. Circulation 54:879

Chatterjee K, Ports T, Rubin S, Massie B, Arnold S, Brundage B, Parmley WW (1979) Sustained beneficial haemodynamic effects during long-term hydralazine therapy in patients with chronic heart failure (Abstr). Circulation [Suppl 2] 28:98

Colucci WS, Wynee J, Holman L, Braunwald E (1980) Chronic therapy of heart failure with prazosin. A randomised double-blind trail. Am J Cardiol 45:337

Elkayam U, Mathur M, Lejemtel T, Frishman W, Stom J, Sonnenblick EH (1979) Comparative haemodynamic effects of oral prazosin and hydralazine in refractory heart failure (Abstr). Clin Res 27:163A

Foster PR, Stein L, Freedman AW, McHenry PL, Statza J (1979) Chronic therapy of congestive heart failure with prazosin (Abstr). Clin Res 27:166A

Franciosa JA, Cohn JN (1978) Sustained haemodynamic effects of nitrates without tolerance in heart failure (Abstr). Circulation [Suppl 2] 28:99

Franciosa JA, Nordstrom LA, Cohn JN (1978) Nitrate therapy for congestive heart failure. JAMA 240:443

Hanley SP, Cowley A, Hampton JR (1980) Danger of withdrawal of vasodilator therapy in patients with chronic heart failure. Lancet I:735

Joye JA, Lee G, Awan NA, De Maria AN, Mason DT (1979) Efficacy of oral prazosin in ambulatory therapy of chronic congestive heart failure (Abstr). Clin Res 27:234A

Kovick RB, Tillisch JH, Berens SC, Branowitz AD, Shine KI (1976) Vasodilator therapy for chronic left ventricular failure. Circulation 53:322

Majid PA, Sharma B, Taylor SH (1971) Phentolamine for vasodilator treatment of severe heart failure. Lancet 2:719

Packer M, Meller J, Medina N, Gorlin R, Herman MV (1980) Haemodynamic evaluation of hydralazine dosage in refractory heart failure. Clin Pharmacol Ther 27:337

Sharma B, Majid PA, Pakrashi BC, Dykes JRW, Taylor SH (1970) Insulin secretion in heart failure. Br Med J 2:396

Silke B, Hendry WG, Taylor SH (1981) Immediate and sustained haemodynamic effects of prazosin during upright exercise in man. Br H J 46:663

Silke B, Lakhani ZM, Taylor SH (1981) Pharmacokinetic and pharmacodynamic studies with prazosin in chronic heart failure. J Cardiovasc Pharmacol 3:329

Taylor SH (1975) Phentolamine in chronic heart failure. In: Taylor SH, Gould A (eds) Phentolamine in heart failure and other cardiac disorders. Huber, Berne Stuttgart Viena

Discussion

Auditorium: What about the pharmacokinetics of drugs in patients with congestive heart failure?

Taylor, Leeds: According to the pharmacokinetics in patients with congestive heart failure, we do not know what is the interplay between absorption, first pass effect, liver clearance and binding to the tissues, to mention just a few of the variables involved in heart failure when any drug is considered.

Auditorium: We have seen the beneficial effects of vasodilators after short-term administration, but what is your opinion as to their value in long-term treatment?

Taylor, Leeds: There is no doubt that a small group of patients do benefit symptomatically in this terminal stage of their disease. However, it is difficult to identify these in the clinic.

Just, Freiburg: You have given us a very critical approach which is likely to bring us into the trough of nihilism of your curve. This would seem rather desirable, since many of us have a certain amount of uncritical optimism, as is usual in the clinic. However, you did not mention the dangers that have arisen from the application of this therapeutic principle. Would you care to give us further information as to the real danger connected with the application of vasodilators in chronic heart failure?

Taylor, Leeds: As you know, this has been studied incompletely. There is no doubt that studies such as those initiated by Cohn in the United States, with large numbers of patients, are the best way of detecting adverse reactions. I was extremely concerned to find recently reports in the literature of a small number of patients treated with prazosin who died when it was stopped. That might be the natural progression of the disease. I think we have to be extremely careful in the description of adverse reactions in these terminal patients.

Cohn, Minneapolis: In your comment you suggested that vasodilators might alter life expectancy. What is your view of the role of vasoconstriction in the progression of the disease from early to moderate heart failure? Is there a positive feedback system which is perhaps activated by the vasoconstriction? If that were true, would you share the hope that early advance in these drugs might alter the history of the disease?

Taylor, Leeds: As far as life expectancy is concerned, I would like to emphasize that four these patients (not NYHA class II, but those with far more chronic and severe heart failure) therapy should be directed to the relief of symptoms. Delay of death is not the primary goal; it may be a very worthwhile secondary goal, but it is not the primary one.

Nitrates in Chronic Heart Failure

Wulf-Dirk Bussmann

Nitrates are excellent agents in the treatment of acute heart failure. Striking clinical and hemodynamic improvement can be achieved in acute pulmonary edema, left heart failure during the acute phase of myocardial infarction, papillary muscle syndrome or rupture of the interventricular septum, hypertensive crisis concomitant with heart failure, and certain types of cardiogenic shock. Isosorbide dinitrate and nitroglycerin have been investigated extensively. Their actions will be reviewed briefly.

An oral dose of 20 mg isosorbide dinitrate invariably reduces left ventricular filling pressure. Cardiac output rises in patients with heart failure and declines moderately when failure is not present (Fig. 1). A similar, though somewhat more pronounced, effect is achieved with 6 mg nitroglycerin per hour (Fig. 2). The sublingual application of nitroglycerin is effective in the emergency of acute pulmonary edema. Systolic, diastolic, and mean pulmonary artery pressure decline by 30%–50% within 3–5 min, and cardiac output rises (Fig. 3). In pulmonary edema resulting from acute myocardial infarction, left ventricular filling pressure declines from 30 to 18 mmHg following intravenous nitroglycerin. Cardiac output increases markedly, and mean arterial pressure falls. This results in significantly lower peripheral resistance (Fig. 4).

Isosorbide 5-mononitrate, a metabolite of isosorbide dinitrate, acts similarly to the parent substance. Here is a good correlation to the plasma levels. A disadvantage is its inability to be absorbed sublingually (Fig. 5). Molsidomine should be mentioned in this context. No difference was seen in terms of its effect on left and right ventricular filling pressure, blood pressure, and cardiac output as compared to nitroglycerin (Fig. 6).

Can the nitrate effect be sustained or at least maintained for several days? This can be shown for intravenous nitroglycerin. A reduction in diastolic pulmonary artery pressure compared to an untreated control group could be demonstrated even after 48 h. The results were obtained in 55 patients with acute myocardial infraction (Fig. 7).

Long-term therapy is more difficult to evaluate. Therapeutic failures are encoutered in end-stage congestive heart failure. Counterregulatory mechanisms represent a problem here (pseudotolerance). A side-effect, at least of arterial vasodilators, is an enhanced propensity for edema.

Endogenous counterregulation can be illustrated well in terms the findings of Thadani et al. (1980). When a high dose of nitrates is given to a standing patient, blood pressure can decline precipitously to 80 mmHg. After one week, counterregulation will have led to a mitigation of this effect, which nevertheless is still present (Fig. 8). Counterregulation is difficult to circumvent.

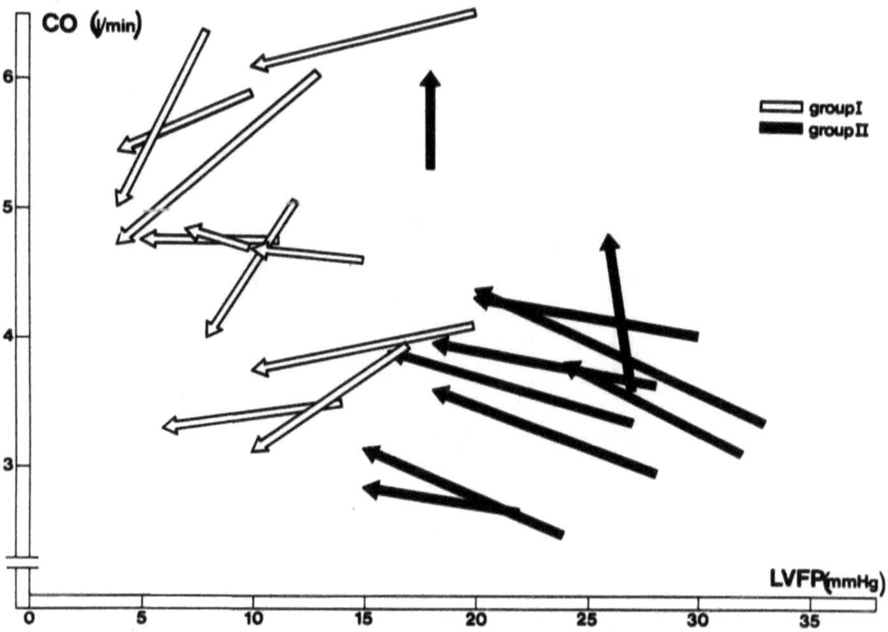

Fig. 1. Effects of oral dose of 20 mg isosorbide dinitrate on left ventricular filling pressure (*LVFP*) and cardiac output (*CO*)

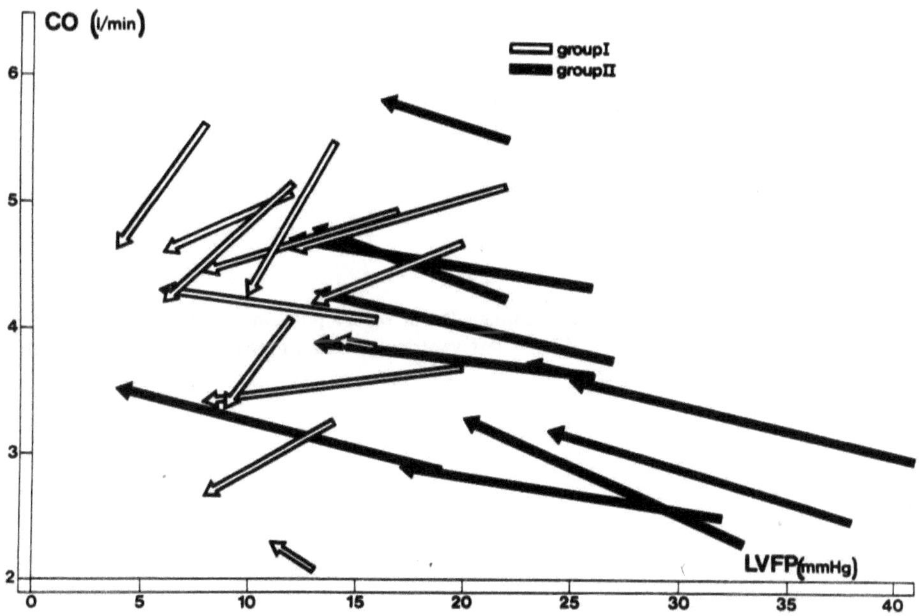

Fig. 2. Effects of nitroglycerin given intravenously in a dose of 6 mg/h on left ventricular filling pressure (*LVFP*) and cardiac output (*CO*)

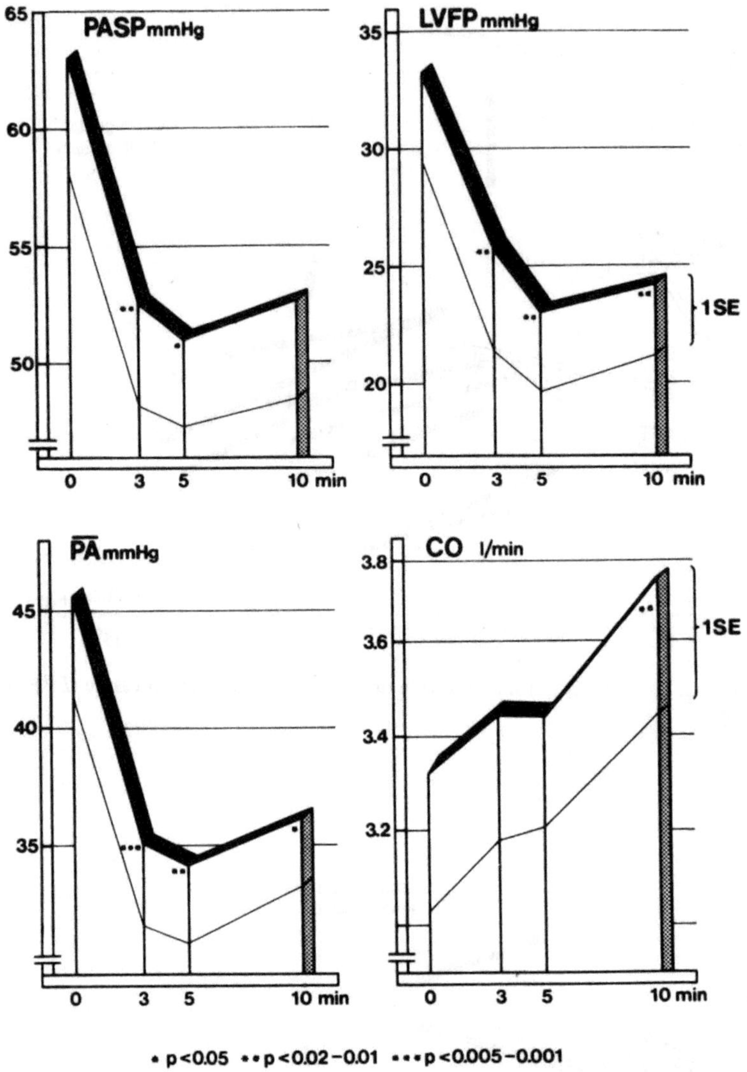

Fig. 3. Effects of nitroglycerin given sublingually in severe pulmonary edema (n = 7). *PASP*, pulmonary artery systolic pressure; *LVFP*, left ventricular filling pressure; *PA*, mean pulmonary artery pressure; *CO*, cardiac output

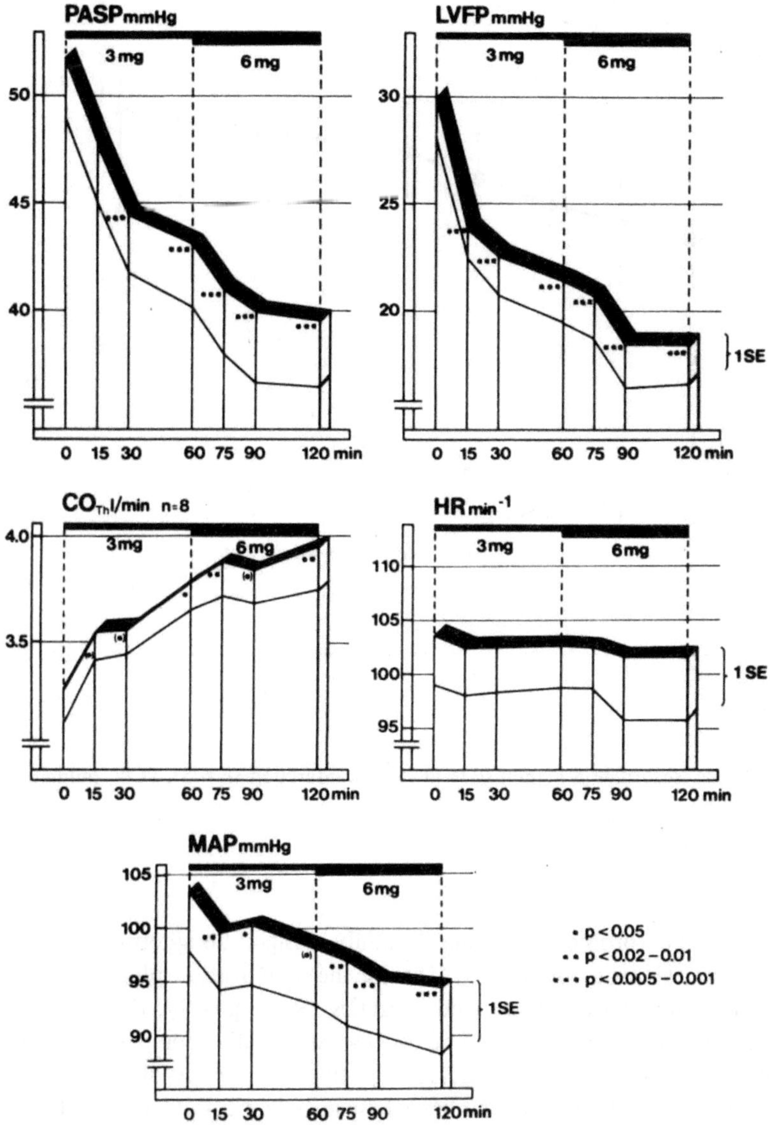

Fig. 4. Effects of nitroglycerin given intravenously in pulmonary edema following acute myocardial infarction (n = 10). *PASP*, pulmonary artery systolic pressure; *LVFP*, left ventricular filling pressure; *CO*, cardiac output; *HR*, heart rate; *MAP*, mean arterial pressure

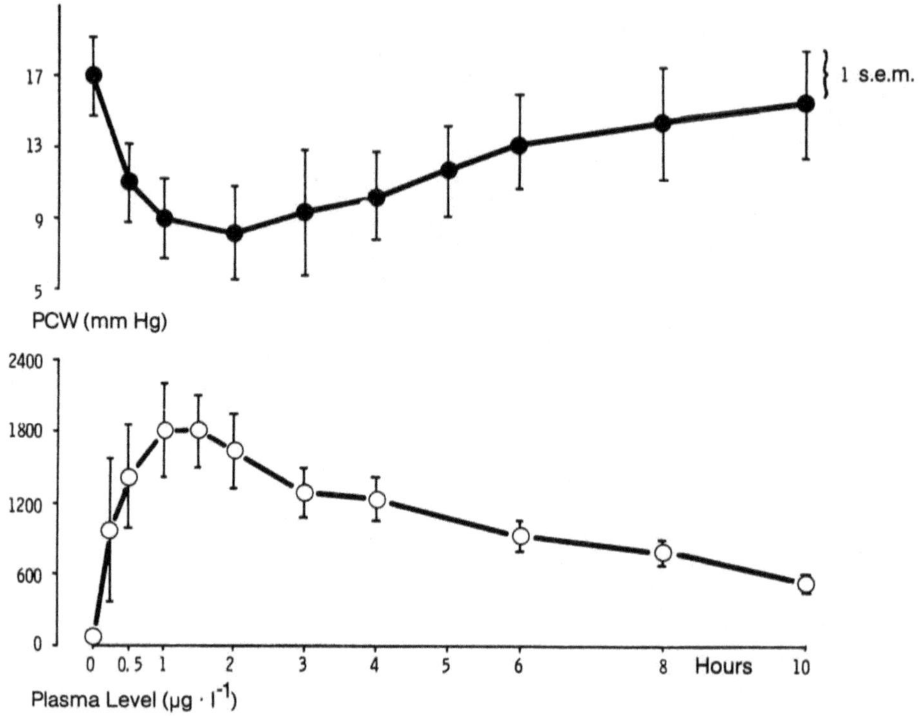

Fig. 5. Effects of 80 mg isosorbide 5-mononitrate in cardiac infarction ($n = 5$). *PCW*, pulmonary capillary wedge pressure

There are also other problems relating to study design. Long-term treatment requires a special clinical protocol (Fig. 9). During the stabilization phase, patients were given a basic regimen of digitalis and diuretics. Subsequently, they received 120 mg isosorbide dinitrate per day for 4–6 weeks. An equally long placebo phase followed. Phentolamine was later investigated, but this will not be discussed here. In addition, the acute effect of 5 mg sublingual isosorbide dinitrate was evaluated during the treatment and placebo phases (Lemke et al. 1978, 1979; Bussmann et al. 1981).

Twenty-three patients with chronic congestive heart failure were treated. Five were in NYHA class II and the remainder were in classes III and IV. In the majority of patients, coronary heart disease was the underlying cause of heart failure, although cardiomyopathies (n = 3) and valvular lesions (n = 6) were also included. Heart volume, which provides an indication of the severity of congestive failure, was increased to over 1200 ml (upper limit of normal, 800 ml). Ejection fraction derived from the angiogram was between 20% and 55%, with a mean of 45% (normal value 75%). The acute administration of sublingual isosorbide dinitrate also provides immediate relief in chronic heart failure (Fig. 10). Heart rate rose moderately, cardiac output remained unchanged, and pulmonary artery pressure declined significantly. During

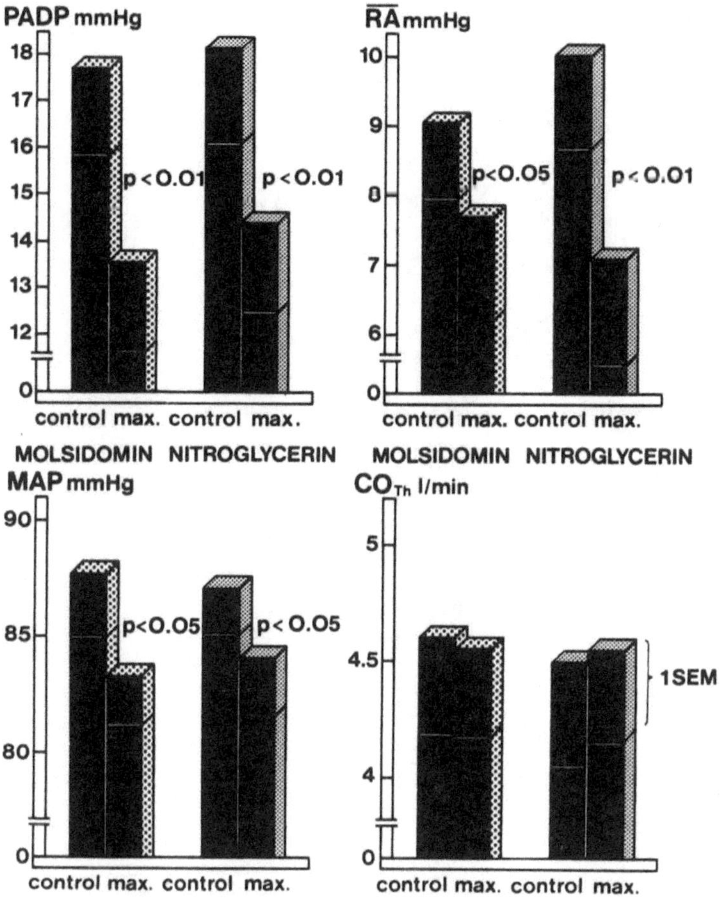

Fig. 6. Comparative effects of molsidomine in an intravenous dose of 8–12 mg and nitroglycerin in a sublingual dose of 1.6 mg ($n = 11$). *PADP*, pulmonary artery diastolic pressure; *RA*, mean right arterial pressure; *MAP*, mean arterial pressure; CO_{Th}, cardiac output

exercise with a work load of 50 W, heart rate actually declined significantly with a distinct increase in cardiac output and a considerable reduction in filling pressure (Fig. 11).

What effect does isosorbide dinitrate have after 4–6 weeks of continuous therapy? The characteristic changes are still evident in attenuated form. A moderate increase in cardiac output and slight though not significant reduction in pulmonary artery pressure are demonstrable (Fig. 12). Patients with high values dropped their filling pressures markedly (Fig. 13).

The findings of Franciosa et al. (1978) and Franciosa and Cohn (1980) are similar: capillary wedge pressure declined significantly during the chronic administration of 160 mg isosorbide dinitrate, but only slightly during the placebo phase. This was a double-blind study. It is important to note that no fundamental clinical improvement occurred, despite the attenuated but still evident hemodynamic effects of isosorbide

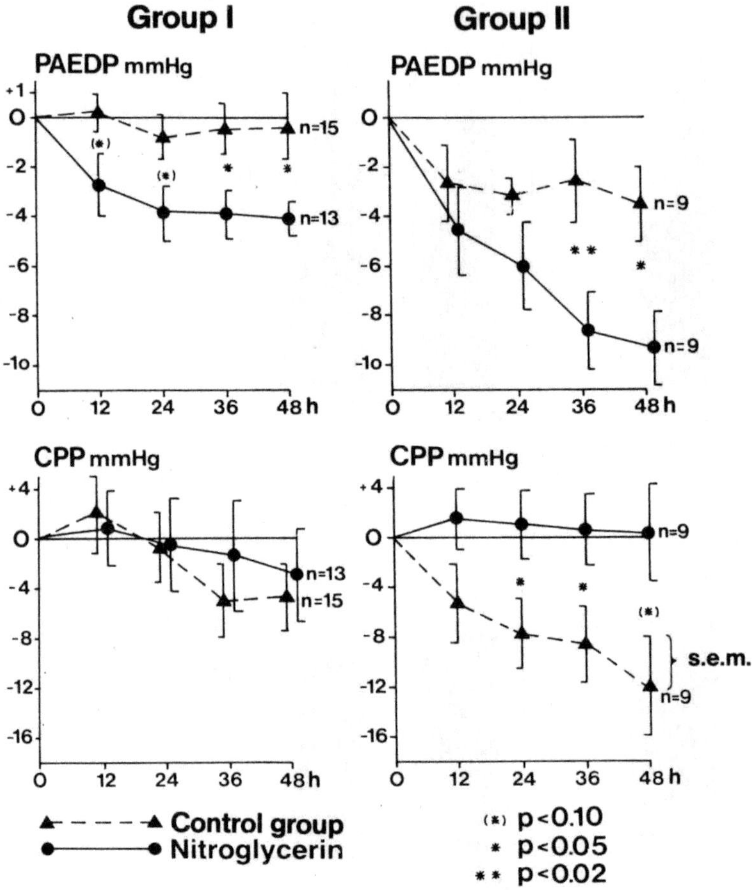

Fig. 7. Effects of intravenous nitroglycerin in acute myocardial infarction. *PAEDP*, pulmonary artery end-diastolic pressure; *CPP*, coronary perfusion pressure. *Group I*, PAEDP < 20 mmHg; *Group II*, PAEDP > 20 mmHg

dinitrate. Functional class remained unaffected. Heart volume did not decline. Findings improved only in individual cases. Franciosa and Cohn (1980) arrived at a similar conclusion.

There are many reasons for the unsatisfactory effects of nitrates in chronic congestive heart failure:

1. Nitrates do not exert an especially great effect on peripheral resistance at the doses used in the study. This accounts for the failure to achieve a marked increase in cardiac output, as is the case with prazosin, for example.
2. Counterregulatory mechanisms apparently play a role in at least partially reversing the venous pooling effect. This cannot be termed tolerance in the true sense.

The antianginal potency of nitrates, for example, is maintained during chronic administration, as Becker was able to demonstrate (Becker et al. 1976).

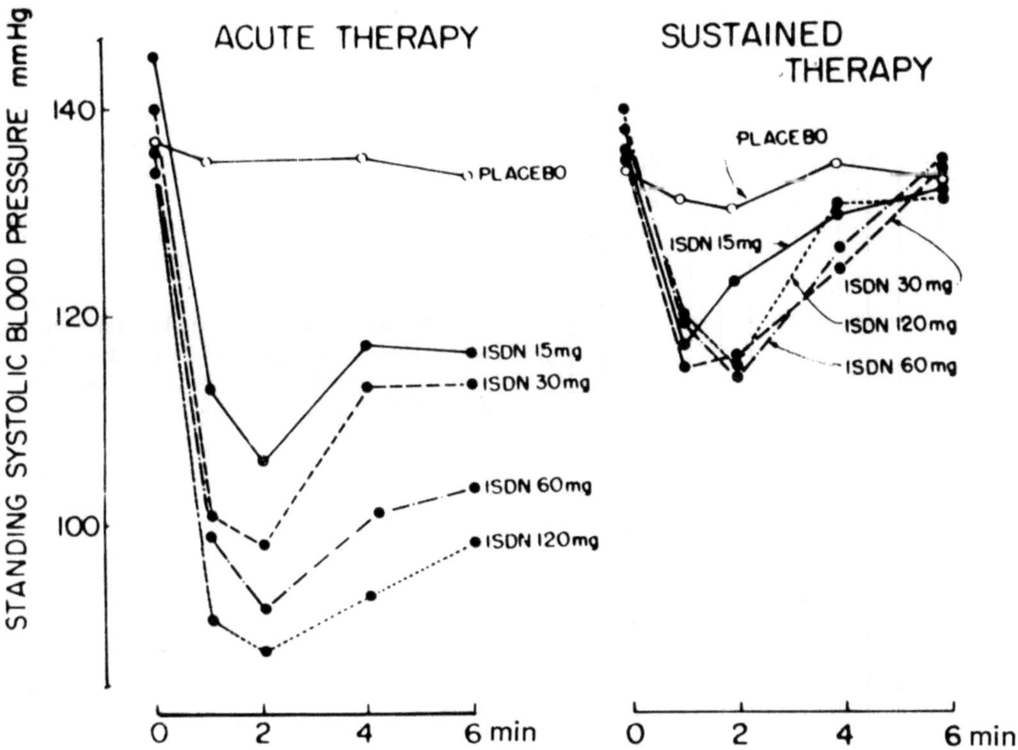

Fig. 8. Counterregulation of effects of nitrates on standing systolic blood pressure. *ISDN*, isosorbide dinitrate

Schneider, a member of our group, recently showed that the dose-response relationship for isosorbide dinitrate in angina pectoris remains demonstrable during chronic therapy (Schneider et al. 1981). An increase in dosage from 5 to 80 and from 30 to 480 mg/day, respectively, led to an increasing reduction in electrocardiographic

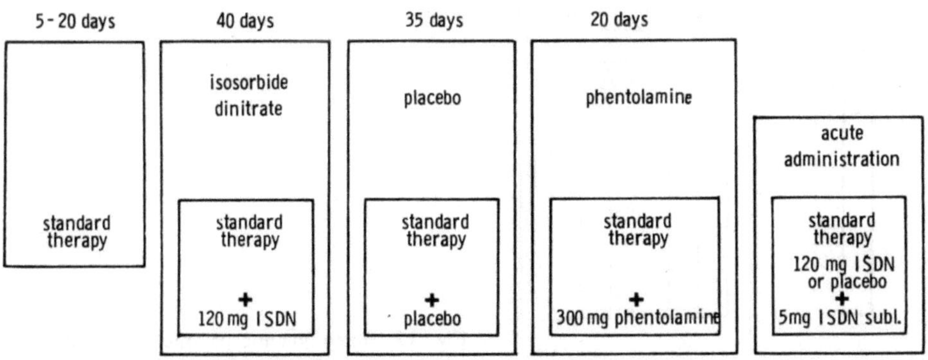

Fig. 9. Clinical protocol for long-term treatment with nitrates. *ISDN*, isosorbide dinitrate

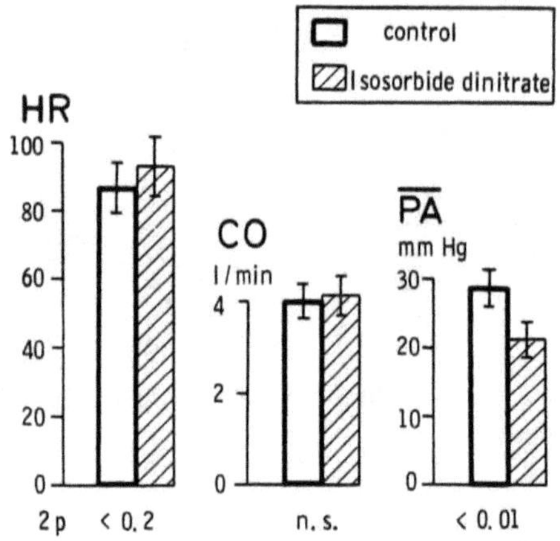

Fig. 10. Acute effects of 5 mg isosorbide dinitrate given sublingually. *HR*, heart rate; *CO*, cardiac output; *PA*, mean pulmonary artery pressure. Subjects at rest (n = 9)

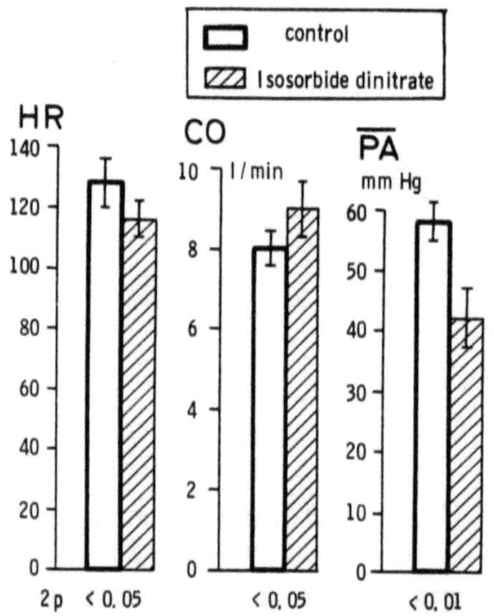

Fig. 11. Acute effects of 5 mg isosorbide dinitrate given sublingually. *HR*, heart rate; *CO*, cardiac output; *PA*, mean pulmonary artery pressure. Subjects under 50 W workload (n = 8)

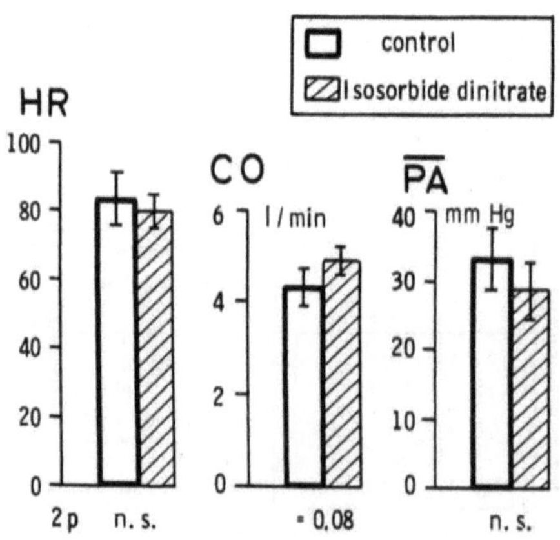

Fig. 12. Hemodynamic parameters in subjects at rest (n = 12) after 4 weeks' isosorbide dinitrate therapy (120 mg/day)

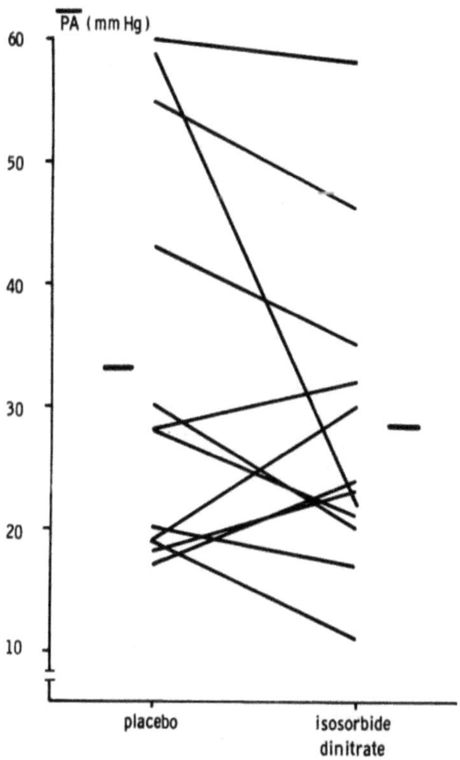

Fig. 13. Effects of isosorbide dinitrate (120 mg/day) for 4 weeks compared with those of a placebo

ST-segment depression during exercise. Pharmacokinetically, this finding suggests an association between increased dosage and increased serum levels in which the concentration of the parent substance, that is, unmetabolized isosorbide dinitrate, rose continuously. One might conclude from these findings that the dose which was employed, 120 mg/day, was too low for patients with chronic heart failure.

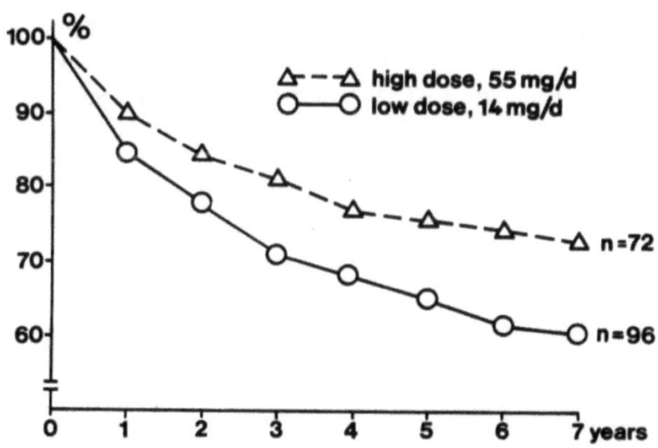

Fig. 14. Comparative 7-year survival rates in coronary heart disease following high and low doses of isosorbide dinitrate (n = 168)

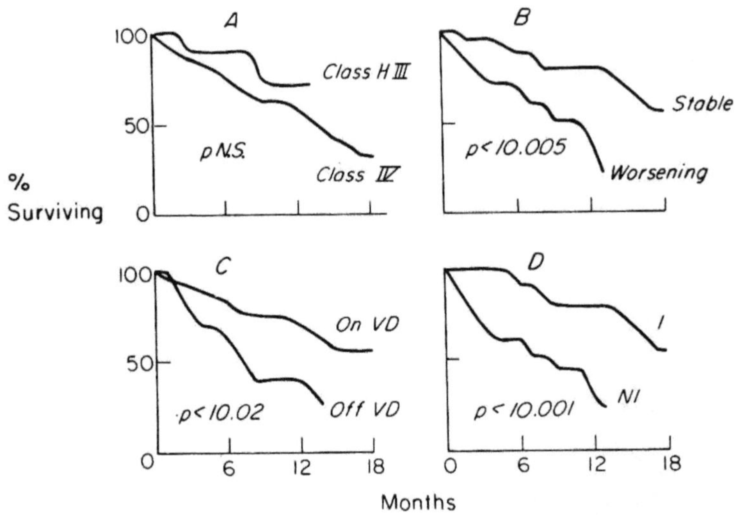

Fig. 15A–D. Percentages of those surviving among responders and nonresponders to therapy for coronary heart disease. *VD*, vasodilators; *I*, initial clinical improvement; *NI*, no improvement

An additional aspect sould be considered. In a recent retrospective study, we were able to demonstrate that patients with coronary heart disease who were treated with high-dose isosorbide dinitrate had a lower 7-year mortality rate than patients given low doses of the same substance (Fig. 14). One might speculate that nitrates ultimately also influence the prognosis in congestive heart failure by as yet unknown mechanisms. Newer findings suggest an inhibition of platelet aggregation (Schafer et al. 1980). These issues remain to be clarified. Nitrates are frequently combined with hydralazine. Massie et al (1981) observed improved survival in patients who initially responded to therapy. Shown are the curves for both responders and nonresponders (Fig. 15).

In conclusion, the role of nitrates in the therapy of congestive heart failure has not yet clearly been defined. First of all, nitrates may have an inadequate effect on the arterial site. Nonetheless, individual patients display a reduction in filling pressure and show clinical improvement. The majority of patients, however, do not obviously benefit from this therapy. It remains to be seen whether more sweeping effects can be achieved by increasing the dosage.

References

Becker HJ, Walden G, Kaltenbach M (1976) Gibt es eine Tachyphylaxie, beziehungsweise Gewöhnung bei der Behandlung der Angina pectoris mit Nitrokörpern. Verh Dtsch Ges Inn Med 82:1208–1210

Bussmann W-D, Lemke R, Kaltenbach M (1981) Chronic isosorbide dinitrate for treatment of refractory heart failure. In: Just H, Bussmann W-D (eds) Therapie der chronischen Herzinsuffizienz mit Vasodilatantien. Verlag edition medizin, Weinheim Derrfield Beach Basel, pp 10–14

Franciosa JA, Nordstrom LA, Cohn JN (1978) Nitrate therapy for congestive heart failure. JAMA 240:433–446

Franciosa JA, Cohn JN (1980) Substained hemodynamic effects without tolerance during long-term isosorbide dinitrate treatment of chronic left ventricular failure. Am J Cardiol 45:648–654

Lemke R, Lippok R, Kaltenbach M, Bussmann W-D (1978, 1979) Orale Langzeitbehandlung der chronischen Linksherzinsuffizienz mit Isosorbiddinitrat und Phentolamin. Z Kardiol [Suppl] 5:16, Z Kardiol 68:82–88

Massie B, Ports T, Chatterjee K, Parmley W, Ostland J, O'Young BS, Haughom F (1981) Long-term vasodilator therapy for heart failure: clinical response and its relationship to hemodynamic measurements. Circulation 63:269–278

Schafer AI, Laxeander RW, Handin RI (1980) Inhibition of platelet function by organic nitrate vasodilators. Blood 55:649–654

Schneider W, Stahl B, Kaltenbach M, Bussmann W-D (1981) Dosis-Wirkungsbeziehungen bei der Behandlung der koronaren Herzkrankheit mit Isosorbiddinitrat (Abstr). Z Kardiol 70:321

Thadani U, Manyari D, Parker JO, Fung H (1980) Tolerance to the circulatory effects of oral isosorbide dinitrate. Circulation 61:526–535

Discussion

Just, Freiburg: Should not a difference be made between patients with coronary artery disease and those with valvular heart disease or primary cardiomyopathy, when drugs like isosorbide dinitrate are considered, as in patients with coronary artery disease the influence of the drug on coronary perfusion and flow dynamics has to be taken into account?

Bussmann, Frankfurt: This point is important, I think. This study included 23 patients with severe heart failure of varying etiology. However, we also separated a group of patients with coronary artery disease, and were not able to observe any essential difference. However, especially using isosorbide dinitrate differences may be observed.

Reindell, Freiburg: The predominant effect of nitrates is a marked venous pooling with a consequent decrease of both right atrial and pulmonary wedge pressure and a reduction of left ventricular volume. If this is true, the heart size must decrease, which has not been observed. Here I see a contradiction!

Bussmann, Frankfurt: I think we have to distinguish between the antianginal effect of a drug and the hemodynamic effect under chronic treatment. I assume that nitrates exert a sustained antianginal effect during long-term treatment. However, I am not sure whether this effect is only due to the increase of the venous capacitance. I think we have to consider a direct effect on the coronary arteries, too, and also the fact that the ventricular volume changes induced by nitrates are too small to be detected by current methods for determination of total heart volume; however, echocardiographic studies have indeed shown a decrease in cavity size. In addition, it must be emphazised that high dosages of nitrates are used. In out study we used 120 mg/day; today we would give 240 or 360 mg/day.

Auditorium: In Monte Carlo, Rudolf (Munich) reported a marked alternation of the effect of isosorbide dinitrate after long-term treatment over 2 years. How do you explain this difference?

Bussmann, Frankfurt: I think we have to distinguish between the 7-year follow-up – which was a retrospective study – and the antianginal effect of nitrates during long-term treatment. Rudolf used isosorbide dinitrate in a slow-release preparation; in our study isosorbide dinitrate was used without slow release – I think this is a difference. At present we do not have the results of the long-term treatment, the results refer to a 1-week period. We are, however, planning a double-blind study in order to determine the long-term effects of isosorbide dinitrate.

The Long-Term Effects of Dihydralazine and Hydralazine on Intractable Cardiac Failure

Ph. Morand, G. Lavigne

The object of this work was to study the long-term effects of the hydrazinophthalazines [hydralazine (Apresoline) or dihydralazine (Nepresol)] using our own clinical studies (38 patients treated with dihydralazine) and a review of the literature. Their short-term effects are well known but there is less information available on their long-term effects (2, 8). It is generally accepted that the pharmacological effects of the two commercially available hydrazinophthalazines are very similar.

Patients

Thirty-eight patients with cardiac insufficiency, belonging to NYHA class IV, were treated for up to 48 months. The average age was 61 years. Nineteen of these patients had nonobstructive cardiomyopathy, 11 had ischemic heart disease, and seven had aortic or mitral valvular cardiopathy. All these patients had intractable cardiac failure as shown by the following average hemodynamic results: cardiac index 1.5 $l/min/m^2$, stroke volume 32 ml; pulmonary wedge pressure 25.5 mmHg (in all cases it was greater than 20 mmHg; systemic vascular resistance was greater than 2100 $dyn/s/cm^{-5}$).

Dosage

The dosage of dihydralazine was 50, 75 or 100 mg 6-hourly, determined on the basis of a target of 50% or more increase in cardiac index over the initial value. To avoid risk of systemic lupus erythematosus (SLE) developing, acetylator status was recorded by the isoniazide test (blood isoniazide level 3 h after oral administration of 300 mg) (1, 8). Slow acetylators were given 50 mg dihydralazine 6-hourly, rapid acetylators 75 or 100 mg 6-hourly. These doses can be exceeded without any negative effects, especially in rapid acetylators.

Packer et al. (9) showed that the desired hemodynamic results could always be obtained if one reached or surpassed 900 mg/day.

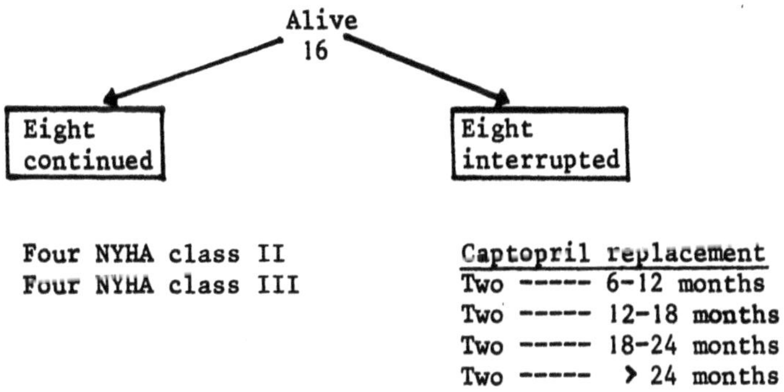

Fig. 1. Long-term follow-up in dihydralazine therapy – survivors

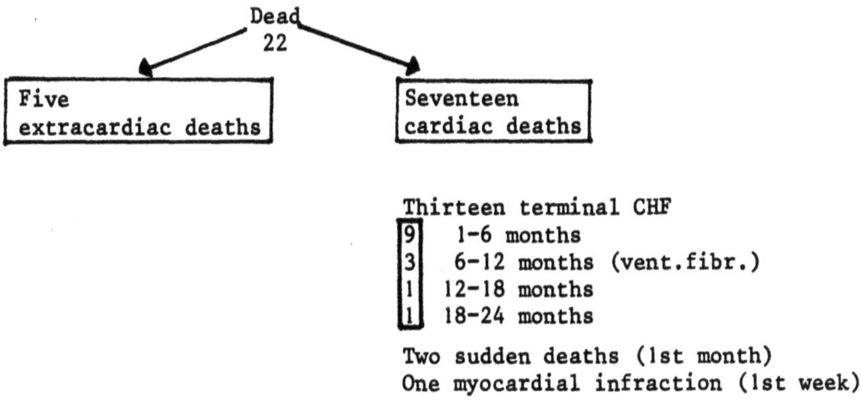

Fig. 2. Long-term follow-up in dihydralazine therapy – nonsurvivors

Results

Clinical

The clinical results are summarized in Figs. 1 and 2. Twenty-two patients died, five from extracardiac causes and the remaining 17 from progressive deterioration due to their heart disease. There were two sudden deaths in the first month and one myocardial infarction in the first week. Of the 16 survivors, eight continued the treatment and showed functional improvement at the beginning of the treatment (for a period of 6–36 months); in eight patients the treatment was changed to captopril.

It is difficult to evaluate the effects of the treatment with dihydralazine and hydralazine on the survival of these patients. We tried nonetheless to do this by comparing the actuarial survival curves of our 19 patients with nonobstructive cardiomyopathy with those established by Connert (3) using conventional digitalis and diuretics treatment. In both series the pulmonary wedge pressure is higher than 20 mmHg. Beyond 6 months, the curve representing the treated series is more favorable (Fig. 3). The survival curves established by Massie et al. (5) for 56 patients with cardiac insufficiency of varying origin and treated with hydralazine and nitrates show a survival rate that is almost identical: 78%, 65%, and 37% at 6 months, 12 months, and 18 months, respectively.

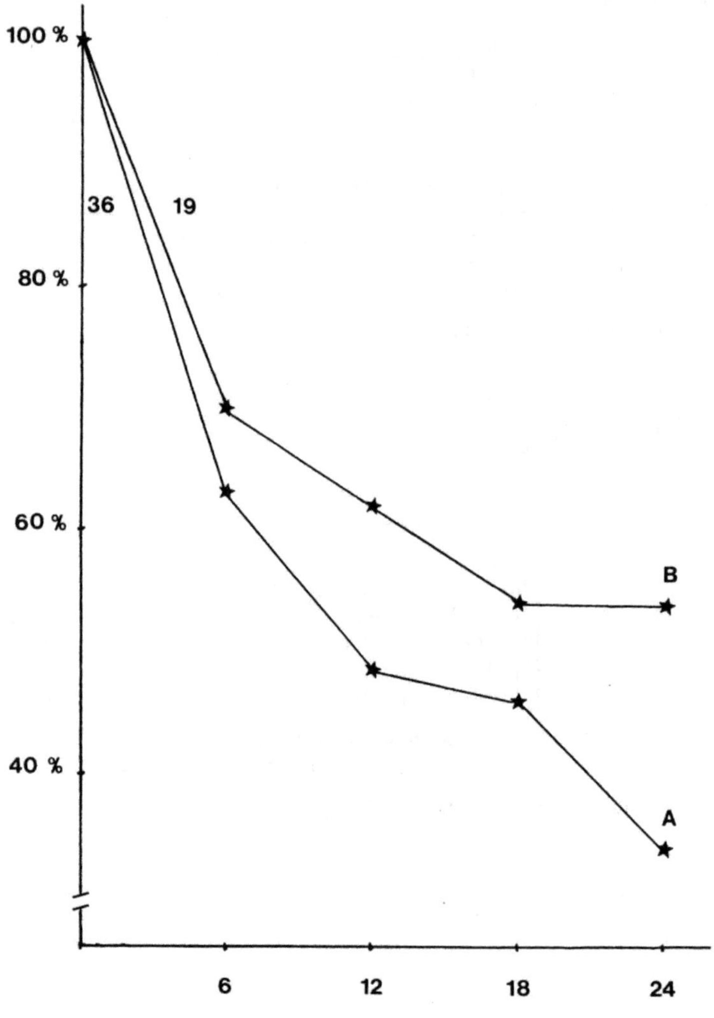

Fig. 3. Survival in nonobstructive cardiomyopathy. Mean pulmonary wedge pressure 20 mmHg. *A*, digitalis and diuretics; *B*, dihydralazine

Hemodynamics

The 80% increase in cardiac output observed by the 2nd day remained at 6–12 and 18 months in spite of a slight decrease at the beginning of the 6th month (Fig. 4). Cardiac frequency markedly decreased in the 1st month and this decrease becomes significant by the 6th month (Fig. 5). Pulmonary wedge and right atrial pressure were also modified but the change remained not significant at the 12th and 18th months. For three patients the decrease in pulmonary wedge pressure was even more marked at the 18th month. It was in these patients, obviously, that the dyspnea was most improved.

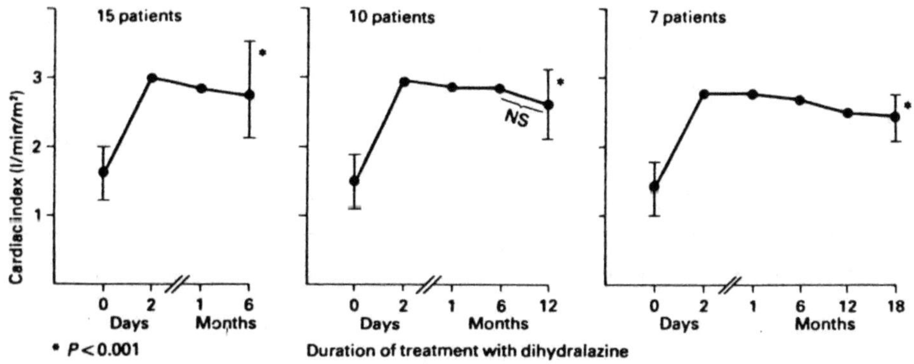

Fig. 4. Cardiac index in heart failure patients treated with dihydralazine and surviving to 6, 12, and 18 months

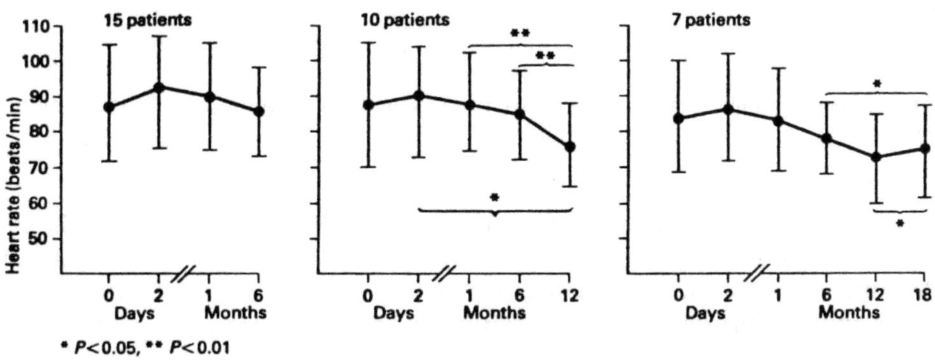

Fig. 5. Heart rate in heart failure patients treated with dihydralazine and surviving to 6, 12, and 18 months (means ± *s*)

The average arterial pressure was not changed even in the long term in this type of patient. The systemic vascular resistance and systolic volume were calculated from the preceeding parameters and vary in the same direction and magnitude. Fig. 6 shows the results obtained in four patients followed up for 2 years; there was a persistent increase in systolic volume, decrease in systemic vascular resistance, and virtually no modification in pulmonary wedge or right atrial pressure. Mathey et al. (7) obtained virtually the same results while studying cardiac output and vascular resistance 6 days and 6 months after the start of the treatment. They found a net increase in cardiac index (60%) associated with a marked drop in systemic vascular resistance from 2000 to 1000 dyn/s/cm^{-5} after 6 months of treatment.

They introduced an important idea concerning renal circulation, in that they found the renal perfusion significantly increased from 300 to 380 ml/mm^{-1} with a decrease in renal vascular resistance of the same magnitude. The effects of vasodilatation on other local flows are less certain; hydralazine supposedly increases coronary and hepatic flow at rest.

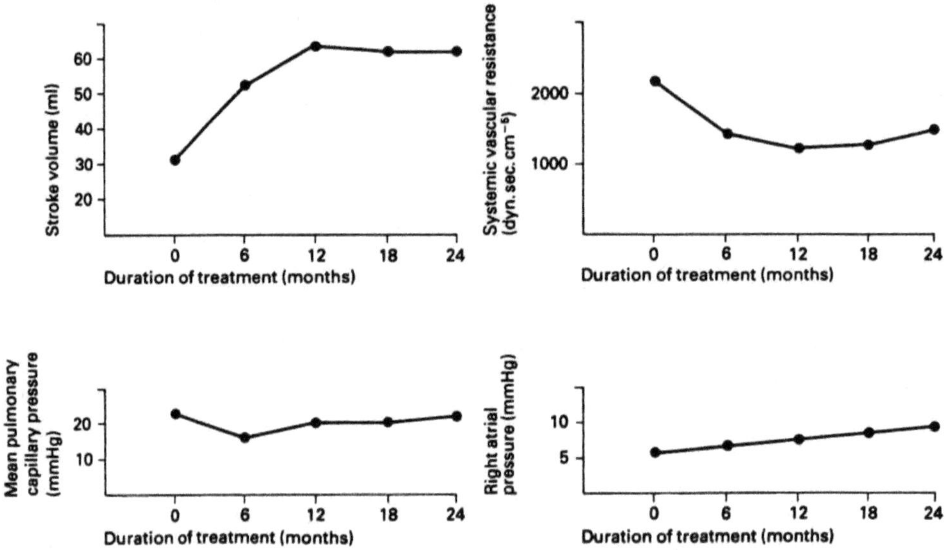

Fig. 6. Hemodynamic changes up to 24 months in four heart failure patients treated with dihydralazine

Eitchett et al (4) studied the effects of stopping hydralazine therapy on the hemodynamic parameters in patients with severe heart failure treated with 300 mg of hydralazine for 4–6 weeks. This interruption was followed by a rapid deterioration and a return to the status quo ante. This is important, and shows that the treatment with hydralazine and dihydralazine must be continuous. The same principle was studied by Massie et al. (5), who constructed survival curves based on different criteria. Survival at 12 months was markedly higher in those patients who continued treatment than in those who discontinued treatment.

In our study, we analyzed survival according to the origin of the cardiac failure for a period of 6 months. Survival is better in nonobstructive cardiomyopathy than it is in vascular or ischemic heart disease (Fig. 7).

The toxicity of hydralazine and dihydralazine is virtually limited to nausea and severe headaches at the start of the treatment, which disappear with continued treatment, and to SLE. The lupus syndrome induced by hydralazine was well studied by Perry (10), who found 44 cases with clinical manifestations among 308 patients treated with 500 mg or more a day. Fourteen of these cases were serious. One should suspect this lupus syndrome when one or more of the following signs or symptoms appear: fever with arthralgia, asthenia, general discomfort, hepatosplenomegaly, or adrenomegaly; the typical specific rash is rare. No patients died due to hydralazine or dihydralazine toxicity.

Perry did not find any specific anatomical renal lesions. Clinical signs and symptoms disappeared with cessation of treatment.

In 12 of our patients, the level of antinuclear antibodies was determined for at least 6 months, and five of these patients had a titer higher than 1/400 (Table 1). Among these five no clinical signs were evident. Treatment was continued, if effective, after

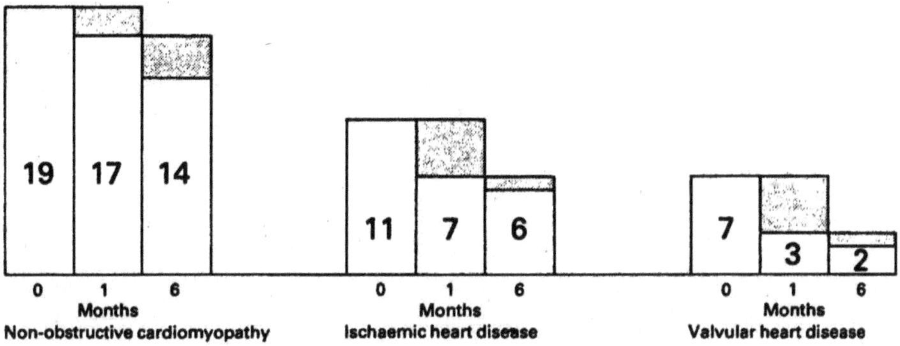

Fig. 7. Survival in 37 heart failure patients treated with dihydralazine, according to etiology. *Shaded areas* represent deaths

this titer was reached without any inconvenience. Of these five patients, three were slow acetylators and two fast acetylators, but it must be remembered that in order to avoid complications in slow acetylators the dosage of dihydralazine was 150 mg/day.

In the literature, we found only one reported case of SLE induced by dihydralazine (11). Reports of angina pectoris or increased risk of myocardial infarction are uncommon (one infarction in our study 1 week after treatment, one worsening angina in Massie's series).

Usually the difference in behavior of the cardiovascular system with respect to hydralazine or dihydralazine is stressed. Depending on whether the heart is insufficient or not, hypotension, tachycardia, and decrease in coronary blood pressure may be observed. These parameters are not altered when the patient with cardiac insufficiency is in a supine position. Therefore hydralazine and dihydralazine are used in cardiac insufficiency irrespective of its origin.

Table 1. Circulating antinuclear antibodies (0–24 months) in 12 patients (no clinical manifestations). *S*, slow acetylator; *F*, fast acetylator; *I*, intermediate. Significant titer ≥ 1/400

Control	M1	M6	M12	M18	M24	
–	–	–	–	1/20	1/20	S
–	–	–	–	1/400		F
–	–	1/400	1/800	1/800	1/800	S
–	–	–	1/20	1/20	1/20	F
–	–	1/20	1/20			S
–	1/100	1/200				F
–	1/20	1/100		1/800		S
–	1/100	1/100				F
1/20	1/100	1/100	1/100	1/100		I
–	–	1/100	1/100	1/100	1/400	F
–	1/200	1/400				S
–	1/200		1/200			F

The recent paper by Massie et al. nonetheless encourages caution in coronary insufficiency, since they showed that in an upright position the cardiac frequency increases; the mean arterial pressure, stroke volume index, and PCW decrease slightly. These conditions favor myocardial hypoxia (6).

Conclusions

Hydralazine and dihydralazine have the same hemodynamic effects in the long term as those noted in the first days of treatment: at least 50% increase in systolic volume and 25% decrease in filling pressure. These effects persist in the long term without tachyphylaxis. No etiology except aortic stenosis is a counterindication. Prudence must be observed in ischemic cardiac insufficiency. Nonobstructive and nonischemic cardiomyopathy are the best indications.

In the long term, hydralazine – or dihydralazine – induced lupus syndrome is possible with doses less than 300 mg but may show no clinical manifestations. The quality of life is certainly improved. It is impossible to affirm that this treatment has an effect on survival, due to the difficulty of comparison with control populations, but it seems probable that it does.

References

1. Boxenbaum HG, Riegelman S (1974) Determination of isoniazide and metabolites in biological fluides. J Pharm Sci 63:1191
2. Chatterjee K, Parmley WW, Massie B, Greenberg B, Werner J, Klausner S, Norman A (1976) Oral hydralazine therapy for chronic refractory heart failure. Circulation 54:879
3. Convert G (1978) Etude prognostique des cardiomyopathies primitives non obstructives. Med Dissertation, Lyon University
4. Eitchett DH, Marin Neto JA, Oakley CM, Goodwin JF (1979) Hydralazine in the management of left ventricular failure. Am J Cardiol 44:303–309
5. Massie B, Ports T, Chatterjee K, Parmley W, Ostland J, O'Young J, Haughom F (1981) Long-term vasodilator therapy for heart failure: clinical response and its relationship to hemodynamic measurements. Circulation 63:269–278
6. Massie B, Kramer B, Gaughom F (1981) Postural hypotension and tachycardia during hydralazine-isosorbide dinitrate therapy for chronic heart failure. Circulation 63:658–664
7. Mathey DM, Hanrath P, Polste J, Witte G, Montz R, Bleifield W (1981) Short term and long term effects of oral hydralazine on left ventricular pump function and renal haemodynamics in chronic left heart failure. In: Rutishauser W (ed) The place of vasodilators in the long-term treatment of intractable heart failure. Hans Huber, Bern Stuttgart Vienna
8. Morand P, Lavigne G, Masson D, Latour F, Alison D (1979) Traitement de l'insuffisance cardiaque chronique sévère par la dihydralazine. Arch Mal Coeur 3:268
9. Packer M, Meller J, Medina N, Gorlin R, Herman MV (1980) Dose requirements of hydralazine in patients with severe chronic congestive heart failure. Am J Cardiol 45:655
10. Perry HM (1973) Late toxicity to hydralazine ressembling systemic lupus erythematosus or rheumatoid arthritis. Am J Med 54:56
11. Wechsler B, Brion NO, Colan C, Ximines H, Godeau P (1979) Lupus érythémateux induit par la dihydralazine. Nouv Presse Méd 45:8

The Use of Hydralazine for Chronic Cardiac Failure: Results and Unanswered Questions

D.G. Mathey

Acute and Chronic Hydralazine Efficacy

It has been shown repeatedly that hydralazine's peripheral arteriolar dilating effect raises cardiac output (Fig. 1) in patients with chronic left heart insufficiency by up to 100%, the mean being 50% (1, 2). While there have been cases of tachyphylaxis reported with treatment involving prazosin, the hemodynamic effect of hydralazine during long-term therapy does not diminish. These observations were confirmed by Chatterjee et al., who determined after 3 months hydralazine treatment a 65% increase in the cardiac index compared to an increase of 56% following an acute hydralazine dose (3).

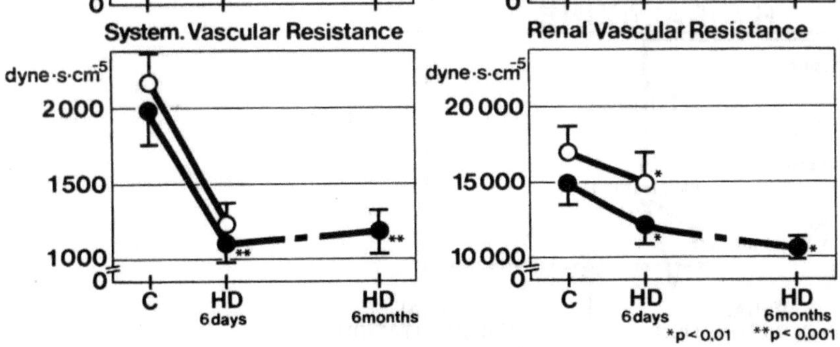

Fig. 1. Acute and long-term effects of hydralazine on left ventricular and renal hemodynamics. *Open circles*, total group of 12 patients; *closed circles*, long-term group of six patients

Organ Blood Supply

The question arising from these results is how the blood volume is distributed in various vascular regions. Since it is difficult in humans to measure the blood supply of an organ, only a few such studies have so far been undertaken. We have attempted to measure the immediate (acute) and long-term effects of hydralazine on renal plasma flow using isotope clearance (2). We found in this investigation that although the cardiac index rose by 50% (Fig. 1), the renal plasma flow did not significantly change following 6 days oral hydralazine treatment. However, after 6 months of treatment the kidney perfusion increased by 32%, although the cardiac index remained constant.

The failure of hydralazine to initially increase the kidney perfusion in proportion to the cardiac output we attribute to the sympathetic alpha-stimulation of the kidney, which is based on the lowered systemic vascular resistance induced by hydralazine (Fig. 2). During chronic hydralazine therapy, a "secondary adjustment" of the baroreceptors, could have developed, leading to decreased sympathetic alpha-stimulation of the kidneys with improved kidney perfusion. In addition, the increased production of intrarenal vasoconstrictors such as angiotensin could be responsible for the failure of renal plasma flow to increase following 6 days oral hydralazine therapy.

In a study recently published by Leier et al. (4) it was shown that after a single oral dose of 75 mg hydralazine the blood supply to the liver and extremities increased almost proportionally to the cardiac output, and that – in contrast to our findings – kidney perfusion also rose drastically (Fig. 3). These studies were carried out strictly during rest.

Hydralazine Efficacy During Exercise

Since chronic left heart insufficiency manifests itself particularly under strain, the pertinent question was whether hydralazine would offer continued hemodynamic improvement during exercise. In accordance with the results of other studies, we actually

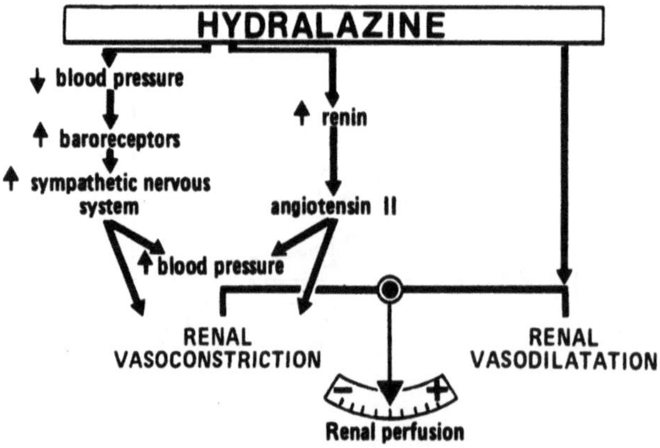

Fig. 2. Possible mechanisms of hydralazine efficacy on kidney perfusion

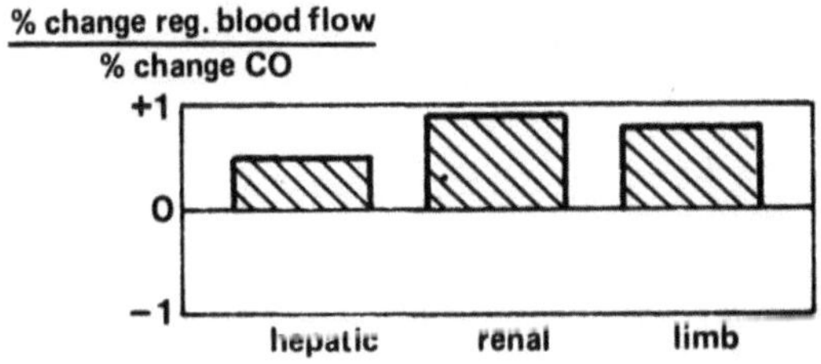

Fig. 3. Relationship between change in organ circulation and change in cardiac output with hydralazine. (Leier et al. 1981)

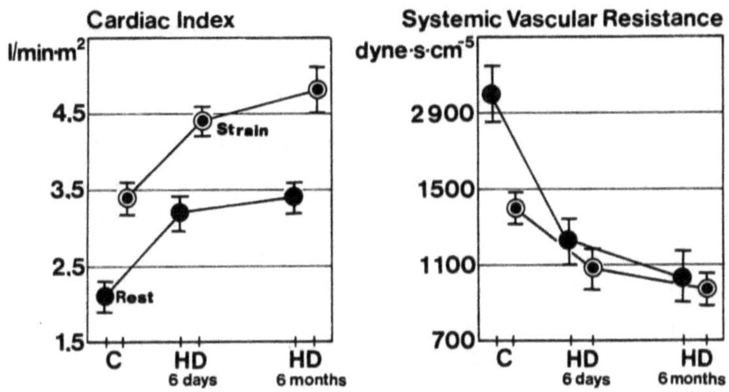

Fig. 4. Marked increase in cardiac index during exercise using hydralazine (*HD*)

observed a considerable increase in cardiac output acutely and chronically, as tested using a bicycle ergometer (Fig. 4).

To summarize the data mentioned so far, it has been established that hydralazine raises cardiac output markedly both at rest and under stress during acute and chronic treatment. It remains questionable whether the organ blood supply is also increased, even though much evidence seems to indicate such an assumption.

There is a wide clinical range in the favorable effects. Only five of 19 patients felt better with oral hydralazine treatment; in 12 patients the symptoms of cardiac insufficiency remained unchanged, and in two the symptoms worsened. In the light of such discrepancy between hemodynamic and clinical results it is necessary to investigate the effect of vasodilators – especially hydralazine – and their role in the prognosis of patients with cardiac insufficiency. The answers to these questions will hinge on whether vasodilators provide a safe method of treatment.

In a double placebo experiment, Franciosa and Cohn (5) found that the maximal load capacity was the same as that of the placebo when a combination preparation of hydralazine and isosorbide dinitrate was employed. Invasive hemodynamic studies were, however, not conducted. These same authors then extended their observations to long-term hydralazine treatment (6). Thirty-two patients were randomly divided into two groups. The hydralazine group received 200 mg daily for 26 weeks; the other was a placebo group. Although in time the load capacity rose significantly, this rise did not vary between the two groups.

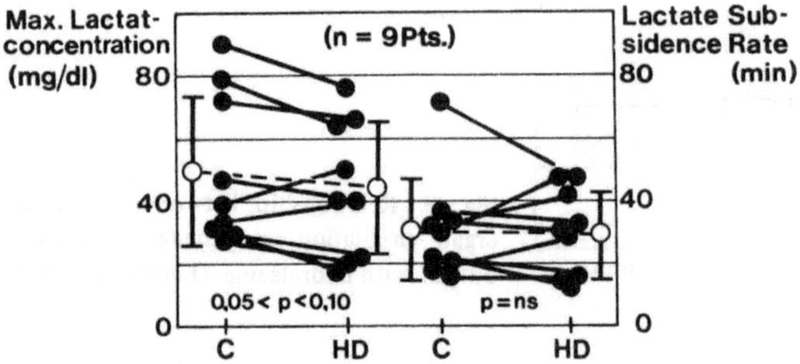

Fig. 5. Effect of hydralazine (*HD*) on maximal serum lactate level during exercise and lactate subsidence rate thereafter. (Rubin et al. 1980)

Congruously with this, Rubin et al. (7) discovered that the maximal serum lactate level during ergometric exercise was not lower with hydralazine than without, although the cardiac capacity did increase by 20% (Fig. 5). In addition, the lactate subsidence rate in the recovery phase immediately following exercise did not change.

These results allow the assumption that the perfusion of skeletal muscle does not increase with hydralazine. Two possible mechanisms should be kept in mind: either the increase in cardiac output does not reach the stressed organs, or the increased circulation is short-circuited at the precapillary level, thus not contributing to the nutritive organ blood supply.

We too have pursued this important question of skeletal muscle circulation during hydralazine treatment; so far, however, without success. We attempted to determine thallium uptake in skeletal muscle by infusion thallium-201 into the femoral artery and then collecting samples from the femoral vein. It seems, however, that thallium uptake in skeletal muscle is not dependent on the circulation. As a consequence, this method proved unsuitable for our purposes. We are now conducting a double blind-placebo study in which we are comparing hydralazine with a placebo. While in a seated position, the patient is gradually stressed until he presents the expected symptoms. On each exercise level the pressures and the cardiac output are recorded. In this way, we hope to be able to correlate capacity and hemodynamics at various load levels.

Prognosis of Cardiac Insufficiency During Hydralazine Treatment

To what degree, if at all, long-term hydralazine treatment influences the prognosis of cardiac insufficiency remains unresolved. Probably only Massie et al. (8) have retrospectively studied the course of hydralazine therapy (Fig. 6).

In 24 of their total of 56 patients there was no initial clinical improvement. The clinical findings of 32 patients improved by one or more stages. There was high mortality in the first 6 months among the 24 patients whose condition did not improve with hydralazine. Nine of these patients died in this period, their clinical condition failing to improve during the course of treatment. Nine of the 32 patients whose condition did improve initially with hydralazine also died, but only two during the first 6 months of treatment. In addition, the condition of the other 23 patients continued

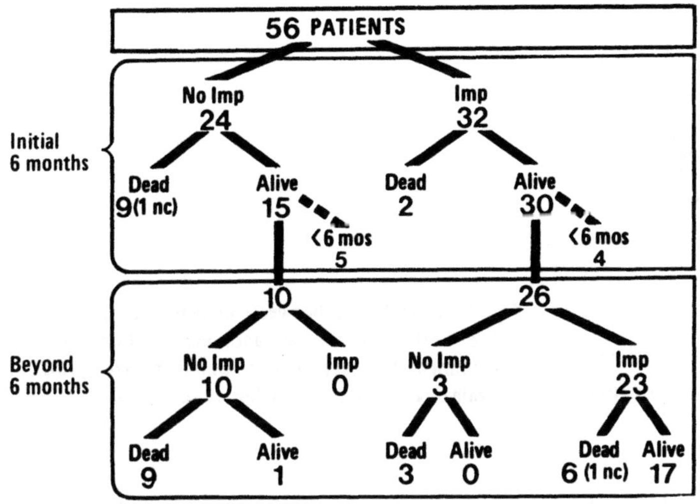

Fig. 6. Clinical results of long-term hydralazine treatment for severe cardiac insufficiency. (Massie et al. 1981)

to improve. The mortality of this group was high, but not comparable to that of the "nonresponder" group. According to the results of this first study, with all the disadvantages inherent in such a retrospective analysis, patients initially helped by hydralazine – even during long-term therapy – did display lasting improvement with lower mortality.

Dosage and Side Effects

For the treatment of cardiac insufficiency, hydralazine is generally given in four daily doses of 25–100 mg. In special cases, doses of up to 1 g and more have been given. Side effects, usually headaches and nausea, were observed in just over half the patients. In one-fifth of the patients these side effects forced termination of the therapy.

Indications

Due to the above side effects and to the yet unanswered questions, we limited the use of vasodilators to patients suffering from cardiac insufficiency of clinical stages III and IV. The etiology of the cardiac insufficiency was of secondary importance, although we found that the increase in cardiac output during exercise was greater in patients with congestive cardiomyopathy than in patients with cardiomyopathy as a result of coronary disease.

Cardiomegaly is an important prerequisite for the use of hydralazine, since only then can a good hemodynamic effect be expected. Packer et al. (9) were able to verify that the increase in the stroke work index correlates well with the left ventricular end-diastolic diameter prior to hydralazine therapy (Fig. 7). They attributed this to the

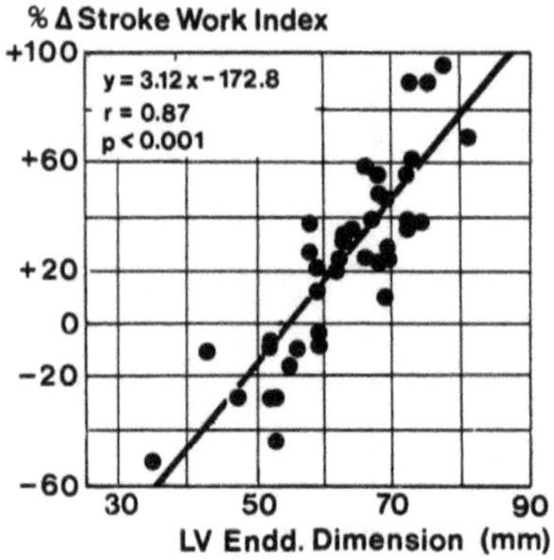

Fig. 7. Correlation between increase in stroke work index with hydralazine and echocardiographically determined left ventricular chamber size prior to hydralazine treatment. (Packer et al. 1980)

fact that the left ventricular wall tension – considered to be a product of the arterial pressure and the left ventricular chamber size – is more strongly reduced in a large ventricle treated with vasodilators than in a smaller one.

It should be mentioned that hydralazine can also be used in the treatment of mitral or aortic insufficiency by decreasing the regurgitation volume in favor of a greater forward stroke volume. Finally, hydralazine is administered for primary pulmonary hypertension. With doses of 4 × 50 mg daily, a decrease in pulmonary vascular resistance of about 30% was observed (10). However, in my opinion the existence of a fixed pulmonary vascular resistance makes such treatment not altogether safe, since the left ventricle would not fill enough to keep up with the dilated periphery, thus perhaps leading to a drop in blood pressure.

It seems important to stress once again that we should attempt to establish the therapeutic mechanism of vasodilators more precisely and to study their effects on patients with cardiac insufficiency before we focus our research on yet another new preparation.

References

1. Chatterjee K, Parmley WW, Massie B, Greenberg B, Werner J, Klausner S, Norman A (1976) Oral hydralazine therapy for chronic refractory heart failure. Circulation 54:879
2. Mathey D, Hanrath P, Polster J, Witte G, Montz R, Bleifeld W (1980) Acute and chronic effects of oral hydralazine on left ventricular pump function and renal hemodynamics in chronic left heart failure. European Heart Journal 1:25
3. Chatterjee K, Ports TA, Brundage B, Massie B, Holly A, Parmely WW (1980) Oral hydralazine in chronic heart failure: sustained beneficial hemodynamic effects. Ann Intern Med 92:600
4. Leier CV, Magorien RD, Desch CE, Thompson MJ, Unverferth DV (1981) Hydralazine and isosorbide dinitrate: comparative central and regional hemodynamic effects when administered alone or in combination. Circulation 63:102
5. Franciosa JA, Cohn JN (1979) Immediate effects of hydralazine-isosorbidedinitrate combination on exercise capacity and exercise hemodynamics in patients with left ventricular failure. Circulation 59:1085
6. Franciosa JA, Weber KT, Kinasewitz GT, West JW, Cohn JN (1980) Long-term hydralazine vs. placebo therapy of chronic heart failure (Abstr). Circulation [Suppl III] 62:994

7. Rubin SA, Chatterjee K, Parmley WW (1980) Metabolic assessment of exercise in chronic heart failure patients treated with short-term vasodilators. Circulation 61:543
8. Massie B, Ports T, Chatterjee K, Parmley WW, Ostlund J, O'Young J, Haughom F (1981) Long-term vasodilator therapy for heart failure; clinical response and its relationship to hemodynamic measurements. Circulation 63:269
9. Packer M, Meller J, Medina N, Gorlin R, Herman MV (1980) Importance of left ventricular chamber size in determining the response to hydralazine in severe chronic heart failure. N Engl J Med 303:250
10. Rubin LJ, Peter RH (1980) Oral hydralazine therapy for primary pulmonary hypertension. N Engl J Med 302:69

Discussion (zu den Seiten 124–138)

Just, Freiburg: The two patients who suffered sudden death in Morand's study, did they undergo autopsy?

Morand, Nice: No, they had had ischemic cardiac disease.

Just, Freiburg: We have seen here a considerable number of patients (5 of 12 cases) with positive antinuclear antibodies in the serum. What consequences did you draw? Did you discontinue therapy, or are you continuing to treat them? And what has happened?

Mathey, Hamburg: We continued the treatment. No clinical substrate or symptoms could be observed.

Cohn, Minneapolis: We were involved in a study that was carried out a year or two ago in the United States, looking at hydralazine with doses up to 300 mg daily as therapy for heart failure, while basic treatment continued with digitalis and diuretics. In that study we were certain that we could see a beneficial effect. There was, however, no statistically significant difference between the placebo-treated and hydralazine-treated groups. The therapy was maintained for a period of 4–6 months. The main reason why the hydralazine group was not significantly improved in comparison to the placebo group was that the placebo group itself did better.

Exercise tolerance, using exercise duration as an endpoint, was improved in both the placebo- and the hydralazine-treated groups, but the difference was not significant. We have evaluated other symptoms and quality of life; every effort was made to define changes in heart size and cardiac function, but we could not find any difference between the two groups. At the moment we are left with the uncomfortable conclusion that hydralazine given alone in the way it was given in the study and in this patient population, involving about 100 cases, does not significantly improve clinical or hemodynamic indices of cardiac performance.

Bussmann, Frankfurt: I would like to consider the side effects of hydralazine. We studied ten patients with congestive cardiomyopathy of NYHA class III–IV. Two patients developed peripheral edema under hydralazine, therefore we stopped the treatment. Now if we use vasodilators in patients with severe congestive heart failure (NYHA III–IV) we increase the dosage of the diuretic.

Mathey, Hamburg: We also had to increase the dosage of diuretics. I suppose this is due to hydralazine, not to the natural course of the underlying heart disease.

Bussmann, Frankfurt: It is surprising that exercise capacity does not change but hemodynamic data improve markedly, as Cohn pointed out. In the individual patient, we may see marked improvement in cardiac output at rest and during exercise, but this patient does not improve his NYHA class. I think this question is so important that we cannot recommend that type of therapy before we know the answer to this question.

Taylor, Leeds: I would like to emphasize one point made by Mr. Cohn. When we carry out investigations like these trials, the patients in the placebo group naturally get far more attention and many show an improvement. Therefore it is important to have properly designed double-blind trials.

I would like to make a comment on what Dr. Mathey said. I think that all of us, as physicians and not as physiologists, concentrate on what the heart is doing. Remember, you cannot improve exercise tolerance without training. Even an athlete will find great difficulties in improving his exercise tolerance. If you initiate a training program on top of your treatment and then it does not improve, then you would make your claim.

Mathey, Hamburg: What we are doing right now is to exercise the patient at different work loads. We measure cardiac output and pulmonary artery pressure, which was not done in the studies by Dr. Cohn, enabling us to correlate hemodynamics and work load. There may be an increase in perfusion which is not going to the capillary bed. This means shunting, and not increased organ perfusion, as we would expect from the high cardiac output.

Brunner, Basel: You pointed out that it would be more important to study the mechanism and the mode of action of vasodilators than to use new drugs. However, side effects of hydralazine are observed frequently. New drugs without deleterious side-effects may therefore be useful.

Mathey, Hamburg: The main question today is the importance of hemodynamic data to patients with heart failure. What does improved hemodynamics mean to the patient? This question does not depend on the drug you use. Looking for new drugs without severe side effects is a worthwhile secondary goal, but not the primary one.

Jahrmärker, Munich: It was argued, that the effects of long-term vasodilation are counterbalanced by reflex mechanisms. Does long-term treatment mean regular or intermittent therapy? Perhaps the latter would be advantageous.

Kersting, Koblenz: I cannot answer this question completely. However, Evans has demonstrated a pronounced acute effect of prazosin with little additional effect after repeated administration. Furthermore, he showed that the control values improved in the course of the study. Therefore I believe that intermittent treatment may be beneficial.

Brunner, Basel: How do vasodilators alter the distribution of cardiac output? Increased need for diuretics suggests that the glomerular filtration fraction decreases. That would imply that the improvement in blood flow results in higher perfusion through shunt.

Mathey, Hamburg: We have measured the effective renal plasma flow, corresponding to the PAH clearence. These measurements suggest an improvement of the nutritive blood flow of the kidney. I think, that the development of edema is independent of the function of the kidney.

Prazosin Therapy in Severe Chronic Congestive Heart Failure

F. Burkart, O. Bertel

Vasodilators are well established as a treatment in acute congestive heart failure and are widely used, especially in intensive care units, in patients with acute myocardial infarction and heart failure (1, 2, 3). For long-term treatment in chronic heart failure, however, the number of different drugs is reduced. Oral application and a prolonged action over several hours are necessary, so that nitroglycerin, sodium nitroprusside, and phentolamine cannot be used in chronic failure. In addition, it is difficult to assess the effect due to the progress of the underlying disease in these critically ill patients. In contrast to hydralazine, which acts predominantly on arteriolar resistance vessels, and to the nitrates, which principally influence venous vessels, prazosin has an effect on both systemic venous capacitance and arteriolar resistance vessels, which can be demonstrated for a period of 6 h following a single oral dose (4). The sustained beneficial effect of prazosin has been established in a large number of publications with encouraging results (5–7). However, rapid attenuation of the acute beneficial effect of prazosin has been reported, and the value of this therapy for chronic treatment has therefore been questioned by some workers (8, 9). We performed a hemodynamically controlled study during 6 months' treatment with oral prazosin in 11 patients with severe left heart failure. The results of repeated catheterizations were demonstrated (10, 11). These data will be presented in some detail before discussing the criticism of attenuation and the questionable long-term effect.

Patients and Methods

Eleven consecutive patients with severe congestive heart failure not responsive to treatment with digitalis and diuretics were selected. Their age was 67.6 ± 6.85 years. Coronary artery disease was diagnosed in six patients either by coronary arteriography or by an ECG with typical transmural myocardial infarction. Three patients had congestive cardiomyopathy documented by left and right heart catheterization and two patients suffered from valvular heart disease. The treatment with digitalis and high doses of frusemide and spironolactone was not changed for at least 1 week prior to control catheterization.

To avoid orthostatic hypotension the first dose of 0.5 mg prazosin was given in the evening, then continued three times daily in increasing doses until the patients complained of dizziness or a marked fall in blood pressure occurred. Thus the dosage ranged from 3 to 20 mg/day.

Cardiac catheterization was performed at the beginning, at 4 and 8 weeks, and after 6 months on unchanged medication. At each control right atrial pressure, pulmonary artery pressure, pulmonary capillary wedge pressure, and mean arterial pressure were measured, and cardiac output, stroke volume index, and peripheral resistance calculated at rest and under exercise (25 W, tilting legs alternately). Results are presented as mean and standard deviation. Student's t-test was used for statistical analysis.

Results

Long-term treatment with prazosin was well tolerated by ten of the 11 patients; it had to be withdrawn in one patient because of dizziness and drop in blood pressure. This patient was therefore excluded from the analysis. Three patients died during the study period (Fig. 1): one died in the 5th week of the treatment a few days after this therapy had been stopped due to lack of improvement in cardiac function; the second died suddenly after having shown a good clinical and hemodynamic result; and the third patient died after 17 weeks from pneumonia. Four patients improved from NYHA class IV to II, four others from class IV to class III, and only two patients did not show any change in functional class.

The hemodynamic data from the ten patients entering the study showed a low cardiac index of 1.95 $l/min/m^2$, an elevated heart rate of 92 beats/min, and a calculated stroke index of 23 ml/m^2. The filling pressure was very markedly elevated (29 mmHg) and the mean arterial pressure was 100 mmHg due to an increased peripheral resistance of 2237 $dyn/s/cm^{-5}$.

Under treatment cardiac index increased significantly after 4 weeks, as did stroke volume index. Both improved significantly further after 8 weeks compared to the

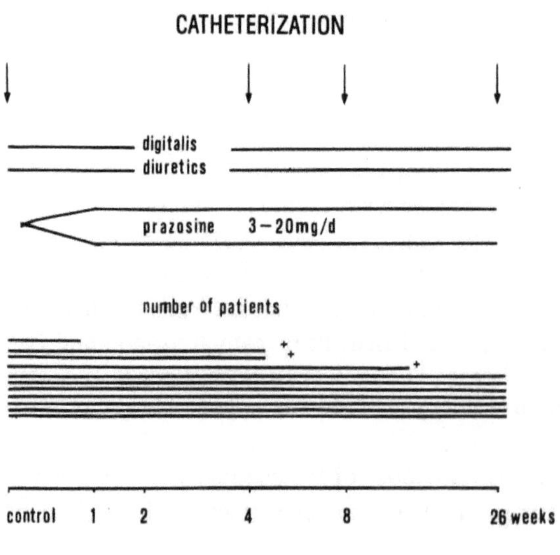

Fig. 1. Study protocol with the number of cardiac catheterizations (*arrows*), prazosin and concomitant therapy, and destiny of the patients in the study (* denotes death)

control value and decreased only slightly after 6 months, still showing a significant change compared to the first values. Heart rate decreased, this change being significant only after 8 weeks, as demonstrated in Fig. 2. The high filling pressure was significantly lower after 4 weeks, and decreased further after 8 weeks and 6 months respectively, as is shown in Fig. 3. Mean arterial pressure was somewhat lower compared to the first value but showed as a whole little change. Due to the increased cardiac index the calculated peripheral resistance was significantly lower in all three measurements under prazosin.

The interpretation of hemodynamic data under exercise is somewhat difficult, as it was not possible for nine out of ten patients to perform ergometry in a supine position at 25 W for 4 min at the beginning of the trial. The legs were therefore tilted upwards, producing an increased venous return and thus a higher preload. For three patients this procedure was not possible because of severe breathlessness even at rest. After 4 weeks seven out of the ten patients could perform supine ergometry, and only one of the three remaining patients could not tolerate the tilting procedure. Finally, after 6 months all seven remaining patients could tolerate either ergometry or volume load. With these changing workloads, the comparison of the mean values is difficult but as exercise increased, the increasing stroke volume index and the decreasing filling pressure as demonstrated in Fig. 4 can be accepted as hemodynamically significant and clinically important improvements.

Discussion

These hemodynamic results suggest a sustained effect of prazosin therapy over a prolonged period. In a similar study Feldman et al. were able to demonstrate a similar beneficial effect on cardiac output after 3 months of treatment (12). Awan et al. were able to show a beneficial effect of this treatment over a period of 12 months (13). Feldman found wide variability in the rise of cardiac output which was not related to the differences in plasma prazosin concentration. In patients with very low filling pressures, this effect seemed to blunt the beneficial effect on cardiac output; therefore, no correlation between systemic vascular resistance in the untreated state and increase in cardiac output could be found. In our series, however, we found an inverse correlation between total systemic vascular resistance and stroke volume index, and the increase in stroke volume index after 4 weeks of treatment was directly correlated both to the decrease in systemic vascular resistance and the reduction of the filling pressure.

We were concerned about the three deaths in this relatively small series and the possibility that the drug itself could have induced this fatal outcome. The history of the patients showed that one had died from pneumonia under hospital observation, in another prazosin had been withdrawn due to lack of improvement; only in one patient who died suddenly could prazosin be discussed as one possible reason for this outcome. On the other hand, prognosis in these severely sick patients is poor, and in all published trials with different vasodilating drugs the dropout number due to death is relatively high. Further studies including control groups will be necessary to elucidate this point.

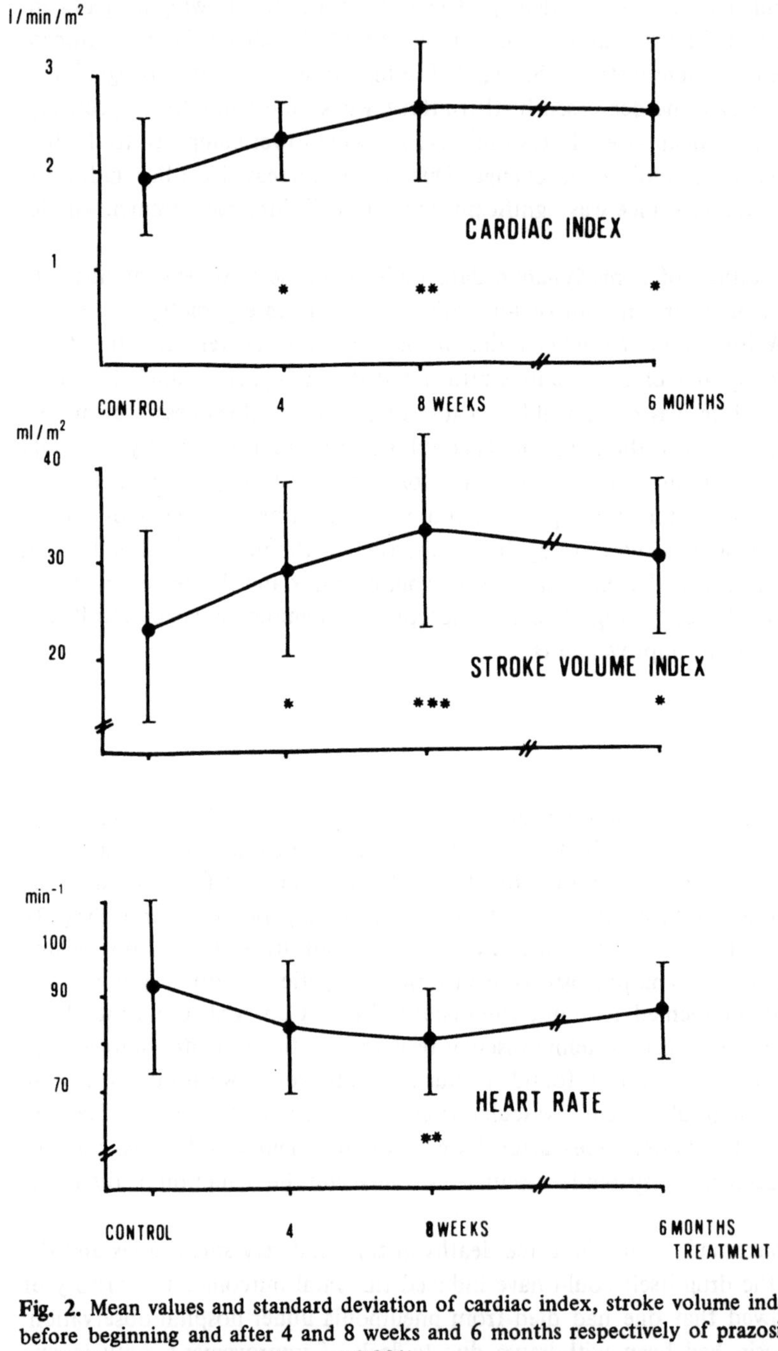

Fig. 2. Mean values and standard deviation of cardiac index, stroke volume index, and heart rate before beginning and after 4 and 8 weeks and 6 months respectively of prazosin treatment. (* = $p < 0.05$, ** = $p < 0.01$, *** = $p < 0.001$)

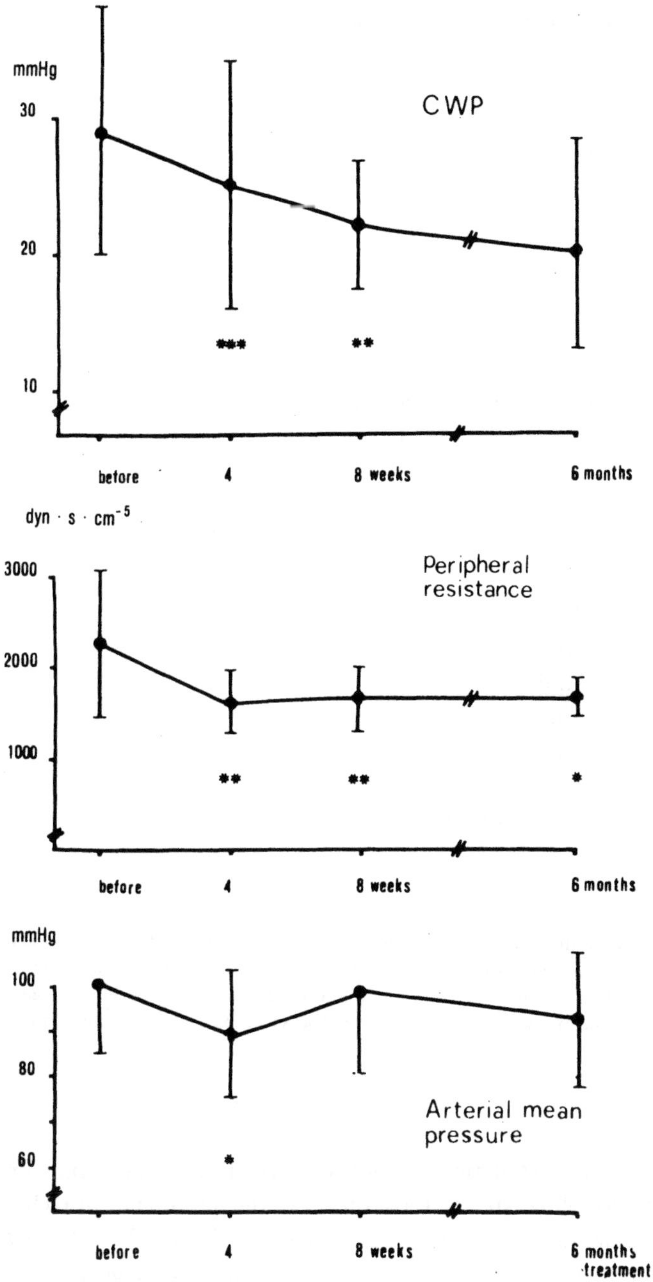

Fig. 3. Mean values and standard deviation of the capillary wedge pressure (*CWP*), the peripheral resistance, and the arterial mean pressure of the patients before beginning and after 4 and 8 weeks and 6 months respectively of treatment with prazosin. (* = $p < 0.05$, ** = $p < 0.01$, *** = $p < 0.001$)

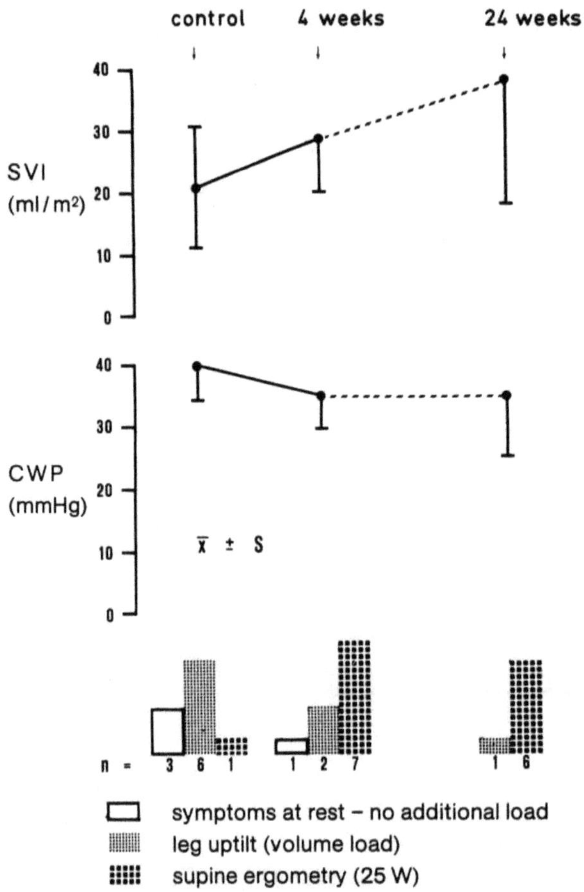

Fig. 4. The behavior of stroke work index (*SWI*) and filling pressure, documented here as capillary wedge pressure (*CWP*), under exercise, which consisted in some tests of tilting the legs, in some of ergometry. Mean values and standard deviation are demonstrated before and after 4 weeks of treatment for ten patients. The values after 24 weeks are given for the seven patients remaining in the study for this length of time

In our study dosage was increased from a very low first dose to the maximum tolerated. This design does not allow an exact assessment of the importance of tachyphylaxis occurring with prazosin in prolonged treatment (8, 9). In these studies decreased hemodynamic response of prazosin in the first 3–4 days could be demonstrated; the authors therefore assumed that no long-term effectiveness was possible with long-term prazosin usage. This same effect of decreased response in repeated prazosin application could be seen in the study of Feldman (12). Awan investigated 16 patients with coronary heart failure receiving prazosin for 12 months for the development of vasodilator tolerance. In six of these patients such tolerance was documented after 7 months, but by increasing the dosage and briefly interrupting prazosin therapy, control of symptoms was maintained in those patients. In this special study forearm blood flow was increased and forearm vascular resistance and venous tone were decreased at the beginning and at the end of the study, but were unchanged after 7 months. The authors concluded that the efficacy of prolonged prazosin treatment could be maintained in such patients by overcoming tolerance with higher effective prazosin dosage or brief prazosin discontinuation. We believe, therefore, together with these authors, that prazosin is a valuable drug in the long-term treatment of severe heart failure, a view which is shared by Awan and Mason in their editorial about this treatment (14).

References

1. Mason DT (ed) (1978) Symposium on vasodilator and inotropic therapy of heart failure. Am J Med 65:101
2. Bertel O (to be published) Vasodilatatoren in der Behandlung der schweren Herzinsuffizienz
3. Bertel O, Baitsch G, Elsässer S, Ritz R, Burkart F (1978) Perorale Nitrattherapie der schweren Herzinsuffizienz nach akutem Myokardinfarkt. Schweiz Med Wochenschr 101:1735
4. Awan NA, Miller RR, De Maria AN, Maxwell KS, Neiman A, Mason DT (1977) Efficacy of ambulatory systemic vasodilator therapy with oral prazosin in chronic heart failure. Circulation 56:346
5. Aronov WS, Donahi DT (1978) Efficacy of trimazosin and prazosin therapy on cardiac and exercise performance in out-patients with chronic congestive heart failure. Am J Med 65:144
6. Straub E (1978) Clinical use of prazosin in congestive heart failure. Circulation 57:200
7. Colucci WS, Wynne J, Holman BL, Braunwald E (1980) Long-term therapy of heart failure with prazosin: a randomized doubleblind trial. Am J Med 45:337
8. Arnold SW, Williams RL, Ports TA, Baugham RA, Bénet LZ, Parmley WW, Chatterjee K (1979) Attenuation of prazosin effect on cardiac output in chronic heart failure. Ann Intern Med 91:345
9. Packer M, Meller J (1978) Oral vasodilator therapy for chronic heart failure: a plea for caution. Am J Cardiol 42:686
10. Bertel O, Burkart F, Bühler FR (1981) Sustained effectiveness of chronic prazosin therapy in severe chronic congestive heart failure. Am Heart J 101:529
11. Burkart F, Bertel O, Bühler FR (1980) Sustained benefit of prazosin therapy in severe congestive heart failure. In: Rutishauser W (ed) The place of vasodilators in the long-term treatment of intractable heart failure. Hans Huber, Bern Stuttgart Vienna
12. Feldmann AC, Ball AM, Winchester MA, Jaillon P, Kates RE, Harrison DC (1981) Beneficial hemodynamic response to chronic prazosin therapy in congestive heart failure. Am Heart J 101:534
13. Awan NA, Lee G, De Maria AN, Mason DT (1981) Ambulatory prazosin treatment of chronic congestive heart failure: development of late tolerance reversible by higher dosage and interrupted substitution therapy. Am Heart J 101:541
14. Awan NA, Mason DT (1981) Oral vasodilator therapy with prazosin in severe congestive heart failure. Am Heart J 101:695

Discussion

Pippig, Gütersloh: How did you titrate the dosage of prazosin? Did you have to change the dosage after the titration period?

Burkart, Basel: No, we did not change the dosage after the titration period. The dose of prazosin was built up to a maximum within the limits of the resultant reduction in blood pressure and the patients' subjective tolerance, exept in the patients with hypotension, for whom we reduced the dosage of diuretics.

Pippig, Gütersloh: What did you tolerate as the lowest blood pressure?

Burkart, Basel: It depended on the control values. Even patients with systolic blood pressure below 100 mmHg may tolerate vasodilator treatment. We acted according to the patients' subjective tolerance and the control values, but not the absolute values.

Hanrath, Hamburg: You have said that you did not change the dosage of diuretics. Did you observe an increase in body weight? We studied 16 patients over a period of 12 months. We found an increase in body weight of about 4%. There was a wide variation in the individual cases, so we had to increase the dosage of diuretics. We also started with a titration period of 4 weeks, but, whereas hemodynamics were sustained after 6 months, there was a pronounced attenuation of the hemodynamics in comparison to the initial improvement.

Burkart, Basel: I think it is very difficult to distinguish between the progression of the disease and the attenuation of vasodilator treatment, especially over a long period of time.

Jahrmärker, München: One patient had severe side-effects. Did you have evidence that this patient was hypovolemic, and what about the hemodynamic control values?

Burkart, Basel: The control values were not too bad; this patient did not belong to the group with the lowest cardiac output values, yet severe hypotension occurred. The same happened when prazosin was given again after withdrawal. It may be that this lack of response was due to severe heart failure with no residual myocardial reserve at all. I think this is possible in an individual case.

Jahrmärker, München. It is because of the differences in natural history – and attenuation of vasodilator treatment – that we need controlled trials. You mentioned that all patients were treated with digitalis and diuretics. Did you use spironolactone as a diuretic drug, too?

Burkart, Basel: It may be that we continued to treat some patients with spironolactone if these patients had taken it before entering the study. We did not start with spironolactone after the beginning of the study.

Jahrmärker, München: We observed similar effects after long-term treatment with spironolactone 10 years ago.

Auditorium: You mentioned so-called tachyphylaxis. I do not think this is true tachyphylaxis. However, could the attenuation of effects be explained by changes in the metabolism of the drug? Are there any data showing that metabolism is changed after long-term treatment with prazosin, e.g., changes in extraction by the liver due to changed liver perfusion, shortening the half-life of the drug?

Burkart, Basel. I have no data on the metabolism of prazosin. In our experience, the so-called tachyphylaxis occurred in patients with initially pronounced vasoconstriction. Prazosin caused increased cardiac output and less peripheral vasoconstriction. If the drug is administered under these circumstances, the effect will be less pronounced. This cannot be called tachyphylaxis.

Auditorium: Is prazosin in your opinion the drug of choice as a vasodilator? What is your personal view?

Burkart, Basel: I prefer prazosin, as the results are comparable to those of the combination of hydralazine and nitrates. Using prazosin the number of tablets can be reduced.

Auditorium: Did your patients die suddenly or though advanced congestive heart failure?

Burkart, Basel: In one case congestive heart failure resulted in sudden death. In one case we don't exactly know.

Early Experience with Captopril in Congestive Heart Failure*

Jay N. Cohn, T. Barry Levine, Gordon Pierpont

The renin-angiotensin system is frequently activated in patients with congestive heart failure. In experimental heart failure this activation appears to be an early phenomenon that abates after sodium retention has led to volume expansion (1). In clinical heart failure, however, activity of the renin-angiotensin system appears to be unrelated to volume status. Plasma renin activity is elevated in some patients who are taking diuretics and in others who have been withdrawn from diuretics (2). Its degree of elevation appears to have only a weak relationship to the severity of the hemodynamic abnormality (3). The most striking correlation with plasma renin activity has been the level of serum sodium. Hyponatremic subjects present almost invariably with exceedingly high plasma renin activity, whereas in those whose serum sodium is within the normal range plasma renin activity is either normal or only modestly elevated (4).

Since arterial vasoconstriction impairs left ventricular pump function in heart failure (5–7), and since vasodilator drugs produce a striking acute hemodynamic improvement in patients with heart failure (8–10), recent interest has focused on the role of the renin-angiotensin system in influencing left ventricular pump function, and on the possible therapeutic effect of drugs that inhibit the renin-angiotensin system. Furthermore, since angiotensin may stimulate catecholamine release (11) and is also a potent stimulator of aldosterone production (12), activation of the renin-angiotensin system may result in diverse responses that might aggravate pump function and contribute to circulatory overload.

The availability of teprotide and saralasin as intravenous agents to antagonize the action of angiotensin II made it possible to assess the role of the renin-angiotensin system in the vasoconstriction of heart failure. Studies of these two agents in patients with heart failure revealed that at least in some subjects a marked vasodilator effect could be produced (13, 14). Furthermore, when subjects with a wide range of plasma renin activity were evaluated a correlation between the degree of elevation of plasma renin activity and the response to these drugs was apparent (15).

With the synthesis of an orally effective converting enzyme inhibitor, captopril (16), chronic therapy for heart failure became a possibility. In early studies with captopril a vasodilator effect involving both the arteriolar and venous beds was apparent. Cardiac output and stroke volume rose and left ventricular filling pressure fell

---

* Supported in part by PHS Grant HL 22977 from the National Heart, Lung and Blood Institute

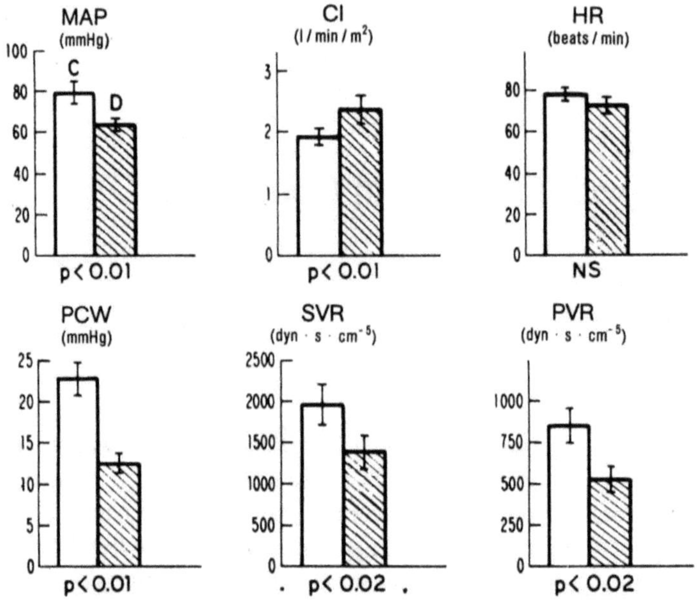

Fig. 1. Hemodynamic response to first dose of captopril in patients with congestive heart failure. Control values (*C*) before and treatment values (*D*) 1.5 h after administration of 25–100 mg captopril. *MAP*, mean arterial pressure; *CI*, cardiac index; *HR*, heart rate; *PCW*, pulmonary wedge pressure; *SVR*, systemic vascular resistance; *PVR*, pulmonary vascular resistance

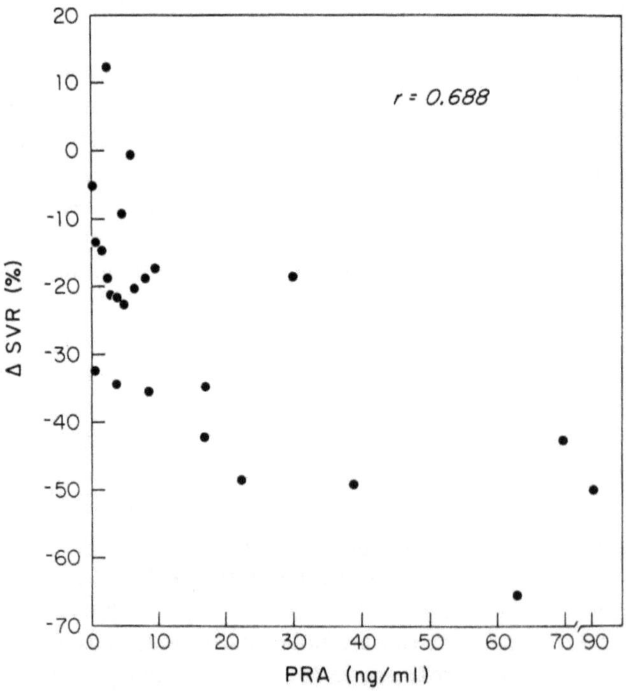

Fig. 2. Correlation between control plasma renin activity (*PRA*) and acute fall in systemic vascular resistance (*SVR*) in response to administration of a single dose of converting enzyme inhibitor. Cohn et al. (15)

in association with a modest fall in arterial pressure (Fig. 1). The hemodynamic response could be correlated with the degree of elevation of plasma renin activity (17) (Fig. 2). The effect of captopril appeared to be sustained and even small doses were effective in producing a fall in pulmonary wedge pressure and a rise in cardiac output. Similar to the responses with other vasodilator drugs given to patients with heart failure, heart rate did not increase in response to captopril but actually tended to fall. Indeed, this absence of tachycardia in response to captopril has also been observed in patients with hypertension, who usually exhibit a reflex tachycardia in response to other vasodilator drugs.

The most provocative issue regarding captopril therapy for heart failure is not its ability to vasodilate the periphery, since many other drugs such as hydralazine and prazosin can accomplish this (18, 19), but whether attacking a specific mechanism of vasoconstriction might have a more salutary effect in restoring regional blood flow to a more normal distribution, in inhibiting the renal sodium retention which characterizes congestive heart failure, and in correcting other abnormalities of the heart failure state that may actually be aggravated by the presence of high circulating angiotensin II levels. Data so far only provide a partial answer to these questions. A sustained hemodynamic response to the drug has been demonstrated in patients who have been maintained on continuous therapy (20). Furthermore, improved exercise tolerance and reduced symptoms of shortness of breath and fatigue have been characteristic in patients maintained on the drug over a period of weeks or months (Fig. 3).

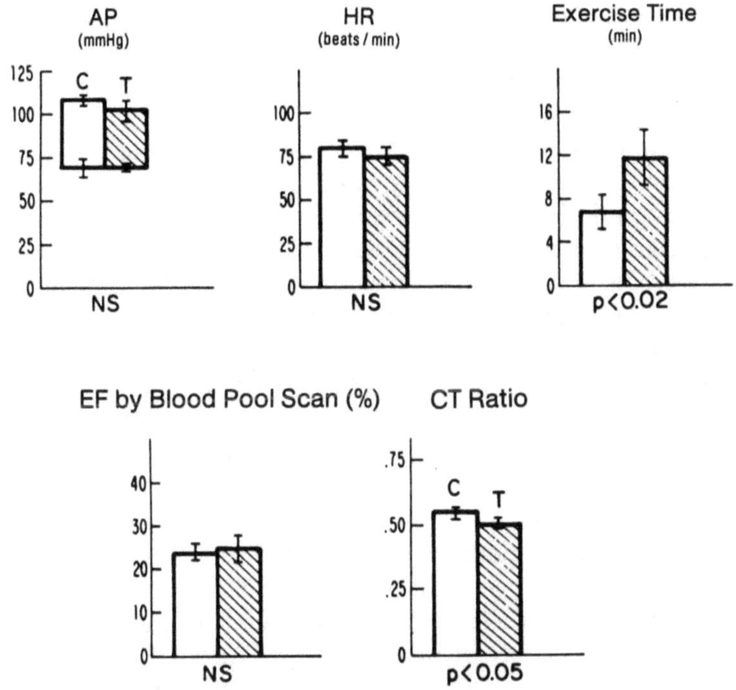

Fig. 3. Responses to chronic captopril therapy in eight patients with congestive heart failure. *C*, control values; *T*, values after 2–5 months of therapy; *AP*, arterial pressure; *CT*, cardiothoracic ratio; *EF*, ejection fraction; *HR*, heart rate

A carefully controlled double-blind trial of captopril therapy versus placebo therapy in patients maintained on treatment with digitalis and diuretics is currently under way.

The influence of captopril therapy on renal sodium excretion and fluid retention in heart failure has not been well delineated. Systemic blood pressure falls considerably in response to captopril in some patients and this hypotensive effect would be expected to have an antinatriuretic effect on the kidney (21), even if captopril were concomitantly reducing renal vascular resistance. On the other hand, the renal vasodilator effect of inhibition of angiotensin II and the reduced levels of aldosterone might be expected to enhance sodium excretion. The net effects of these opposite influences on renal function have, as expected, produced disparate results. In some patients a dramatic improvement in renal function has accompanied the hemodynamic response to the drug. In others renal deterioration has occurred at the onset of therapy, probably at least in part because of the fall in systemic blood pressure. In a controlled 3-day trial of captopril therapy in patients who were maintained on metabolic balance, subjects with heart failure revealed no significant change in sodium excretion, but some did exhibit a fall in creatinine clearance apparently related to the degree of reduction of systemic blood pressure (22).

Toxic effects of captopril remain a problem with some patients. Although albuminuria and nephrotoxicity have been reported in patients with hypertension, this complication has not to our knowledge occurred in patients with congestive heart failure. Rashes have been reported in as many as 15% of patients with hypertension treated with captopril, and rashes have also occurred in patients with congestive heart failure. The rash appears to be dose-related, and at least in some patients has been controlled by reduction of dose or even by reinstitution of a similar dose. We have observed one patient who developed a striking maculopapular erythematous eruption after taking captopril for several months. Reinstitution of even tiny doses of captopril thereafter resulted in prompt recrudescence of the rash.

Preliminary and uncontrolled experiences with captopril have suggested not only that it is an effective vasodilator in some patients, but also that the clinical response to the drug may be even more dramatic than the hemodynamic response to acute administration. Whether this clinical response reflects a progressively improving hemodynamic response to the drug with sustained administration cannot be determined. Even in patients with low plasma renin activity who have not exhibited a very dramatic acute response, chronic improvement has been noted during sustained therapy. Another possibility is that the high angiotensin II levels in some patients with heart failure exert additional subtle effects on the state of well-being, and that captopril is effective in reversing these.

If captopril or other converting enzyme inhibitors prove to be safe and effective agents for use in congestive heart failure, it will be necessary to determine whether these agents should be employed specifically in patients with high plasma renin activity or whether they should be employed in all patients with heart failure. Furthermore, this problem identifies the need for an understanding of the mechanism by which renin activity is stimulated in congestive heart failure. The most striking correlate in our hands has been the serum sodium concentration. Thus a low serum sodium seems to be closely correlated with a high plasma renin activity. Whether the hyponatremia proceeds the hyperreninemia, or whether the high renin itself is a factor in producing hyponatremia, has not yet been determined. The place of angiotensin-converting enzyme inhibitor therapy in the management of heart failure remains to be determined. This exciting new area of therapeutics increases our need for an understanding of the hemodynamic and hormonal interrelationships in congestive heart failure and how they interact to contribute to the progressive circulatory deterioration that characterizes the syndrome. The high mortality rate in this disease demands that such studies be carried out without delay.

References

1. Watkins L Jr, Burton JA, Haber E, Cart JR, Smith FW, Barger AC (1976) The renin-angiotensin-aldosterone system in congestive failure in conscious dogs. J Clin Invest 57:1606
2. Vandongen R, Gordon RD (1970) Plasma renin in congestive heart failure in man. Med J Aust 1:215
3. Levine TB, Cohn JN, Vrobel T, Franciosa JA (1979) High renin in heart failure: a manifestation of hyponatremia. Trans Assoc Am Physicians 92:203–207
4. Levine TB, Franciosa JA, Vrobel T, Cohn JN (to be published) Hyponatremia as a marker for high renin heart failure. Br Heart J
5. Cohn JN (1973) Vasodilator therapy for heart failure: the influence of impedance on left ventricular performance. Circulation 48:5–8
6. Ross J Jr (1976) Afterload mismatch and preload reserve: a conceptual framework for the analysis of ventricular function. Prog Cardiovasc Dis 18:255
7. Cohn JN, Franciosa JA (1981) Effect of systemic vasoconstriction on left ventricular function: implications regarding hypertension and heart failure. In: Strauer BE (ed) The heart in hypertension. Springer, Berlin Heidelberg New York, pp 307–311
8. Guiha NH, Cohn JN, Mikulic E, Franciosa JA, Limas CJ (1974) Treatment of refractory heart failure with infusion of nitroprusside. N Engl J Med 291:587
9. Miller RR, Vismara LA, Williams DO, Amsterdam EA, Mason DT (1976) Pharmacological mechanism for left ventricular unloading in clinical congestive heart failure: differential effects of nitroprusside, phentolamine and nitroglycerin on cardiac function and peripheral circulation. Circ Res 39:127
10. Cohn JN, Franciosa JA (1977) Vasodilator therapy of cardiac failure. N Engl J Med 297:254
11. Zimmerman BG, Gomer SK, Liao JC (1972) Action of angiotensin on vascular adrenergic nerve endings; facilitation of norepinephrine release. Fed Proc 31:1344
12. Boyd GW, Adamson AR, James VHT, Peart WS (1969) The role of the renin-angiotensin system in the control of aldosterone in man. Proc R Soc Med 62:1253
13. Gavras H, Flessas A, Ryan TJ, Brunner HR, Flaxon DP, Gavras I (1977) Angiotensin II inhibition treatment of congestive cardiac failure in a high-renin hypertension. JAMA 238:880
14. Curtiss C, Cohn JN, Vrobel T, Franciosa JA (1978) Role of the renin-angiotensive system in the systemic vasoconstriction of chronic congestive heart failure. Circulation 58:763–770
15. Cohn JN, Mashiro I, Levine TB, Mehta J (1979) Role of vasoconstrictor mechanisms in the control of left ventricular performance of the normal and damaged heart. Am J Cardiol 44: 1019–1022
16. Ferguson RK, Turini GA, Brunner HR, Gavras H, McKinstry DN (1977) A specific orally active inhibitor of angiotensin-converting enzyme in man. Lancet I:775
17. Levine TB, Franciosa JA, Cohn JN (1980) Acute and long-term response to an oral converting-enzyme inhibitor, captopril, in congestive heart failure. Circulation 62:35–41
18. Franciosa JA, Pierpont G, Cohn JN (1977) Hemodynamic improvement after oral hydralazine in left ventricular failure: a comparison with nitroprusside infusion in 16 patients. Ann Intern Med 86:388

19. Miller RR, Awan NA, Maxwell KS, Mason DT (1977) Sustained reduction of cardiac impedance and preload in congestive heart failure with the antihypertensive vasodilator, prazosin. N Engl J Med 297:303
20. Dzau VJ, Colucci WS, Williams GH, Hallenberg NK (1980) Sustained effectiveness of converting enzyme inhibitor in patients moribund with congestive heart failure. New Engl J Med 302:1373–1380
21. Hollander W, Judson WE (1958) Electrolyte and water excretion in arterial hypertension. II. Studies in subjects with essential hypertension after antihypertensive drug treatment. Circulation 17:576
22. Pierpont GL, Francis GS, Cohn JN (to be published) Effect of captopril on renal function in patients with congestive heart failure. Br Heart J

Discussion

Auditorium: Looking back at these patients who responded with a marked hypotension, was there any way of predicting that they might have done so?

Cohn, Minneapolis: A patient with a low serum sodium would be a candidate for a marked hypotensive response and would have high renin values. Neither right atrial pressure nor application and dosage of diuretics could allow prediction of such complications.

I think the renin is responding to something other than volume in heart failure. One patient developed very impressive hypotension during chronic therapy, and he was on diuretics. He had responded more to diuretics, and we were able to restore the blood pressure very easily just by giving him 1 liter of sodium chloride solution: he gained 1 kg, which was enough to restore perfectly normal pressure. I do not want to imply that volume and blood pressure are unrelated in this group. I would agree with you completely that if the patient becomes hypotensive the important thing is to stop the diuretic, and that does moderate the response. But in looking at the baseline of these patients I do not think it is the lack of volume that made them hyporeninemic, but I think that the volume extension will improve the hemodynamics in part because this drug, as we pointed out yesterday, controls right atrial pressure quite a lot. It is venodilator. In part, hypotension may be related to the reduction of filling pressure, so if one gives volume one will restore the hemodynamics quite independently of the renin-angiotesin system.

Auditorium: Some authors have reported hypochloremia. We have not seen this. What is your experience?

Cohn, Minneapolis: One patient became hypochloremic. It only occurred in that one patient.

Just, Freiburg: May I ask a question concerning the patients who developed a rather severe degree of hypotension? This can surely only happen if together with the falling resistance there is no compensatory increase in stroke volume and peripheral perfusion. We know this from the effect of dioxazide. The hemodynamic measurements here show a very large increase in cardiac output. If this effect is inhibited through β-blockade, for example, then the fall in blood pressure may be deleterious to the patient. What did you find in your patients with the drastic fall in blood pressure due to captopril? Did they have a concomitant increase in stroke volume?

Cohn, Minneapolis: I do not think it makes any difference. The cardiac output does go up, and the patients seem very well. After captopril many of the patients feel very much better than they should on the basis of the hemodynamic improvement.

Zelis, Hershey: You pointed out that captopril is a venodilator.

Cohn, Minneapolis: It is an arterial dilator, too. The only thing that surprises me is that it is also a venodilator. The first research study I ever did, 20 years ago, was to infuse angiotensin into normal people and measure the venoconstriction, and it does not constrict veins. I do not understand why captopril reduces venous tone. It is definitely an arterial dilator.

Zelis, Hershey: What do you think about the mode of action of captopril?

Cohn, Minneapolis: I think the acute response is due to competition. Whether the delayed response has something to do with the kinins or other subtle vasoconstrictor stimuli or whether the decrease in sympathetic neural transmission is due to the interaction of angiotensin II and norepinephrine is very difficult to answer from our knowledge at present.

Meinertz, Mainz: We have seen two patients in whom sudden death occurred 1–3 days after beginning captopril therapy. These two patients had severe terminal coronary artery disease. They experienced a marked fall in blood pressure and felt very bad. And 2–3 days after the beginning of therapy sudden death occurred in one patient from ventricular fibrillation.

Cohn, Minneapolis: I think this is why we need a control trial.

Auditorium: The effect on veins could be mediated indirectly by vasopressin by reducing angiotensin II. I would therefore suggest looking for vasopressin: we know that vasopressin has very important effects on veins.

Cohn, Minneapolis: Yes, we have measured vasopressin levels before and after captopril: they do not change.

Acute Effect of Nifedipine on Left Ventricular Performance at Rest and During Exercise in Patients with Left Ventricular Dysfunction

Peter Hanrath, Peter Kremer

Summary

In order to determine the acute hemodynamic effect of nifedipine at rest and during a standardized supine bicycle exercise test (3 min, 50 W), 14 patients with left ventricular dysfunction due to coronary artery disease (ten) and congestive cardiomyopathy (four) were studied before and 60 min after taking 30 mg nifedipine sublingually.

At rest (R) and during exercise (E) nifedipine produced a significant increase in left ventricular systolic performance in terms of stroke volume index (R: 33 ± 6 to 38 ± 4 ml/m^2, $p < 0.005$; E: 32 ± 5 to 37 ± 6 ml/m^2, $p < 0.005$) and cardiac index (R: 2.9 ± 0.4 to 3.6 ± 0.5 l/min/m^2, $p < 0.001$; E: 4.1 ± 0.7 to 4.9 ± 0.9 l/min/m^2; $p < 0.001$) due to a marked reduction in systemic vascular resistance (R: 1517 ± 246 to 1129 ± 247 dyn · s · cm^{-5}, $p < 0.001$; E: 1170 ± 176 to 908 ± 129 dyn · s · cm^{-5}, $p < 0.01$). At rest, nifedipine produced a probably reflex-transmitted increase in heart rate (R: 89 ± 11 to 95 ± 10 min^{-1}, $p < 0.005$; E: 126 ± 13 to 131 ± 11 min^{-1}, N.S.) due to a significant decrease in mean arterial blood pressure (R: 100 ± 11 to 90 ± 11 mmHg, $p < 0.001$; E: 112 ± 12 to 104 ± 8 mmHg; $p < 0.05$). Thus the pressure-rate product at rest and during exercise remained unchanged after nifedipine application (R: 8.8 ± 1.1 · 10^3 to 8.5 ± 1 · 10^3 mmHg · min^{-1}, N.S.; E: 14 ± 1.6 · 10^3 to 13.7 ± 1.5 · 10^3 mmHg · min^{-1}, N.S.).

Pulmonary artery pressure did not change at rest, but systolic (75 ± 21 to 66 ± 22 mmHg, $p < 0.001$), mean (53 ± 12 to 45 ± 15 mmHg, $p < 0.005$), and end-diastolic (37 ± 8 to 31 ± 10 mmHg, $p < 0.005$) pulmonary artery pressure dropped significantly during exercise, probably due to a shift in the left ventricular pressure-volume relationship. The findings of this study indicate that acute hemodynamic improvement at rest and during exercise in patients with coronary artery disease and congestive cardiomyopathy can be achieved by sublingual application of nifedipine. This beneficial effect is caused by the peripheral as well as by the myocardial action of nifedipine.

Nifedipine is a drug with a strong calcium flux inhibiting effect, which has proved useful in the treatment of angina pectoris (1–3). This beneficial effect of nifedipine is primarily based on a vasodilating action on coronary arteries, resulting in an increased coronary blood flow even in ischemic areas (4, 5), and secondarily due to its vasodilating action on the arteriolar vessels, which lowers the arterial blood pressure, thus reducing the oxygen demand (5, 6). Apart from these facts, the agent has also proved to have a protecitve effect upon ischemic myocardium, as demonstrated in several biochemical and morphological studies (7–9). With regard to its peripheral and myocardial action, it was the purpose of the present study to evaluate the acute hemodynamic effects of nifedipine on left ventricular performance at rest and during exercise in patients with left ventricular dysfunction due to coronary artery disease or congestive cardiomyopathy.

Patients

The acute hemodynamic effect of nifedipine on cardiovascular dynamics was evaluated in 14 patients. The average age of these patients (12 male, two female) was 48 ± 9 years. The etiology of moderate left heart failure was coronary artery disease proved by selective angiography in ten patients, and congestive cardiomyopathy without significant narrowing of the coronary arteries in four patients.The average ejection fraction – determined from monoplane (RAO 30°) left ventricular angiograms – was 34% (range: 23%–49%).

All patients had clinical signs of left heart failure (NYHA class II–III) with exertional dyspnea as the predominant symptom. All patients with coronary artery disease had suffered myocardial infarction in their history. Two of the ten patients with coronary artery disease and two of four patients with congestive cardiomyopathy had additional mitral incompetence, proved by angiography. The basic therapy given to these patients was digitalis, diuretics, and nitrates. These drugs were withdrawn at least 12 h before the hemodynamic study was performed.

Hemodynamic Measurements

In all patients pulmonary artery pressure was recorded with a Swan-Ganz catheter. Cardiac output was measured in triplicate at each stage using the thermodilution technique. Arterial pressure was monitored by sphygmomanometer in all patients. After the hemodynamic baseline measurement at rest had been taken in 12 of the 14 patients, a standardized supine bicycle exercise stress test of 3 min with a workload of 50 W was performed. When all hemodynamic parameters had reached their control values 30 min later, all patients received 30 mg nifedipine sublingually. Sixty min after application of nifedipine, the same measurements at rest and during exercise as in the control state were repeated. Earlier observations in our laboratory, as well as those reported by others (10), have shown that maximal action of nifedipine involving hemodynamic changes occurs after 60–90 min.

Derived hemodynamic variables were calculated as follows: Cardiac index (CI) (l/min/m^2) = cardiac output (l/min/m^2 body surface area)
Stroke volume index (SVI) (ml/m^2) = cardiac index/heart rate
Mean arterial pressure (AP) (mmHg) = (SAP + 2 DAP)/3 where SAP is systolic and DAP is diastolic arterial pressure

Table 1. Acute hemodynamic response (*A*) to nifedipine in the individual patient at rest (*R*) and during exercise (*E*), compared to control values (*C*)

Pat. no.	Age/sex	Diagnosis		HR (min^{-1})		MAP (mmHg)		SPAP (mmHg)		MPAP (mmHg)		EDPAP (mmHg)		SVI (ml/m^2)		CI ($l/min \cdot m^2$)		SVR ($dyn \cdot s \cdot cm^{-5}$)		PVR ($dyn \cdot s \cdot cm^{-5}$)	
				C	A	C	A	C	A	C	A	C	A	C	A	C	A	C	A	C	A
1	66/M	CAD	R	74	80	113	107	30	28	22	17	17	13	36	40	2.8	3.2	1614	1338	71	50
			E	125	130	115	113	95	80	68	50	50	40	33	37	4.1	4.9	1108	932	173	82
2	54/M	CAD	R	80	85	115	113	25	25	16	15	12	9	35	41	2.8	3.5	1840	1458	64	77
			E	121	135	120	123	48	32	35	21	25	15	39	51	4.7	6.9	1157	794	96	39
3	42/M	CAD	R	100	110	95	83	55	50	53	35	25	15	38	38	3.8	4.1	1085	874	321	164
			E	155	150	103	98	78	60	58	48	40	35	31	37	4.8	5.5	936	769	210	102
4	42/F	CAD	R	93	98	90	78	35	32	25	23	18	13	35	44	3.2	4.3	1358	879	106	113
			E	–	–	–	–	–	–	–	–	–	–	–	–	–	–	–	–	–	–
5	51/M	CAD	R	98	110	102	93	70	68	45	46	30	30	27	30	2.7	3.3	1665	1220	245	210
			E	120	135	103	98	105	100	68	70	40	40	30	31	3.6	4.2	1230	1005	334	308
6	44/M	CCM	R	79	90	88	87	35	34	20	24	10	17	32	41	2.5	3.7	1600	1105	182	89
			E	–	–	–	–	–	–	–	–	–	–	–	–	–	–	–	–	–	–
7	33/M	CCM	R	105	100	90	77	52	52	38	38	32	30	25	38	2.6	3.8	1263	742	84	77
			E	120	120	105	98	68	60	48	41	39	32	23	33	2.7	3.9	1424	922	122	85
8	52/M	CCM	R	78	83	100	97	27	26	18	18	13	13	31	33	2.5	2.7	1662	1543	77	70
			E	120	110	108	110	71	68	55	45	38	33	38	42	4.6	4.6	891	907	140	99
9	44/M	CAD	R	105	105	95	83	55	55	40	35	25	23	28	43	2.9	4.5	1394	825	194	101
			E	140	135	108	98	72	73	58	57	38	42	30	43	4.2	5.8	971	643	180	98
10	48/M	CCM	R	84	100	93	85	46	46	31	30	18	17	30	36	2.5	3.6	1959	1183	114	147
			E	120	140	120	105	75	70	48	48	32	30	31	36	3.8	5.1	1315	848	175	145
11	44/F	CAD	R	95	100	110	103	20	20	15	15	8	10	32	39	3.1	3.9	1725	1268	110	62
			E	135	145	130	107	40	35	30	26	20	15	38	37	5.1	5.4	1238	951	95	98
12	38/M	CAD	R	82	86	122	93	32	30	27	22	12	15	46	40	3.8	3.5	1457	1220	179	92
			E	105	130	128	102	55	40	45	25	35	15	40	37	4.2	4.8	1365	971	107	95

Table 1 (continued)

Pat. no.	Age/sex	Diagnosis		HR (min^{-1}) C	A	MAP (mmHg) C	A	SPAP (mmHg) C	A	MPAP (mmHg) C	A	EDPAP (mmHg) C	A	SVI (ml/m^2) C	A	CI ($l/min \cdot m^2$) C	A	SVR $dyn \cdot s \cdot cm^{-5}$) C	A	PVR ($dyn \cdot s \cdot cm^{-5}$) C	A
13	60/M	CAD	R	77	82	93	87	40	45	30	32	20	20	39	37	3.0	3.0	1283	1243	138	171
			E	125	122	108	103	80	75	52	50	35	35	27	32	3.4	3.9	1329	1129	209	164
14	59/M	CAD	R	98	100	90	80	78	76	53	50	40	32	29	36	2.8	3.6	1333	914	193	206
			E	130	125	90	97	110	100	67	62	45	38	27	31	3.5	3.9	1075	1021	263	253
	48 + 9		R	89	95	100	90	43	42	31	29	20	18	33	38	2.9	3.6	1517	1129	148	116
				±11	±10	±11	±11	±17	±17	±13	±11	±9	±8	±6	±4	±0.4	±0.5	±246	±247	±75	±53
				$p < 0.005$		$p < 0.001$		N.S.		N.S.		N.S.		$p < 0.005$		$p < 0.001$		$p < 0.001$		$p < 0.05$	
			E	126	131	112	104	75	66	53	45	37	31	32	37	4.1	4.9	1170	908	175	131
				±13	±11	±12	± 8	±21	±22	±12	±15	±8	±10	±5	±6	±0.7	±0.9	±176	±129	±72	±77
				N.S.		$p < 0.05$		$p < 0.001$		$p < 0.005$		$p < 0.005$		$p < 0.005$		$p < 0.001$		$p < 0.01$		$p < 0.001$	

CAD, coronary artery disease; *CCM*, congestive cardiomyopathy; *HR*, heart rate; *MAP*, mean arterial pressure; *SPAP*, systolic pulmonary artery pressure; *MPAP*, mean pulmonary artery pressure; *EDPAP*, end-diastolic pulmonary artery pressure; *SVI*, stroke volume index; *CI*, cardiac index; *SVR*, systemic vascular resistance; *PVR*, pulmonary vascular resistance

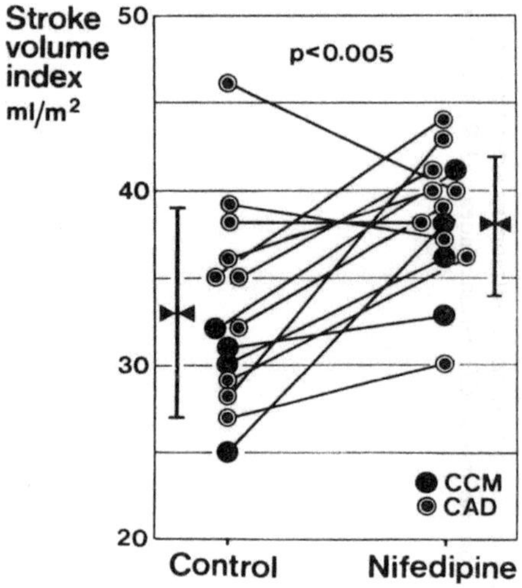

Fig. 1. Changes in stroke volume index at rest from control to 60 min after nifedipine application (n = 14). *CCM*, congestive cardiomyopathy; *CAD*, coronary artery disease. (Hanrath et al. 1982)

Systemic vascular resistance (SVR) $(\text{dyn} \cdot \text{s} \cdot \text{cm}^{-5}) = \text{AP}/(80\ \text{CO})$
Pulmonary vascular resistance (PVR) $(\text{dyn} \cdot \text{s} \cdot \text{cm}^{-5}) = \text{MPAP} - \text{EDPAP}/(80\ \text{CO})$, where MPAP is mean pulmonary artery pressure and EDPAP is end-diastolic pulmonary artery pressure

Statistical analysis was performed using paired Student's t-test.

Results

The acute hemodynamic response to nifedipine in the individual patient at rest and during exercise is listed in Table 1. Sixty min after sublingual administration of 30 mg nifedipine, the stroke volume index improved in 11 of 14 patients at rest, with an average increase of 18% (33 ± 6 to 38 ± 4 ml/m^2; $p < 0.005$). Cardiac index increased from 2.9 ± 0.4 to 3.6 ± 0.5 l/min/m^2 ($p < 0.001$). The individual change in stroke volume index at rest is illustrated in Fig. 1.

The relatively strong increase in cardiac output (25%) at rest is caused by a significant increase in heart rate (89 ± 11 to 95 ± 10 min^{-1}, $p < 0.005$). Nifedipine produced a marked reduction in systemic vascular resistance (1517 ± 246 to 1129 ± 247 dyn $\cdot$ s $\cdot$ cm^{-5}, $p < 0.001$), resulting in a decrease in mean arterial pressure (100 ± 11 to 90 ± 11 mmHg, $p < 0.001$). The pressure rate product (8.8 ± 1.1 $\cdot 10^3$ to 8.5 ± 1 $\cdot 10^3$ mmHg $\cdot$ min^{-1}) remained unchanged at rest. Systolic (SPAP), mean (MPAP), and end-diastolic (EDPAP) pulmonary artery pressure (SPAP: 43 ± 17 to 42 ± 17 mmHg, N.S.; MPAP: 31 ± 13 to 29 ± 11 mmHg, N.S.; EDPAP: 20 ± 9 to 18 ± 8 mmHg, N.S.) did not change at rest.

Under a standardized exercise stress test which was performed in 12 patients, a significant increase in stroke volume after nifedipine administration (32 ± 5 to 37 ± 6 ml/m^2, $p < 0.005$) was observed. The individual response is illustrated in Fig. 2. This rise in stroke volume index was accompanied by a comparable increase in cardiac index (4.1 ± 0.7 to 4.9 ± 0.9 l/min/m^2, $p < 0.001$). Heart rate remained

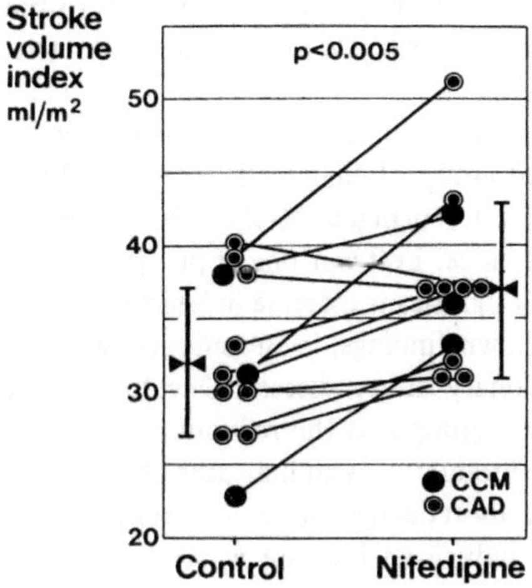

Fig. 2. Changes in stroke volume index during exercise from control to 60 min after nifedipine application (n = 12). *CCM*, congestive cardiomyopathy; *CAD*, coronary artery disease. (Hanrath et al. 1982)

unchanged before and after nifedipine during exercise (126 ± 13 to 131 ± 11 min^{-1}, N.S.). The improvement in systolic performance was related to a significant decrease in systemic vascular resistance (1170 ± 176 to 908 ± 129 dyn · s · cm^{-5}, $p < 0.01$), which was accompanied by a significant fall in mean arterial pressure (112 ± 12 to 104 ± 8 mmHg, $p < 0.05$). After nifedipine the pressure-rate product during exercise remained unchanged compared to the control measurement before treatment (14 ± 1.6 · 10^3 to 13.7 ± 1.5 · 10^3 mmHg · min^{-1}). This was due to a more pronounced increase in heart rate in six of the 12 patients during ergometry.

Pulmonary artery pressures were more significantly affected after nifedipine during exercise than in the control state. Systolic pulmonary artery pressure was 75 ± 21 mmHg before and 66 ± 22 mmHg ($p < 0.001$) after nifedipine therapy. Mean and end-diastolic pulmonary artery pressure decreased from 53 ± 12 to 45 ± 15 mmHg ($p < 0.005$) and from 37 ± 8 to 31 ± 10 mmHg ($p < 0.005$), respectively. Pulmonary vascular resistance decreased significantly from 175 ± 72 to 131 ± 77 dyn · s · cm^{-5} ($p < 0.001$) during exercise. Nifedipine-related side-effects were not observed during this acute intervention study.

Discussion

The present data confirm recent findings by Klugman et al. (11) showing nifedipine to be a potent vasodilating agent which may improve systolic performance at rest in patients with congestive cardiomyopathy. These authors observed a 54% increase in stroke volume on an average estimated from left ventricular angiograms after sublingual application of 20 mg nifedipine. Although we used a higher dose of nifedipine, the average increase in stroke volume index at rest was not as strong. This fact seems more likely to be due to the difference in patient population than to a possible negative inotropic effect. A negative inotropic effect of nifedipine has been documented in in vitro studies (12) after intracoronary application in animal experiments, but not after intravenous application of nifedipine in a therapeutic dosage (13). Thus the observed improvement in systolic performance seems to be related primarily to the vasodilating effect of nifedipine caused by a reduced calcium influx into the arterial smooth muscle cells.

In contrast to our data and those of Klugmann et al. are recent observations by Matsumoto et al. (14), who reported in a mixed group of eight patients with mild to moderate left heart failure no improvement in left ventricular stroke volume despite a significant reduction in systemic vascular resistance. Left ventricular pump function in terms of cardiac index was improved due to a significant increase in heart rate. This increase in heart rate, confirmed at rest in our own findings, is an undesirable effect of nifedipine, particularly in patients with coronary artery disease. Due to the fact that the tachycardia was only mild in our study group and the reduction of arterial pressure was relatively strong, the pressure rate product – as an indicator of myocardial oxygen consumption – remained unchanged. By reducing the vascular resistance to left ventricular ejection, vasodilators generally induce an increase in stroke volume. As a result, arterial pressure reduction is usually modest, as was also reported for hydralazine therapy (15).

Many patients with left ventricular dysfunction do not have symptoms of limited cardiac reserve at rest, but such symptoms can occur during mild to moderate exertion. Thus an important therapeutic objective of any vasodilating drug is the improvement of cardiac performance during exercise. In the present study we observed a significant improvement of left ventricular systolic pump function in terms of stroke volume and cardiac index, with no increase in heart rate compared to the control state. Similar observations have so far not been reported in the literature. Although the trend of the entire group revealed a significant improvement in systolic pump function during exercise, two of the 12 patients showed no increase in stroke volume index. This may represent a severe degree of myocardial dysfunction with a lack of cardiac reserve.

Nifedipine has been reported to have no venous pooling effect (16, 17). This finding is also confirmed by our results. No significant decrease in end-diastolic pulmonary artery pressure at rest was observed. With regard to the absence of a venous pooling effect it is likely that the significant reduction of pulmonary pressure during exercise after nifedipine application may be related to a downward shift of the left ventricular pressure-volume relationship. This may be caused by a relief of exercise-induced ischemia during nifedipine therapy, as has resently been shown in clinical as well as experimental studies (18, 19).

The present study indicates that acute sublingual nifedipine therapy in patients with moderate left heart failure due to coronary artery disease or congestive cardiomyopathy is associated with:

1. A beneficial effect on stroke volume and cardiac output at rest as well as during exercise
2. A reduction in pulmonary artery pressure during exercise.

The beneficial effect of nifedipine on left ventricular systolic and diastolic performance is probably related to the peripheral as well as to the myocardial action of this agent. Further clinical studies are needed to document the long-term effect of such a therapy.

References

1. McIlwraith GR, Kidner PH, Oram S (1980) Effect of Nifedipine on exercise tolerance in angina pectoris. Br Heart J 44:335
2. Moshowitz RM, Piccini RA, Nacarelli GV, Zelis R (1979) Nifedipine therapy for stable angina pectoris. Preliminary results of effects on angina frequency and treadmill exercise response. Am J Cardiol 44:811
3. Previtali M, Salerno JA, Tavazzi R, Ray M, Medici A, Chimienti M, Specchia G, Bobber P (1980) Treatment of angina at rest with nifedipine: a short term controlled study. Am J Cardiol 45:825
4. Engel HJ, Wolf R, Handeshagen H, Lichtlen PR (1980) Different effects of nitroglycerin and nifedipine on regional myocardial blood flow during pacing induced angina pectoris. Eur Heart J [Suppl B] 1:53
5. Henry PD, Schuchleib R, Clark RE, Perez JE (1979) Effect of nifedipine on myocardial ischemia: analysis of collateral flow, pulsatile heart and regional muscle shortening. Am J Cardiol 44:817
6. Vater W, Schlossmann K (1976) Effects of nifedipine on the hemodynamics and the oxygen consumption of the heart in animal experiments. In: Jatene AD, Lichtlen PR (eds) Proc of the 3rd international Adalat symposium. Exerpta medica, Amsterdam, p 33
7. Clark RE, Christlieb IY, Henry PD, Fischer AE, Nora JD, Williamson JR, Sobel BE (1979) Nifedipine: a myocardial protective agent. Am J Cardiol 44:825
8. Selwyn AP, Welman FK, Fox K, Morlock P, Pratt T, Klein M (1979) The effects of nifedipine on acute experimental myocardial ischemia and infarction in dogs. Circ Res 44:16
9. Nayler WG (1980) The pharmacological protection of the ischemic heart, the use of calcium and beta-adrenoceptors antagonists. Eur Heart J [Suppl B] 1:5
10. Beck OA, Wolf R, Hochrein H (1977) Die hämodynamische Wirkung von Nifedipin bei kompensierten und dekompensierten Koronarkranken. Münch Med Wochenschr 119:39
11. Klugmann S, Salvi A, Camerini F (1980) Hemodynamic effects of nifedipine in heart failure. Br Heart J 43:440
12. Kaufmann L (1977) Differenzierung verschiedener Kalziumantagonisten. Münch Med Wochenschr 119:6
13. Verdouw PD, Ten Cate FJ, Hugenholtz P (1980) Effect of nifedipine on segmental myocardial function in the anesthetized pig. Eur J Cardiol 63:209
14. Matsumoto S, Ito T, Sada T, Tahahashi M, Su K, Ueda A, Okata F, Sato M, Sekine J, Ito Y (1980) Hemodynamic effects of nifedipine in congestive heart-failure. Am J Cardiol 46:476
15. Chatterjee K, Parmley WW, Massie B, Greenberg B, Wiener J, Klausen S, Norman A (1976) Oral hydralazine therapy for chronic refractery heart failure. Circulation 54:879
16. Mostbeck A, Partsch H, Peschel L (1976) Investigations on peripheral blood distribution. In: Jatene AD, Lichtlen PR (eds) 3rd international Adalat symposium. New therapy of ischemic heart disease. Exerpta medica, Amsterdam, p 91
17. Lichtlen PR, Engel HJ, Amende J, Rafflenbeul W, Simon R (1976) Mechanisms of various antianginal drugs. Relationship between regional flow behaviour and contractility. In: Jatene AD, Lichtlen PR (eds) 3rd international Adalat symposium. Exerpta Medica, Amsterdam, p 14
18. Henry PD, Schuchleib R, Davis J, Weiss ES, Sobel BE (1977) Myocardial contracture and accumulation of mitochondrial calcium in ischemic rabbit heart. Am J Physiol 233:H677
19. Lorell BH, Turi Z, Grossman W (1980) Modification of left ventricular response to pacing tachycardia by nifedipine in patients with coronary artery disease (Abstr) Circulation 62II: 992
20. Hanrath P, Kremer P, Bleifeld W (1982) Influence of nifedipine on left ventricular dysfunction at rest and during exercise. Eur Heart J 3
21. Polese et al. (1979) Am J Med 30:825
22. Klugmann et al. (1980) Eur Heart J [Suppl] 1:49

Discussion

Auditorium: Nifedipine is a calcium slow-channel blocking agent, so its effect on the peripheral circulation must be taken into account. This drug antagonizes its own beneficial effect (reduction of afterload) with its negative inotropic effect. Without the direct effect on the myocardium, the improvement might have been much greater.

Hanrath, Hamburg: This is an important point. There are experimental and clinical studies showing that nifedipine does not exert negative inotropic effects when given intravenously within the therapeutic range. In this context, I would like to mention the study by Verdouw (13). However, given intracoronarily nifedipine shows negative inotropic effects in animal experiments as well as in clinical studies. Nifedipine was therefore given to patients with coronary artery disease and impaired ventricular function. I think in this context, the combination of direct myocardial and peripheral effects of the drug merits clinical interest.

Löllgen, Freiburg: When did you repeat the exercise test? Can you exclude the possibility of an effect due only to repetition, considering the minor changes of pulmonary artery and wedge pressure? Such small changes may be observed when exercise tests are repeated after a short time.

Hanrath, Hamburg: We performed the control measurements at rest and on exercise, and repeat these measurements 60–90 min after administration of the drug. After such an interval the maximal effect can be expected, according to our own experience and the reports in the literature.

Bechthold, Mainz: There have been reports in the literature demonstrating an impairment of ventricular function, e.g., pulmonary edema, after administration of nifedipine to patients with severe congestive heart failure. Did you observe impairment at all in those patients with heart failure?

Hanrath, Hamburg: No, as shown in the figures, two patients showed a decrease in stroke volume index; furthermore, in two patients pulmonary wedge pressure increased after administration of nifedipine.

Auditorium: Nifedipine has even been used in the treatment of acute pulmonary edema, demonstrating that removal of pulmonary edema is possible by afterload (21). I do not think the negative inotropic effect of nifedipine plays an essential role in the application of that drug in clinical conditions.

Hanrath, Hamburg: As far as I know, in Frankfurt nifedipine has even been administered in patients with acute myocardial infarction.

Bussmann, Frankfurt: These studies were performed 4–5 years ago, showing effects of nifedipine as beneficial as in your study: decrease in arterial blood pressure and increase in cardiac output even in patients with congestive heart failure. However, an increase in heart rate was always observed, and this is a typical effect of nifedipine. I think, therefore, that nifedipine might not be the optimal drug in this situation. I think that verapamil – reducing heart rate – would be better.

Hanrath, Hamburg: I think the influence of nifedipine on heart rate depends on the dosage used. In one study, published in the American Journal of Cardiology (14), heart rate increased after administration of 20 mg nifedipine in patients with cardiomyopathy. However, there was an Italian study which did not report an increase in heart rate (22).

Just, Freiburg: I do not think the acute and chronic effects of the drug can necessarily be compared.

Bussmann, Frankfurt: The increase in heart rate is caused by an activation of baroceptors via decrease of arterial blood pressure. I doubt whether this is true during chronic treatment.

Scholz, Hannover: It is certainly clear that the so-called calcium antagonists are negative inotropic agents. If this negative inotropic effect cannot be seen in humans, one must take into account the fact that this negative inotropic effect is counterbalanced by vasodilatation, which causes an activation of baroceptors and thereby an activation of the sympathetics system. If a β-blocking agent is added, the heart rate and ventricular function will decrease strongly. Some time ago, these agents were used only in combination with digitalis. Today we know that a clinical dosage levels the reflex effect overrides the direct negative inotropic effect. Now we apply calcium antagonists in patients with congestive heart failure. I call for caution, especially in view of a general recommendation.

Hanrath, Hamburg: In clinical practice we are successfully treating many patients with coronary artery disease with β-blocking agents, in addition to calcium antagonists in high dosages, without deleterious effects.

Auditorium: The success of such a treatment can be questioned. Furthermore, we cannot today assess the influence on life expectancy.

Greef, Düsseldorf: There are many differences between the different calcium antagonists. Nifedipine, for example, is more potent than verapamil in the peripheral vasculature. When higher doses of nifedipine are used, the peripheral effect becomes pronounced, whereas the reduction in contractility will be observed later than with verapamil. I think, then, that only nifedipine can be combined with β-blocking agents.

Auditorium: This is what we do in the clinic.

Greef, Düsseldorf: Why do you measure hemodynamics 30–90 min after administration of the drug? These data do not give evidence of the efficiency of calcium antagonists in the treatment of congestive heart failure.

Hanrath, Hamburg: I agree, but we wanted to demonstrate the maximal hemodynamic effect of the drug. This occurs after 60–90 min. We have to carry out studies to evaluate the long-term effect of nifedipine. Furthermore, calcium antagonists predominantly affecting peripheral vasculature may be developed in the future. These agents may be much more interesting in the treatment of congestive heart failure.

Treatment of Congestive Heart Failure with Dipyridamole

U. Sigwart, M. Grbic

In addition to treatment with digitalis and diuretics, therapy of congestive heart failure with vasodilating agents has proved to be effective. Unfortunately, most of the available arteriolar vasodilators are far from ideal.

For some time, we have been treating congestive heart failure with dipyridamole, a well-known substance which has been employed for more than 25 years in the treatment of cardiovascular disease, with varying indications. In Goodman and Gilman's handbook "The Pharmacological Basis of Therapeutics" (1) we find the following remarks:

"Dipyridamole is similar to papaverin in many of its pharmalogical properties. In therapeutic doses dipyridamole produces a slight alteration of systemic blood pressure and coronary blood flow. The drug does decrease coronary vascular resistance and increases coronary blood flow and oxygen tension in coronary sinus blood. Dipyridamole appears to act predominantly on small resistance vessels."

In another chapter on antithrombotic and thrombolytic agents, we find:

"Dipyridamole is a vasodilator that, in combination with warfarin, inhibits embolization from prosthetic valves and, in combination with aspirin, prolongs the survival of platelets in patients with thrombotic diseases. Dipyridamole by itself has little or no clinical effect."

These contradictory statements within one handbook are astonishing and evoke a certain curiosity. Since dipyridamole is rather nontoxic and has been used in high doses in many women during pregnancy to prevent thromboembolic disease associated with prosthetic heart valves without ever showing teratogenic or other important side effects, it appears justified to give dipyridamole in equally high doses as a vasodilator.

The use of dipyridamole as a vasodilator has suffered from the fact that the drug was advertised some 25 years ago as an agent for the treatment of angina pectoris. Subsequently, reports appeared in the literature documenting a higher frequency of anginal attacks after treatment with dipyridamole. This was attributed to a typical coronary steal effect, as found with most arterial vasodilators in the presence of organic coronary stenosis. The same symptomatology is known for hydralazine, prazosin, captopril, and other drugs.

In order to test the effectiveness of dipyridamole in treatment of congestive heart failure, we studied its effects acutely in 12 patients and later in a total of 19 patients during chronic treatment. All patients had congestive cardiomyopathy with NYHA

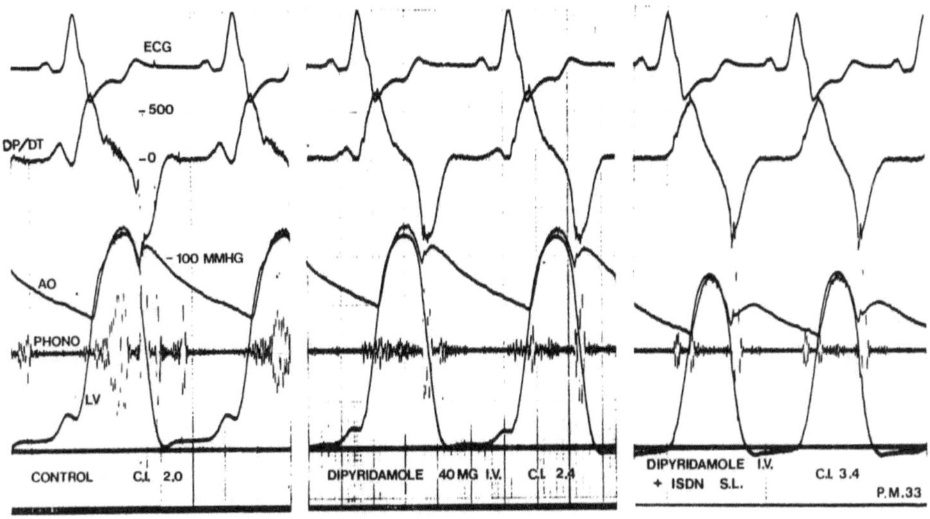

Fig. 1. Original recording of ECG, left ventricular pressure, aortic pressure, phonocardiogram, and first derivative of left ventricular pressure before and after injection of 40 mg dipyridamole and sublingual administration of 5 mg ISDN. Note the disappearance of a mitral incompetence documented with the intracardiac phonocardiogram

class III–IV dyspnea and had already been treated conventionally, for example with digitalis and diuretics. The acute study was performed during routine diagnostic cardiac catheterization. Cardiac output was measured with a Swan-Ganz thermodilution cathether in the pulmonary artery, using a Roche automatic injection pump which injects ice-cold saline in a triggered mode starting at the peak of the R wave. For left heart catheterization we employed a double micromanometer catheter (Millar 771) which permits simultaneous and very accurate measurement of left ventricular and aortic pressure (Fig. 1).

After baseline measurements, 40 mg dipyridamole was injected in two equal parts with a 1-min interval. Fig. 2 depicts the behavior of the most important parameters. The cardiac index increased by about 30% after application of dipyridamole, whereas the peripheral resistance dropped. Since the heart rate failed to change significantly, the increase in cardiac index must have been due to an augmentation of left ventricular stroke volume. Interestingly, left ventricular and aortic systolic pressure decreased only insignificantly and no change was observed in left ventricular contractility parameters.

The additional administration of nitrates [5 mg isosorbide dinitrate (ISDN) sublingually] significantly reduces the left ventricular end-diastolic pressure, which is not effected by dipyridamole, without affecting the left ventricular stroke volume.

Fig. 3 shows the left ventricular function curve: there is an upward shift after dipyridamole, with insignificant displacement to the left; additional ISDN displaces the curve significantly to the left by reducing the left ventricular filling pressure.

There were seven other patients in whom no acute hemodynamic investigation was performed in the catheterization laboratory. Upon discharge from the hospital, all

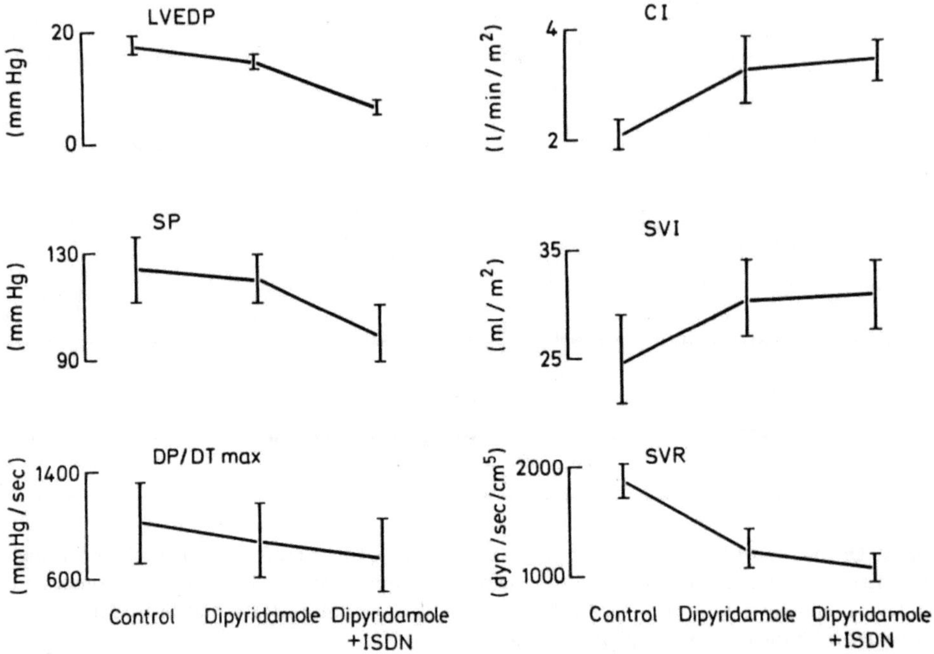

Fig. 2. Hemodynamic parameters before and after dipyridamole and ISDN. *LVEDP*, left ventricular end-diastolic pressure; *SP*, systolic pressure; *dp/dt max*, maximum velocity of pressure rise in left ventricle; *CI*, cardiac index; *SVI*, stroke volume index; *SVR*, systemic vascular resistance

patients were treated daily with 600 mg dipyridamole, divided into four doses of 150 mg, and 160 mg ISDN, divided into four doses of 40 mg. With regard to side effects headache occurred almost invariably in the initial phase, with decreasing intensity, and in three cases slight gastrointestinal disorders were noted but did not necessitate a change of treatment. During the follow-up study, an average improve-

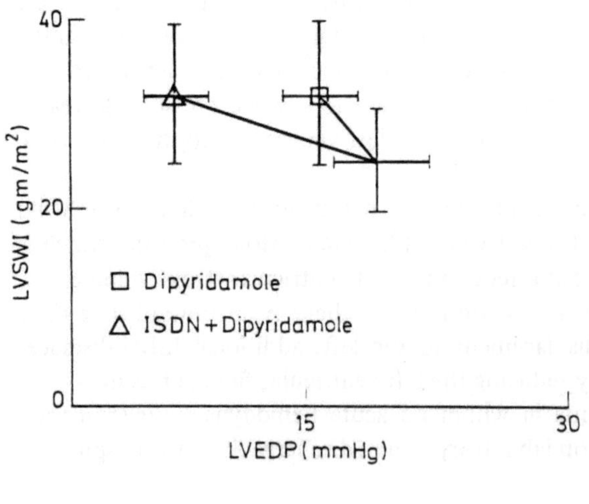

Fig. 3. Normalization of the left ventricular function curve in congestive heart failure after administration of dipyridamole and ISDN

ment of 1.4 NYHA dyspnea classes was found, in conjunction with a reduction of heart size of some 16%.

Conclusion

Dipyridamole appears to be an effective vasodilator in the treatment of congestive heart failure, with only minor side effects. Judging from our initial experience, combined treatment with nitrates is recommended to lower the preload, which is not significantly effected by dipyridamole. Further clinical studies are indicated in order to test the long-term efficacy of dipyridamole and to investigate the development of tolerance to it.

Reference

Gilman AG, Goodman LS, Gilman A, Mayer SE, Melmon KL (eds) (1980) The pharmacological basis of therapeutics, 6th edn, Macmillan, New York

Discussion

Scholz, Hannover: Would you agree that all so-called coronary vasodilators can now be used in this context?

Sigwart, Lausanne: Yes, if these agents have less severe side effects than the currently used vasodilators.

Audience: Dipyridamole has effects on myocardial metabolism. Do you have an idea of the negative effects of the drug when given in chronic therapy?

Sigwart, Lausanne: As far as I know, dipyridamole is quite nontoxic in the doses employed clinically. The actions of dipyridamole seem to be linked, at least in part, to the metabolism and transport of adenosine and adenine nucleotides; in particular, dipyridanole inhibitis the uptake of adenosine by endothelial and other cells. Adenosine, which is released from the hypoxic myocardium, is a (coronary) vasodilator and appears to be an important signal for the autoregulation of (coronary) blood flow.

Audience: You probably know that dipyridamole is used during thallium scintigraphy in order to provoke myocardial ischemia. What do you think about this possible adverse effect?

Sigwart, Lausanne: This side effect can also be observed employing hydralazine or other vasodilators. However, selection of patients is recommended, e.g., if you treat patients with critical coronary stenosis, the coronary steal effect may play an important role. In these cases, one should either not use dipyridamole or try to titrate the dosage until symptoms appear, then use a lower dosage.

Audience: You cannot perform coronary angiography in each patient?

Sigwart, Lausanne: I think that in the clinic we are able to distinguish cardiomyopathy and coronary artery disease in the majority of cases.

Audience: As expected, dipyridamole causes an increase of cardiac output and concomitantly a comparable reduction of afterload. Would you agree that dipyridamole predominantly effects the resistance vessels?

Sigwart, Lausanne: Yes, I agree.

Therapeutic Agents with Combined Inotropic and Vasodilating Properties

H. Just, H. Drexler, M. Irmer

The mechanism of action and the therapeutic effectiveness of vasodilating substances are well known. Both arteriolar and venous dilatation tend to reduce ventricular dimensions and thereby improve contractile geometry (22, 23). Ventricular radius and thereby circumference determine the force of contraction for any given work. According to the law of Laplace, shortening distance decreases with increasing chamber radius for ejection of a given stroke volume. Since work is the product of force and distance, the former has to increase at any given workload if the chamber enlarges. Therefore with increasing ventricular dimensions even the healthiest contractile element will inevitably reach the point of failure. On the other hand, reduction of chamber size will lessen the workload and allow even subnormally performing myocardium to maintain sufficient shortening.

Therefore vasodilatation should be the first step in the treatment of congestive heart failure. Inotropic stimulation will then provide a most useful adjunct. With lesser degrees of failure, inotropic stimulation alone may improve cardiac performance, reducing sympathetic tone and thereby allowing vascular resistance to fall. Ultimately a reduction in chamber size should be achieved.

In the intact circulation, reduction of arteriolar systemic resistance will reduce aortic pressure. The baroreceptor reflex is activated: heart rate increases and sympathetic stimulation enhances contractility. In the failing heart, increased central blood volume and filling pressure will imediately proved sufficient ventricular filling for adequate augmentation of stroke volume. Blood pressure will therefore be maintained, provided there are no obstacles in the central circulation such as valvular stenosis, severe pulmonary hypertension, or inability fo the heart to respond.

With insufficient venous blood supply, severe myocardial damage, or defective baroreceptor response, additional inotropic stimulation may be necessary to maintain cardiac output and thereby circulatory stability. Serious complications of vasodilator treatment frequently result from this very mechanism, i.e., inadequate secondary increase of stroke volume or cardiac output, in response to venous and/or arteriolar dilatation. Additional inotropic stimulation would serve to prevent such complications.

Substances combining vasodilating and inotropic properties will therefore be ideally suited for the treatment of cases of cardiac insufficiency. However, such drugs, although in use and well known for many years, have shown drawbacks, such as arrhythmogenic properties, poor reabsorption, or unfavorable pharmacokinetic properties. Since an urgent need is felt, and better means for hemodynamic and pharmacologic evaluation are now available, older substances should be reevaluated and new compounds studied.

Digitalis glycosides, rated highly by clinicians experienced in the treatment of heart failure, have recently fallen into disrepute because of their vasoconstrictive actions. However, arteriolar constriction is seen only with very rapid digitalization, and the effect abates with repeated administration, and in the failing heart is overridden by a pronounced secondary indirect vasodilatation with improvement of cardiac output. The point to be emphasized here is the venoconstriction due to digitalis. This effect is a rather complex one in that through activation of splanchnic sphincter veins it leads to splanchnic blood pooling, with subsequent reduction of central blood volume. This effect occurs early after administration and provides in reality an unloading effect for the right and the left heart with reduction in chamber size. Nevertheless, digitalis glycosides will not be discussed at any further here.

Methylxanthines

Theophylline and aminophylline are used in different preparations and have been administered to patients with heart failure for many years. Variable reabsorption, and therefore reduced bioavailability, as well as short duration of action, have led to disappointment. However, preparations with more favorable pharmacokinetic properties have been and are being developed. The usefulness of this group of drugs should therefore be reassessed. The mode of action of the xanthines is complex. A direct inotropic action occurs through competitive inhibition of cyclic nucleotide phosphodiesterase and probably also through a direct influence on intracellular calcium flux. Sympathetic and glycoside receptors are not affected by the xanthines.

The vasodilatation is effected through a direct, though ill-defined action upon vascular smooth muscle. With small doses, a centrally mediated vasoconstrictor stimulus is operative at the same time. In therapeutic doses, the net effect is dilatory. Diuretic and bronchodilatory actions are well known.

Aminophylline (5, 13, 18) in doses of 300–500 mg augments cardiac output by approximately 25%, stroke volume by 30%, and stroke work by 40%. Total peripheral resistance falls by 25%–45%. The significance of this effect is seen in a reduction of cardiac volume of 1%–3%. An improvement of left ventricular ejection fraction by 5% has been described with long-term administration (13).

ARL-115 BS

This drug is related to the xanthines structurally as well as in its mode of action. Its mechanism of action and clinical effectiveness have been studied extensively only recently; a detailed account is given elsewhere in this volume (see pp 185 ff).

This substance has both a better bioavailability and a longer half-life than theophylline or aminophylline. Its arrhythmogenic properties seem to be less pronounced. Clinical evaluation is therefore urgently needed.

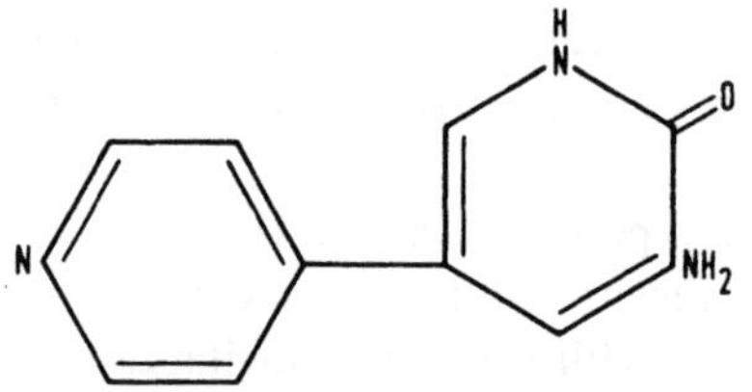

Fig. 1. Chemical structure of amrinone

Amrinone

This bipyridine derivative (Fig. 1) has recently been introduced as an inotropic drug. The data published thus far have, however, also shown a marked vasodilating effect which cannot be considered secondary. Perorally, amrinone is well reabsorbed. Its duration of action reaches 7 h. It is therefore suitable for chronic treatment. Accordingly, it has been tested in chronic refractory heart failure (2, 8, 9). Its site of action has not been defined precisely, but is different from that of the sympathetic receptors and does show no relation to the digitalis glycosides.

The inotropic and vasodilating effects are remarkable. At doses of 2.5 mg/kg body weight, cardiac minute volume is augmented by 40%–69%. Total peripheral resistance falls by approximately 40%. The right and left heart filling pressures are reduced by 20%–30%. A pronounced arteriolar and venular dilatory action must therefore be assumed (8), augmenting the effects of the dominant inotropic stimulation. Heart rate has remarkably remained almost constant in the studies published thus far. Coronary perfusion increased slightly, which is related to the vasodilating properties. Myocardial oxygen consumption fell, a highly desirable effect which is rather unusual in this group of drugs with both vasodilating and inotropic properties. This phenomenon is related to a reduction in ventricular wall tension (9).

So far a more extensive evaluation of this promising substance has been hampered by serious side effects such as thrombo- and leukopenia.

Beta-sympathomimetic Substances

Beta-stimulation with inotropic (beta-1) and/or vasodilating actions has long been used in the treatment of heart failure, especially in the acute situation and in shock. The classic substances are isoproterenol and its optical isomer orciprenaline. Of course adrenaline itself and the now widely employed synthetic derivative dobutamine exhibit inotropic and vasodilating properties. In the discussion here, however, we shall confine ourselves to drugs suitable for application in chronic heart failure. Only those which show a sufficiently reliable reabsorption and a duration of action of at least several hours will be considered.

Orciprenaline is supplied in a galenically retarded form with a pharmacokinetic half-life of approximately 2 h. Besides this undesirable property, orciprenaline, as well as isoproterenol, shows marked chronotropic action, a strong arrhythmogenic capacity, and a pronounced tendency to augment myocardial oxygen consumption.

Myocardial necrosis can be induced with high doses. These substances are therefore deemed unsuitable for chronic use in heart failure.

Variations of the basic molecular structure have led to substances such as dobutamine with lesser chronotropism and a relative dissociation of beta-1- and beta-2-stimulatory properties. However, the drug is still rapidly inactivated by the catechol-O-methyltransferase (COMT) enzyme, reducing half-life to less than 2 min after intravenous infusion.

Further attempts at modification of the molecule have led to replacement of the meta-phenol group, with the subsequent stability of the molecule toward the COMT enzyme resulting in satisfactory reabsorption and sufficient duration of action. Those substances were synthesized toward beta-2 activity: fenoterol (16), salbutamol (15), pirbuterol (14), and terbutaline were developed for treatment of bronchospastic conditions or for tokolysis (fenoterol). They exhibit eight- to tenfold lesser beta-1 activity than isoproterenol, but have similar beta-2-stimulating power.

Only recently have these compounds been evaluated in the treatment of chronic heart failure (7, 17, 20, 21). At doses of 2.5 mg/kg body weight total peripheral resistance is lowered by 20%–30%. With higher doses a greater reduction is brought about (Fig. 2). In patients with chronic refractory failure, heart rate remained almost unchanged (!). Cardiac pumping action was markedly improved with an increase if

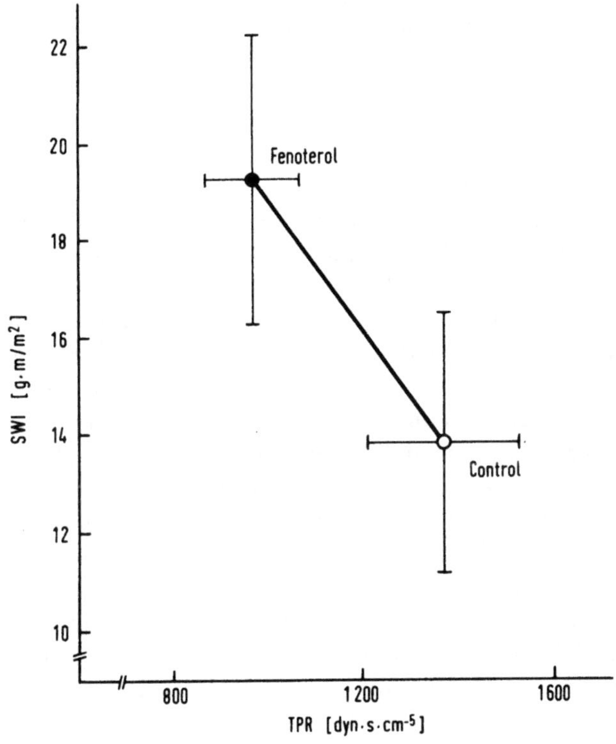

Fig. 2. Response of total peripheral resistance (*TPR*) and stroke work index (*SWI*) to fenoterol in seven patients with severe refractory heart failure. It can be seen that the pronounced reduction of TPR is accompanied by a quantitatively similar increase of SWI. From Irmer et al. (7)

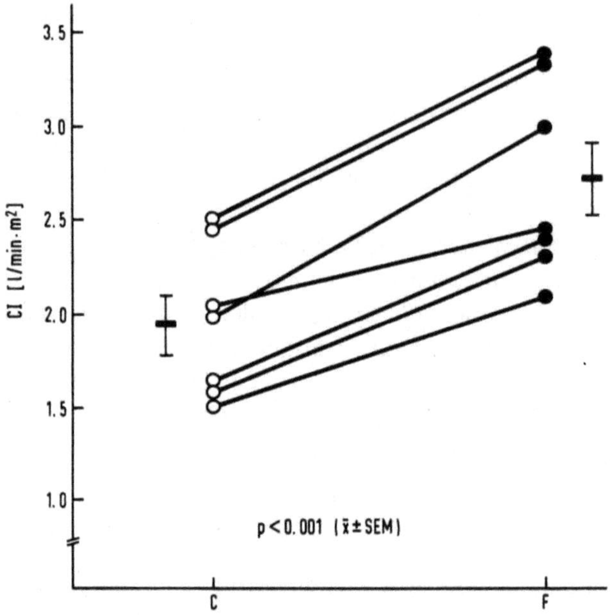

Fig. 3. Response of cardiac index to i.v. fenoterol in patients with refractory heart failure. Increment of cardiac index (*CI*) is similar to that of SWI and stroke volume because heart rate changed only slightly. *C*, control values; *F* after fenoterol. From Irmer et al. (7)

stroke volume of 30% and cardiac output of 40% (Figs. 2 and 3). Beta-2-stimulation, i.e., vasodilatation cannot alone be responsible for this marked response. A beta-1 effect must have been operative at the small dose employed (Fig. 4). Likewise, impressive augmentation of cardiac pumping could be measured with long-term treatment in patients with chronic refractory heart failure (17).

The closely related substance, salbutamol (3, 11, 19), exhibits similar actions at doses of 0.5–2 mg/kg. Its vasodilating power has been related to that of nitroprusside (17). Both drugs in this study showed a reduction of total peripheral resistance by 21% with subsequent comparable augmentation of cardiac output. This observation has to be interpreted as evidence of a pronounced dominance of the beta-2 effect, since nitroprusside has no inotropic power of its own. In patients recovering from open-heart surgery, left atrial pressure fell by 40% with both drugs. Augmentation of heart rate was here less pronounced with salbutamol than with nitroprusside!

In another study (19) patients with chronic refractory failure due to congestive cardiomyopathy received 0.5 mg/kg per minute salbutamol. Cardiac index rose by 105%, stroke volume index by 100%. Heart rate rose only by 8% in the face of a reduction of peripheral resistance by 55%! The authors of this very careful hemodynamic and angiocardiographic study attempted to separate the inotropic and vasodilating properties of salbutamol, at the same time studying myocardial compliance. They observed that both vasodilatation (dominant) and inotropism were present. Myocardial compliance was found to be improved, i.e., stiffness was reduced; in fact

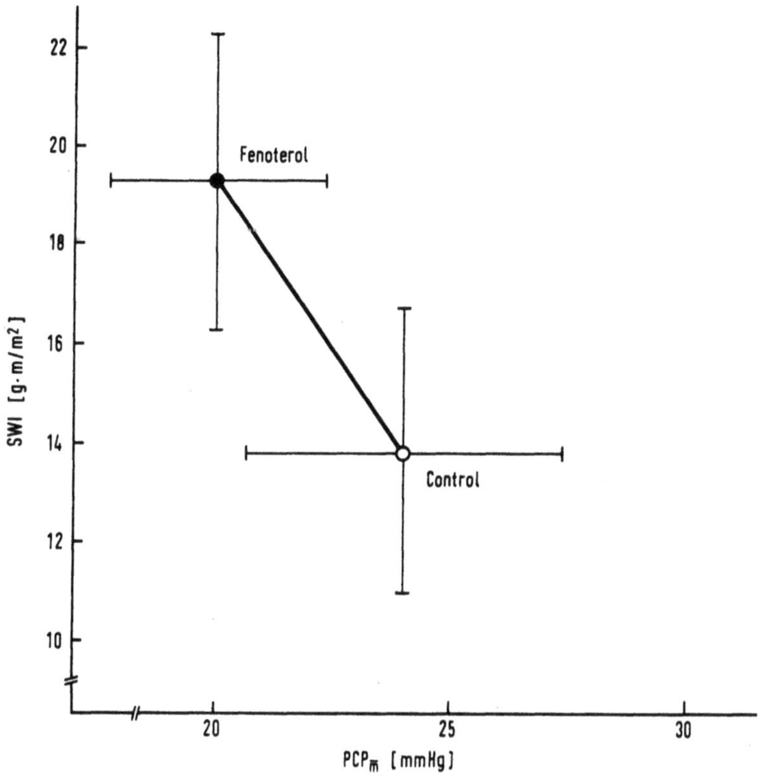

Fig. 4. Response of pulmonary capillary wedge pressure (*PCP_m*) and stroke work index (*SWI*) to i.v. fenoterol in patients with refractory heart failure. The inotropic effect is indicated by the marked rise of SWI in the face of falling PCP_m. From Irmer et al. (7)

compliance returned to normal in all cases studied. The cause and significance of this effect remain so far undetermined but seems highly important and woth pursuing.

In animal experimental studies, cardiac efficiency has been determined under salbutamol and orciprenaline (15). It could be demonstrated that salbutamol had a tendency to improve the efficiency of energy utilization, whereas orciprenaline reduced this value. Dogs with nonfailing hearts also showed a lesser degree of augmentation of heart rate with salbutamol than with orciprenaline. The difference was, however, abolished after vagosympathectomy and therefore reflex mechanisms must be postulated. The authors conclude that with salbutamol the vasodilating effect is more pronounced and the inotropic stimulation therefore meets with reduced output impedance and ventricular dimensions, favoring a more economic contractile performance.

In spite of the thus far reported favorable results achieved with beta-stimulants in heart failure, these substances do not seem to be optimal. Side effects are frequent and not necessarily harmless. Palpitation, ectopic arrhythmias, restlessness, agitation, and gastrointenstinal disturbances are frequent. In addition, for all of these drugs a more or less marked loss of effectiveness is seen with long-term administration.

At the present time, all drugs combining inotropic and vasodilating properties have to be considered far from ideal for the treatment of chronic heart failure. Until newer, more suitable drugs are available, the combination of vasodilators with inotropic drugs will remain the treatment of choice.

References

1. Awan KA, Evenson MK, Needham KE, Evans TO, Hermanovich J, Taylor CR, Amsterdam E, Mason DT (1981) Hemodynamic effects of oral pirbuterol in chronic severe congestive heart failure. Circulation 63:96
2. Benotti JR, Grossman W, Braunwald E, Carabello BA (1980) Effects of amrinone on myocardial energy metabolism and hemodynamics in patients with severe congestive heart failure due to coronary disease. Circulation 62:28
3. Bourdillon PDV, Dawson JR, Foale RA, Timmis AD, Poole-Wilson PA, Sutton GC (1980) Salbutamol in treatment of heart failure. Br Heart J 43:206
4. Drexler H, Löllgen H, Just H (1981) Short and long term effects of hydralazine and combined hydralazine-prenalterol therapy in severe chronic congestive heart failure. Klin Wochenschr 59:647
5. Goodman LS, Gilman A (1965) The pharmacological basis of therapeutics, 3rd edn. Macmillan, New York, p 356
6. Gattiker H, Rothlin M (1971) Vergleich der Wirkungen des neuen Sympathomimetikums Th 1165 a mit jenen von Isoprenalin auf den Kreislauf herzoperierter Patienten. Arzneimittelforsch 21:1191
7. Irmer M, Wollschläger H, Just H (1981) Behandlung der schweren Herzinsuffizienz mit dem Beta-Stimulator Fenoterol. Klin Wochenschr 59:639
8. Lejemtel TH, Keung E, Sonnenblick EH, Ribner HS, Matsumoto M, Levis R, Schwartz W, Alousi AA, Davolos D (1979) Amrinone: a new non-glycosidic, non-adrenergic cardiotonic agent effective in the treatment of intractable myocardial failure in man. Circulation 59:1098
9. Jentzer JH, LeJemtel TH, Kirk ES, Sonnenblick EH (1979) The beneficial effect of amrinone on myocardial oxygen consumption during acute left ventricular failure. Circulation [Suppl 2] 59/60:30
10. Just H (1978) Dobutamin. Springer, Berlin Heidelberg New York
11. Lal S, Savidge RS, Davies DM, Ali MM, Soni V (1972) Intravenous salbutamol and cardiogenic shock. Lancet I:853
12. Mason DT (1978) Symposium on vasodilator and inotropic therapy of heart failure. Am J Med 65:101
13. Matthay R, Berger H, Davies R, Lohe J, Zaret B (1979) Prolonged improvement in cardiac performance by oral long-acting theophylline in chronic obstructive pulmonary disease. Circulation [Suppl 2] 59/60:106
14. Moore PS, Constantine JW, Barth WE (1978) Pirbuterol: a selective beta$_2$-adrenergic bronchodilator. J Pharm Exp Ther 207:410
15. Nayler WG, McInnes J (1972) Salbutamol- and orciprenalin-induced changes in myocardial function. Cardiovasc Res 6:725
16. O'Donnell SR (1970) A selective beta-adrenoceptor stimulant (Th 1165 a) related to orciprenaline. Eur J Pharmacol 12:35
17. Poole-Wilson PA, Lewis G, Angerpointer T, Malcolm AD, Williams BT (1977) Hemodynamic effects of salbutamol and nitroprusside after cardiac surgery. Br Heart J 39:721
18. Rutherford JD, Vatner SF, Braunwald E (1981) Effects and mechanism of action of aminophylline on cardiac function and regional blood flow distribution in conscious dogs. Circulation 63:378
19. Sharma B, Goodwin JF (1978) Beneficial effect of salbutamol on cardiac function in severe congestive cardiomyopathy: effect on systolic and diastolic function of the left ventricle. Circulation 58:449

20. Tse TF, Yu D, Wang RYC, Wong P (1978) Effect of Berotec in congestive cardiac failure. Eighth World Congress of Cardiology, Tokyo (Abstr), p 1269
21. Warrell DA, Robertson DG, Newton Homes H, Connolly ME, Patterson JW, Beillin LJ, Dollery T (1970) Comparison of cardiorespiratory effects of isoprenaline and salbutamol in patients with bronchial asthma. Brit Med J 1:65
22. Zelis R, Flaim SF, Moskowitz RM, Nellis SH (1979) How much can we expect from vasodilator therapy in congestive heart failure? Circulation 59:1092 (editorial)
23. Just H, Bussmann WD (1981) Therapie der chronischen Herzinsuffizienz mit Vasodilatantien. Edition Medizin, Weinheim Deerfield Beach Basel

Hemodynamic Effects of Combined Inotropic and Vasodilator Drugs in Severe Congestive Heart Failure: Hydralazine and Prenalterol

H. Drexler, H. Löllgen, H. Just

Cardiac performance can be improved by positive-inotropic agents, by vasodilators, or by a combination of both. In addition to digitalis and diuretics, vasodilatation has become generally accepted as a therapeuic principle in the treatment of severe congestive heart failure. Furthermore, combined treatment with vasodilators and inotropic agents has had beneficial short-term effects. For instance Miller et al. (1) described a greater augmentation of cardiac performance after adding dopamine to nitroprusside in patients with congestive heart failure.

Sodium-nitroprusside causes a pronounced decrease of left ventricular filling pressure and a slight augmentation of cardiac output. Dopamine predominantly enhances cardiac output, with only little effect on left ventricular filling pressure. Combined treatment with dopamine and sodium-nitroprusside increases the beneficial action of each agent. Therefore, combined treatment was considered to be a promising approach also in the long-term therapy of severe congestive heart failure.

Depending on baseline hemodynamics, hydralazine, prazosin, nitrates, etc. can be used as vasodilators in this context; however, apart from digitalis, positive-inotropic agents for oral use are scarce and have not been investigated as to their value in the management of chronic congestive heart failure. Amrinone, pirbuterol and prenalterol are available for clinical investigation and first experiences have been reported (2–5).

Recently, the beneficial effects of combined treatment with hydralazine and amrinone have been reported and compared with the hemodynamic effects of these agents given alone (6). Amrinone (300 mg/day) has been shown to be effective in increasing cardiac output and lowering left ventricular filling pressure. Hydralazine predominantly causes an augmentation of cardiac output, with only sligh reduction of left ventricular filling pressure. The hemodynamic effects of low-dose amrinone can be amplified by simultaneous administration of a vasodilator such as hydralazine. It must be emphasized, that all these patients continued to take their usual doses of digitalis and diuretics throughout the study. In addition to the beneficial effects on resting dynamics, exercise capacity was improved in five of eight patients. The increase of exercise capacity was sustained after three weeks treatment. The improvement in exercise tolerance with the combination of amrinone and hydralazine seemed to be greater than that reported with the use of vasodilators alone in patients with similar

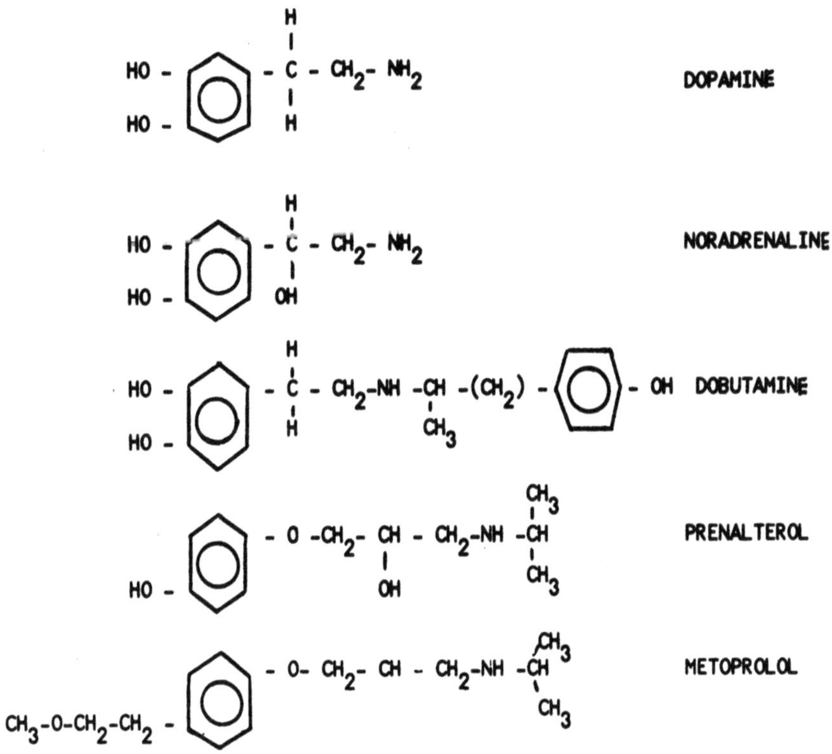

Fig. 1. Molecular structure of prenalterol and other β-adrenoceptor agonists

functional disability (6–9). However, amrinone's mechanism of action is uncertain, and in particular, higher doses may have severe side effects, e.g., thrombocytopenia (10).

In contrast, in molecular structure prenalterol is similar to the catecholamines (Fig. 1) and exerts its effects via β_1-receptors (11), therefore having positive-inotropic and positive-chronotropic effects.

The hemodynamic effects of prenalterol (after oral administration) are represented in Fig. 2. In six patients with severe congestive heart failure a pronounced augmentation of cardiac index was observed 60 min after ingestion (+40%). Pulmonary wedge pressure was slightly decreased in four of six patients. Heart rate was unchanged in all but one patient; however, the increase of heart rate in this patient after acute administration could not be observed after long-term treatment over a 4-week period, although augmentation of cardiac index was maintained.

No studies have yet evaluated the long-term effects of any combined vasodilator and inotropic therapy in chronic congestive heart failure. Therefore we have studied ten patients with severe chronic congestive heart failure in order to assess the short- and long-term effects of combined hydralazine-prenalterol therapy compared to hydralazine treatment alone. The protocol was as follows:

1. Baseline measurement, including heart rate, systemic blood pressure, pulmonary wedge pressure, pulmonary arterial pressure, and cardiac output (thermodilution).

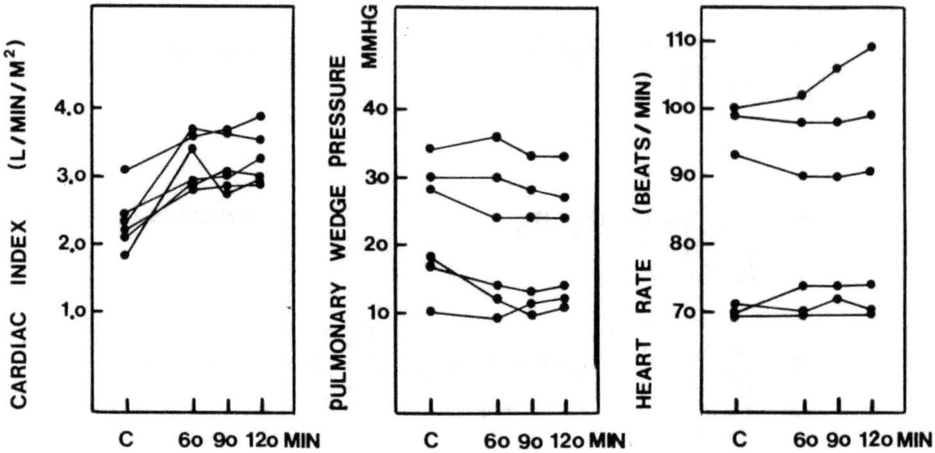

Fig. 2. Effect of prenalterol on cardiac index, pulmonary wedge pressure, and heart rate in six patients with severe heart failure 60, 90, and 120 min after oral administration of 20 mg prenalterol. *C*, control values

Exercise testing on bicycle ergometer, beginning at 25 W for 3 min (n = 6). Oral administration of 100 mg slow-release hydralazine, with measurement of hemodynamics 30, 60, and 90 min after administration and at exercise (limited by symptoms or high pulmonary wedge pressure).

2. Three to four weeks treatment with 2 × 100 mg slow-release hydralazine.
3. Measurement of hemodynamics and additional administration of 5 mg prenalterol 30 min after administration of 100 mg hydralazine.
4. Three to four weeks treatment with 2 × 100 mg slow-release hydralazine and 4 × 5 mg prenalterol with measurement of hemodynamics after that period.

The hemodynamic effects of hydralazine are given in Figs. 3 and 4, which show augmentation of cardiac output with pronounced decrease of total peripheral resistance. These effects were maintained after 3–4 weeks treatment. Additional administration of prenalterol caused further augmentation of cardiac index and stroke volume index, these effects also being maintained after long-term treatment (Fig. 4).

In the group as a whole, there was no essential change in heart rate and systemic blood pressure. Long-term administration of hydralazine did not decrease pulmonary wedge pressure, whereas prenalterol did, slightly but significantly ($p < 0.05$; Fig. 3).

Hemodynamic responses were assciated with an improvement of NYHA functional class (Fig. 5).

In five patients, exercise testing could be performed. Chronic hydralazine treatment improved maximal exercise capacity in one patient, and chronic combined treatment increased exercise tolerance in three of five patients (Fig. 6). Figure 7 shows exercise hemodynamics after 3 min at a workload of 25 W. In three of five patients, cardiac output increased after hydralazine treatment alone. A decrease of pulmonary wedge pressure was observed in three patients after chronic combined treatment.

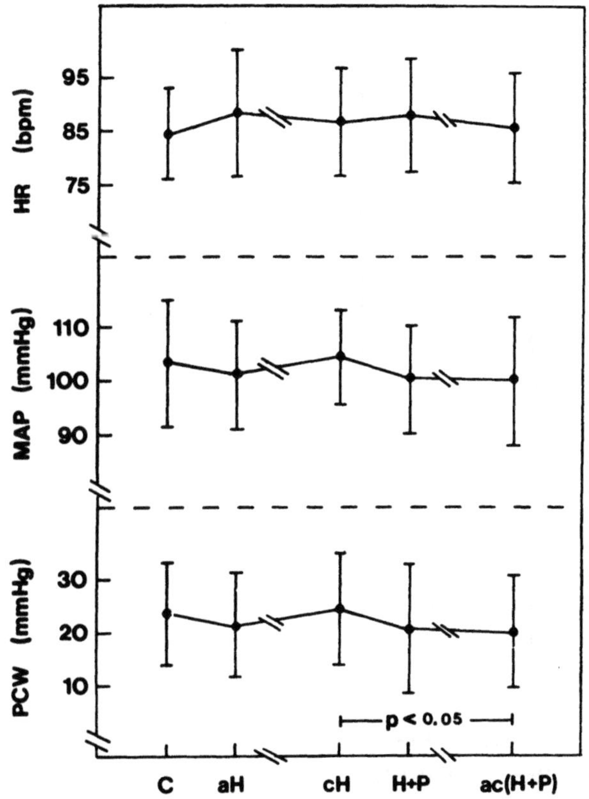

Fig. 3. Effect of hydralazine and combined treatment with hydralazine and prenalterol on heart rate (*HR*), mean systemic arterial pressure (*MAP*), and pulmonary capillary wedge pressure (*PCW*). *C*, Control values; *aH*, 90 min after acute administration of 100 mg hydralazine; *cH*, hemodynamics after 3–4 weeks treatment with 200 mg hydralazine; *H + P*, 90 min after acute administration of 100 mg hydralazine and 60 min after administration of 5 mg prenalterol, following 3–4 weeks treatment with 200 mg hydralazine; *ac(H + P)*, hemodynamics after 3–4 weeks treatment with 200 mg hydralazine and 20 mg prenalterol, 90 min after repeated hydralazine and 60 min after prenalterol administration

Discussion

Although the results of combined vasodilator and cardiotonic treatment are quite preliminary at present (6, 12), the data suggest that positive-inotropic agents such as amrinone or prenalterol may enhance the efficacy of vasodilators in the treatment of chronic heart failure.

The main limitation of cardiac performance in chronic congestive heart failure is cardiac output. Vasodilators increase cardiac output, but may not further enhance exercise-induced vasodilatation, rather increasing flow in other organs. In contrast, inotropic agents increase cardiac output without changing the distribution of flow, thereby enhancing the flow to skeletal muscle. Although no definite conclusion can be drawn regarding the small number of patients studied at exercise, the data obtained would support this hypothesis. In addition, augmented exercise tolerance has been

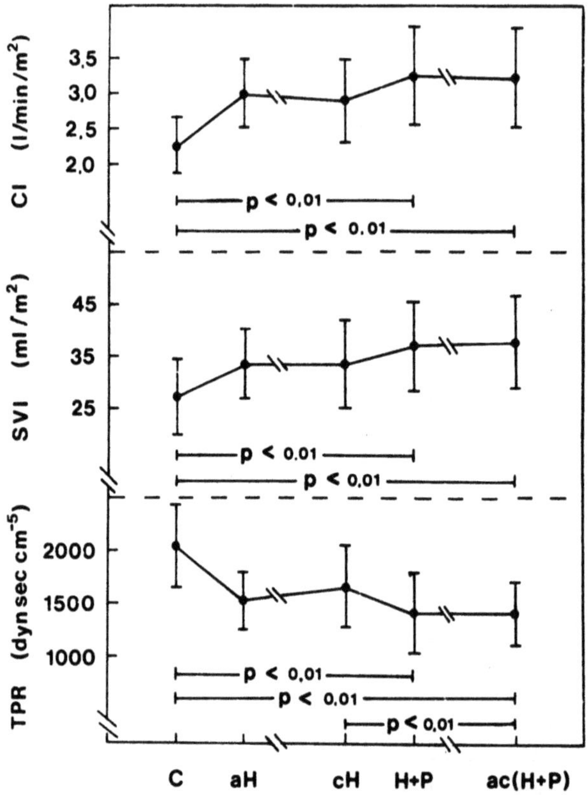

Fig. 4. Effect of hydralazine and combined hydralazine and prenalterol treatment on cardiac index (*CI*), stroke volume index (*SVI*), and total peripheral resistance (*TPR*). Other abbreviations as in Fig. 1

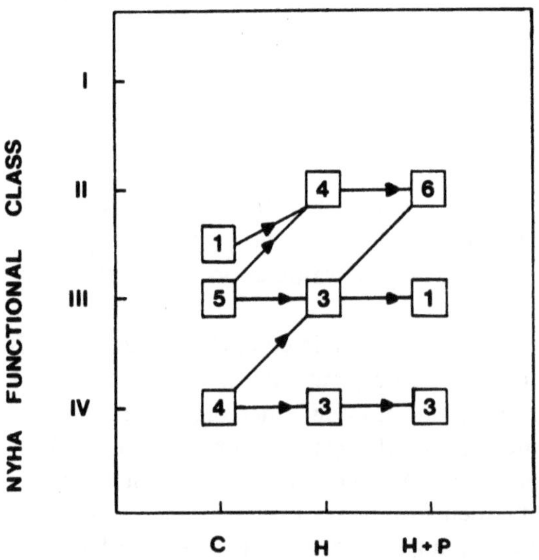

Fig. 5. NYHA functional class before treatment (*C*), after treatment with hydralazine (*H*), and after combined treatment (*H + P*)

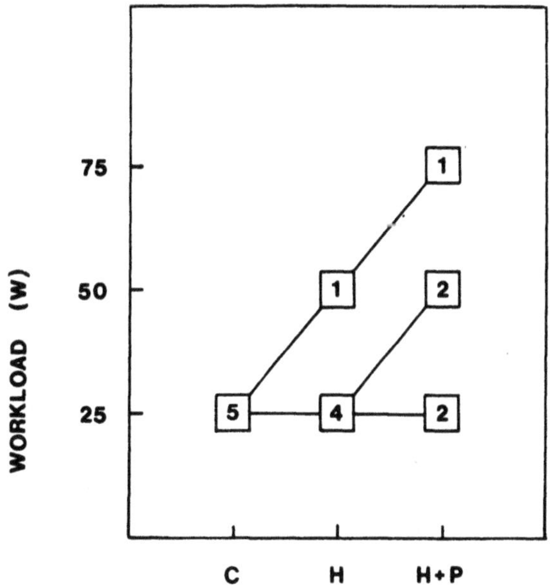

Fig. 6. Stepwise-increasing exercise testing before treatment (*C*), after chronic hydralazine (*H*), and after chronic combined therapy (*H + P*)

translated into an improvement of symptoms as demonstrated by changes in NYHA class. Several studies suggest long-term benefit from vasodilator treatment, e.g., hydralazine in patients with chronic heart failure (13, 14). Our results from the 3-week treatment with hydralazine, are consistent with this, but a randomized double-blind trial is needed to confirm these observations.

Furthermore, additional administration of positive-inotropic agents, such as amrinone (6) or prenalterol (our study), leads to further improvement of hemodynamics. During ambulatory therapy combining hydralazine and prenalterol, considerable improvement in NYHA functional class was demonstrated. Subjective benefits were accompanied by objective documentation of enhanced left ventricular function by hemodynamics at rest and in some patients at exercise.

Thus hydralazine and prenalterol caused sustained augmentation of left ventricular function. It is emphasized that although the hemodynamic and clinical effects of prenalterol in combination with hydralazine are encouraging, precise definition of the therapeutic benefit of such combinations awaits further trials. It is remarkable, though that our initial experience with prenalterol did not disclose any augmentation of ventricular ectopic activity.

Conclusions

Prenalterol may enhance the effectiveness of hydralazine therapy in congestive heart failure (Fig. 8). Long-term oral administration of hydralazine and prenalterol can produce sustained improvement.

Under exercise, inotropic agents such as prenalterol and amrinone may be more beneficial than vasodilators.

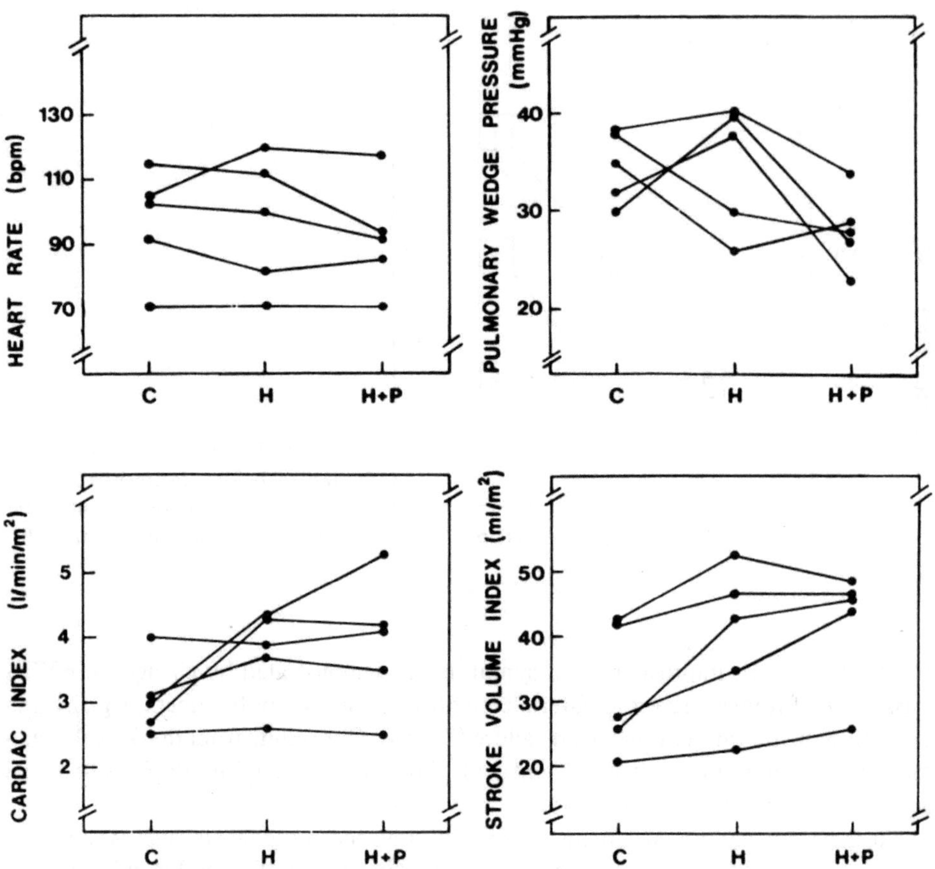

Fig. 7. Exercise hemodynamics for five patients (workload 25 W for 3 min) before treatment (*C*), after treatment with hydralazine (*H*), and after combined hydralazine-prenalterol therapy (*H + P*). Measurements were taken at peak exercise

References

1. Miller RR, Wan NA, Joye JA, Maxwell KS, DeMaria AN, Amsterdam EA, Mason DT (1977) Combined dopamine and nitroprusside therapy in congestive heart failure. Circulation 55:881–884
2. Benotti JR, Grossman W, Braunwald E, Davolos DD, Alousi AA (1977) Hemodynamic assessment of amrinone. N Engl J Med 299:1373–1377
3. Awan NA, Evenson MK, Needham KE, Evans TO, Hermanowich J, Taylor CR, Amsterdam EA, Mason DT (1981) Hemodynamic effects of oral pirbuterol in chronic severe congestive heart failure. Circulation 63:96–101
4. Hutton I, Murray RG, Boyes RN, Rae AP, Hillis WS (1980) Hemodynamic effects of prenalterol in patients with coronary heart disease. Br Heart J 43:134–137
5. Kupper W, Kuck KH, Sonntag F, Bleifeld W (1980) Effects of a new β_1-agonist, prenalterol, on myocardial metabolism and left ventricular function in patients with chronic heart failure. Circulation [Suppl 3] 62:298

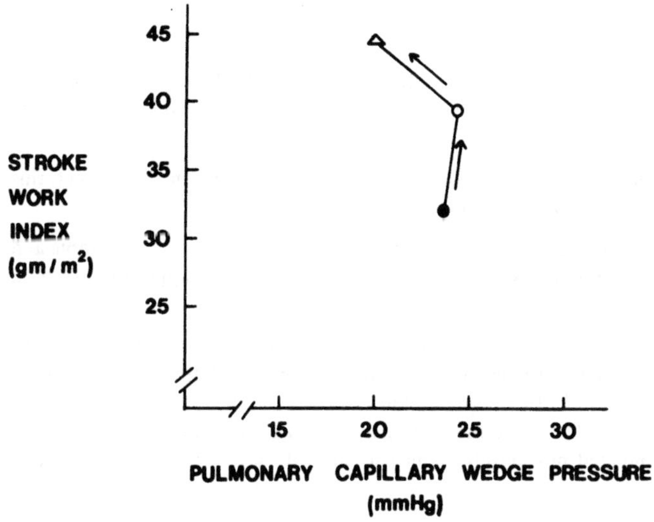

Fig. 8. Stroke work index and pulmonary capillary wedge pressure during control period (•), with hydralazine alone (○), and with combined treatment (△). Hydralazine shifts the Frank-Starling relationship upwards, and additional administration of prenalterol causes a further shift upwards and to the left

6. Siegel LA, Keung E, Siskind SJ, Forman R, Feinberg H, Strom J, Efstathakis D, Sonnenblick EH, LeJemtel TH (1981) Beneficial effects of amrinone-hydralazine combination on resting hemodynamics and exercise capacity in patients with severe congestive heart failure. Circulation 63:838–843
7. Franciosa JA, Cohn JN (1979) Immediate effects of hydralazine-isosorbide dinitrate combination on exercise capacity and exercise hemodynamics in patients with left ventricular failure. Circulation 59:1085–1091
8. Rubin SA, Chatterjee K, Ports TA, Gelberg HJ, Brundage BH, Parmley WW (1979) Influence of short-term oral hydralazine therapy on exercise hemodynamics in patients with severe chronic heart failure. Am J Cardiol 44:1183–1189
9. Goldman SA, Johnson LL, Escala E, Cannon PJ, Weiss MB (1980) Improved exercise ejection fraction with long-term prazosin therapy in patients with heart failure. Am J Med 68:36–41
10. Wynne J, Malacoff RF, Benotti JR, Curfman GD, Grossman W, Homan BL, Smith TW, Braunwald E (1980) Oral amrinone in refractory congestive heart failure. Am J Cardiol 45:1245–1249
11. Carlsson E, Dahlöf CG, Hedberg A, Persson H, Tangstrand B (1977) Differentiation of cardiac chronotropic effects of β-adrenoceptor agonists. Naunyn Schmiedebergs Arch Pharmacol 300:101–105
12. Drexler H, Löllgen H, Just H (1981) Short- and long-term effects of hydralazine and combined hydralazine-prenalterol therapy in severe chronic congestive heart failure. Klin Wochenschr 59:647–654
13. Chatterjee K, Ports T, Rubin S, Massie B, Arnold S, Brundage B, Parmley WW (1978) Sustained beneficial hemodynamic effects during long-term hydralazine therapy in patients with chronic heart failure. Circulation [Suppl] 28:57–58
14. Massie B, Chatterjee K, Parmley WW (1979) Vasodilator therapy for acute and chronic heart failure. In: Goodwin JF, Yu NP (eds) Progress in cardiology, Vol 8. Lea & Febiger, Philadelphia, pp 197–234

Discussion

Scholz, Hannover: Amrinone seems to be a specific phosphordiesterase inhibitor, differing from theophylline only in its much poorer solubility. Why is theophylline not mentioned in this symposium?

Drexler, Freiburg: Actually, AR-L 115 is just a modification of theophylline. In this respect, this subject will be mentioned in the course of this symposium.

Just, Freiburg: The methylxanthines do indeed play an important role, as they exert inotropic and vasodilating effects. There are some derivatives, which are useful for repeated application. I think we have to keep that in mind.

Auditorium: In the first instance, it seems fascinating to combine β_1-agonists and vasodilators. However, it must be considered that the sympathetic system is stimulated almost maximally in patients with congestive heart failure. In this respect, I am surprised about the additional effect of the β_1-stimulating agent prenalterol. The additional administration can only be efficient, I think, if the sympathetic system is stimulated moderately, as in patients with less severe congestive heart failure.

Drexler, Freiburg: These ten patients represent a group with severe congestive heart failure, indicated by an average cardiac index of less than 2.5 liters/min/m^2. In a group of patients with similar baseline hemodynamics, prenalterol caused a pronounced augmentation of cardiac output, although plasma norepinephrine ranged from 400 to 700 mg/ml; therefore, although plasma catecholamines are evaluated, the β_1-agonist exerts a positive effect. In part, this may be due to the different effects of, e.g., norepinephrine and prenalterol on the peripheral vasculature. Norepinephrine antagonizes its positive inotropic effect by a strong arterial vasoconstriction, whereas prenalterol seems to have no essential effects on the resistance vessels.

Auditorium: You reported that exercise tolerance increased in some cases. How did you determine improvement of exercise tolerance? Which guidelines did you use; dyspnea, fatigue, heart rate, or hemodynamics?

Drexler, Freiburg: These results are from only five patients, therefore no fundamental conclusions can be drawn. The improvement of exercise capacity in three of these five patients was indicated by increased exercise time and improved exercise hemodynamics. Exercise testing was performed until the patients' symptomatic maximum was reached or until pulmonary wedge pressure exceeded 40 mmHg.

Greef, Düsseldorf: At what time after ingestion of the drug did you measure the hemodynamics?

Drexler, Freiburg: After obtaining baseline hemodynamics, 100 mg hydralazine were given, followed by measurements of hemodynamics 30, 60, and 90 min after administration of the drug. After 3–4 weeks treatment, hemodynamic measurements were repeated 12 h after the last dose and again 30, 60, and 90 min after administration of the drug. Prenalterol was given 30 min after the application of hydralazine, as indicated by the different pharmacokinetics of the two drugs.

Auditorium: I wonder, why does a drug linke a β_1-stimulating agent show sustained efficiency after chronic treatment?

Drexler, Freiburg: In this study, prenalterol was given combined with hydralazine, therefore the single effect of prenalterol is difficult to determine. However, in another study, when prenalterol was given singly, this agent exerted an augmentation of cardiac output of about 40%–50%. After 4 weeks treatment an attenuation of 10%–15% compared to the acute effect was noted, but nevertheless a significant improvement could be observed.

AR-L 115 BS, a Substance with Inotropic and Vasodilating Effects

M. Schlepper, J. Thormann, W. Kramer

AR-L 115 BS [2-(2methoxy-4-methylsulfinyl-phenyl) imidazo (4,5–6) pyridine] (Thomae) (Fig. 1) bears no structural relationship to any of the other well-known inotropic substances. The spectrum of its effects, however, seems to be very similar to that of amrinone (1–6) (Table 1). The two substances have the following effects in common:

1. An increase in contractility as measured by dp/dt_{max} and echocardiographic parameters
2. An augmentation of cardiac output and ejection fraction
3. A decrease in left ventricular end-diastolic pressure (i.e., preload)
4. A mild reduction of mean arterial pressure (i.e., afterload)
5. A failure to be influenced by β-blocking agents when applied in therapeutic dosages
6. An effect on heart rate that seems to be related to the reduction of arterial pressures, i.e., there is no direct influence on sinus node automaticity.

The above effects are seen in patients with congestive heart failure or significant coronary artery disease and in healthy volunteers; in the case of patients with congestive heart failure, these effects are produced even when the standard therapeutic program is administered, such as digitalis, diuretics, and vasodilating drugs (2, 5, 7).

Comparison of results regarding myocardial oxygen consumption in the presence of increased pump function and inotropic action is limited, since investigations of the two substances were carried out in groups with different underlying cardiac disease. However, for the increase in myocardial pump function resulting from both substances an appropriate oxygen consumption could not be elicited (5, 8, 20).

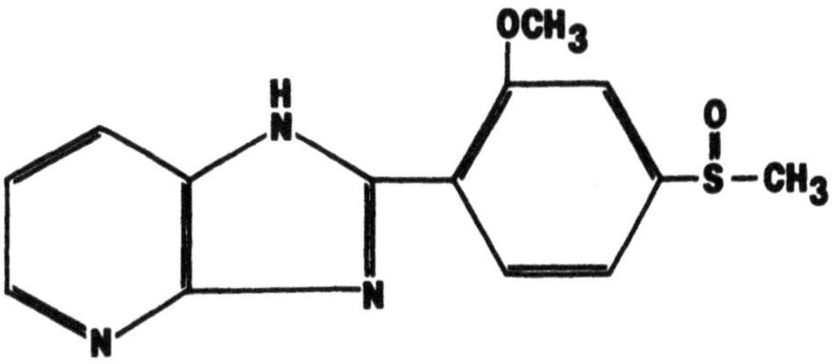

Fig. 1. Chemical structure of AR-L 115 BS [2-(2-methoxy-4-methylsulfinyl-phenyl)-imidazo(4,5-b) pyridine]

Table 1. Comparison of results obtained with AR-L 115 BS and amrinone. *HR*, heart rate; *AOMP*, aortic pressure mean; *LVFP*, left ventricular filling pressure; *CI*, cardiac index; *LVSWI*, left ventricular stroke work index; *SVR*, stroke-volume ratio; *CSF*, coronary sinus flow; *MVO_2*, myocardial oxygen consumtpion; *EF*, ejection fraction; *TH*, fractional posterior wall thickening

	Amrinone		AR-L 115 BS		AR-L 115 BS
Clinical diag.	Congestive heart failure (cor. art. disease)		Congestive heart failure (cong. cardiomyop.)		Ischemic heart disease (cor. art. dis.)
Refs.	(1), (2), (5)	(3), (4)	(7), (16)		(8), (20)
Dosage	IV 1.8–3.5 mg/kg body wt.	Orally 50–300 mg	IV 3 mg/kg body wt.	Orally 200 mg every 8 h	IV 2 mg/kg body wt.
HR	N.S.	N.S.	+18%	N.S.	+19%
AOMP	N.S.	– 6%	– 8%	N.S.	–10%
LVEP	–24%	–25%	–25%	–29%[a]	–30%
CI	+54%	+41%	+23%	–31%[a]	+29%
LVSWI	+76% (0)	(EF + 50%)	(0)	(EF + 35%)	N.S.
SVR	–34%	–30%	0	0	–25%
dp/dt_{max}	+42%	0	0	(% Δ TH + 33%)	+27%
CSF	–17%	0	0	0	+39%
MVO_2	–30%	0	(–9%)	0	+34%

[a] Swan-Ganz

This report summarizes our results with AR-L 115 BS in healthy volunteers, in patients with congestive cardiomyopathy during intravenous and chronic (oral) therapy (7), and in two groups of patients with coronary artery disease (8). These results are compared to those obtained with amrinone and in animal experiments. In investigations with drugs exhibiting vasodilating and positive inotropic effects, one has to differentiate between these properties. Apparently, inotropic parameters such as ejection fraction and mean circumferential fiber-shortening rate are influenced by pre- and afterload alterations, while fractional shortening is independent, and pre-ejection period (PEP) and left ventricular ejection time (LVET) are influenced in such a way that as net effect minimal or no changes of total electromechanical systole may result (9, 10). Furthermore, investigational methods used depend on the underlying cardiac disease: echocardiographic examination will provide reproducible control of inotropic effects in patients with global involvement of the myocardium, as in congestive cardiomyopathy, but not in patients with regional myocardial disease, such as coronary artery disease.

Examination of Healthy Volunteers

The aim of the study was: (a) to quantify the effects of AR-L 115 BS in correlation with the serum levels; (b) to assess side-effects, and (c) to investigate possible blocking effects on AR-L 115 BS actions by the calcium antagonist verapamil.

Ten volunteers were investigated, eight of whom were subjected to a second trial in which verapamil was applied as an additional drug. Apart from heart rate and blood pressure, end-diastolic diameter and volume were assessed, as well as ejection fraction and fractional shortening. Rate-corrected total electromechanical systole was derived from electrocardiographic (ECG) and phonocardiographic records (paper speed 100 mm/s) (9, 10). Disturbances in color perception were quantified using the Farnsworth-Munsell 100-hue test and the Lanthony test (11).

Following an initial bolus injection of 20 mg, AR-L 115 BS was infused in increasing dosages (1.4, 2.8, and 4.2 mg/min) over a period of 15 min (see Fig. 2). In the second study verapamil 0.15 mg/kg body wt. was given intravenously in the 23rd min, i.e., during the phase of the infusion of 2.8 mg/min AR-L 115 BS. Since serum levels of the substance did not differ significantly at any time, with or without verapamil, their mean values were used for presentation (Fig. 3). The serum levels show a continuous increase up to a maximal value of 2400 ng/ml in the 45th min, and a decrease when the infusion was finished. At the end of the second study (i.e., 105th min) the serum level was 1200 ng/ml.

Ejection fraction (Fig. 3) increased by an average of 26% (from 66% to 83%); the difference in ejection fraction – 80% at the 30th min and 83% after the application of a higher dosage – was no longer significant. At the end of the investigation (105th min) the ejection fraction was still significantly greater than in controls (66% vs 71%). The injection of verapamil interrupts the tendency of the ejection fraction to increase. In the 45th min ejection fraction values are significantly different from those of the first trial; at this point, however, they are no longer different from controls. Fractional shortening and corrected mechanical systole (see Fig. 4) show quite a similar course

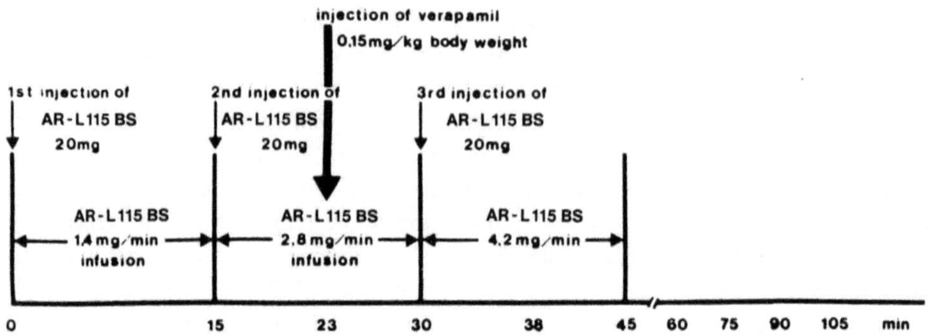

Fig. 2. Schematic protocol of the study in healthy volunteers

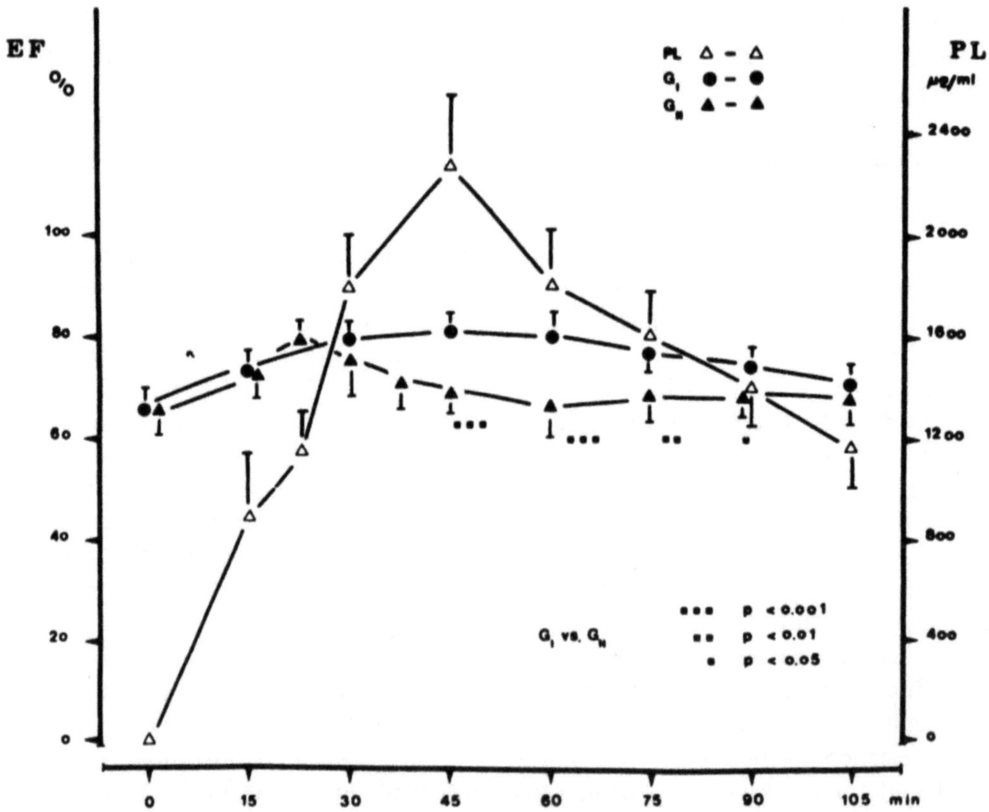

Fig. 3. Changes in ejection fraction (*EF*) and plasma levels (*PL*) after administration of AR-L 115 BS during the two investigational courses in healthy volunteers. G_I, group I; G_{II}, group II

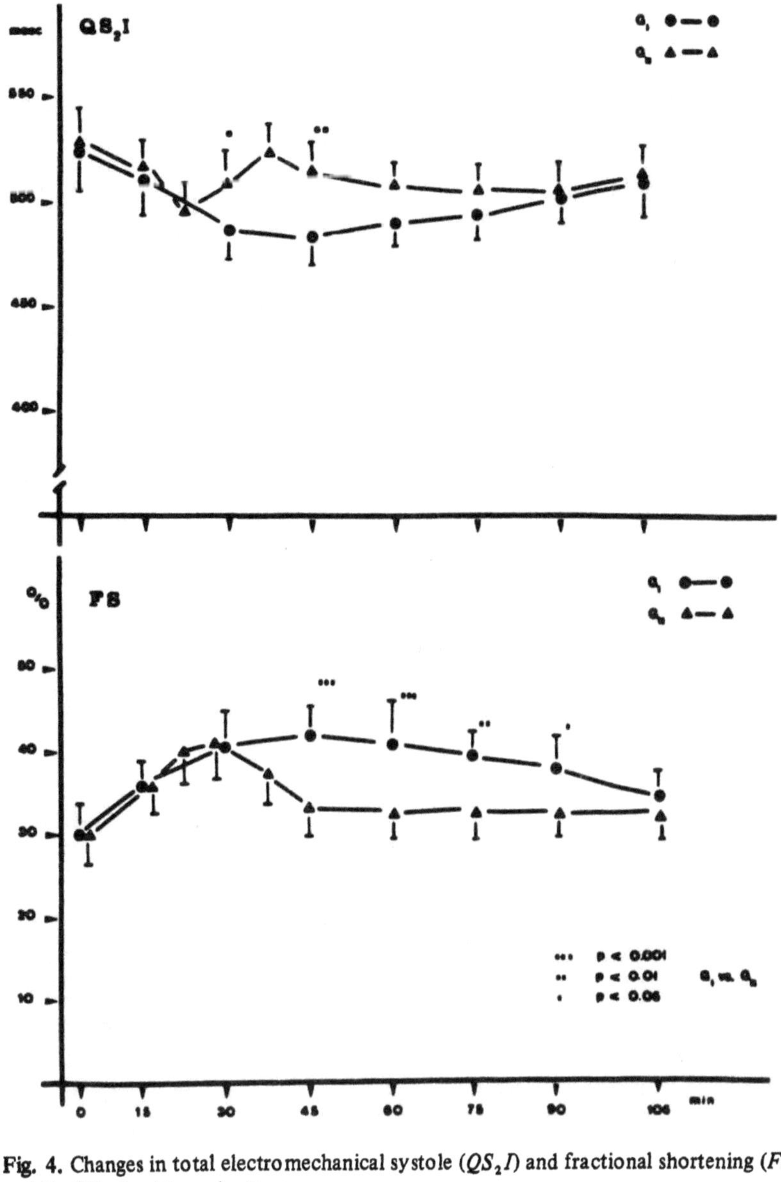

Fig. 4. Changes in total electromechanical systole (QS_2I) and fractional shortening (FS) in groups I and II of the healthy volunteers

as regards both the effects of AR-L 115 BS and the additional application of verapamil.

Heart rate and blood pressure measured in a supine position were not influenced by AR-L 115 BS. However, in orthostasis in the 30th min (not shown in Fig. 5), systolic but more predominantly diastolic blood pressure showed an obvious reaction. Verapamil induced a decrease of systolic and diastolic blood pressure immediately after injection, and an increase of heart rate, most likely by reflex mechanisms.

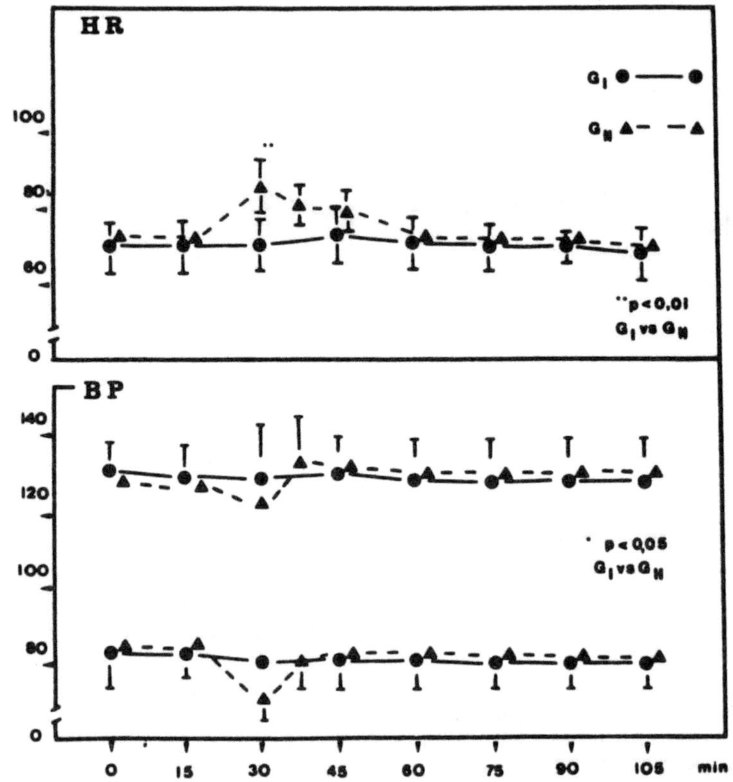

Fig. 5. Changes in heart rate (*HR*) and blood pressure (*BP*) in healthy volunteers after AR-L 115 BS (group I = G_I) and additional verapamil administration (group II = G_{II})

These findings allow the following conclusions. AR-L 115 BS shows positive inotropic effects even in healthy hearts; their maximum occurs at serum levels of 1200–1800 ng/ml. A further increase in serum levels does not result in a further increase in positive inotropic action. This is in accordance with findings of Hauff et al. (12), who, in patients with congestive heart failure, did not find further therapeutic improvement when increasing serum levels of AR-L 115 BS above a certain level.

Verapamil blocks the positive inotropic effect in spite of a fall in blood pressure. A true positive inotropic effect, independent of pre- and afterload reduction, is represented by the changes in fractional posterior wall shortening and total electromechanical systole. These findings have been corroborated by animal experiments (13).

Positive inotropic effects of the drug amrinone have been presented for the systolic time interval only (6). Oral application (0.3–3.5 mg/kg body wt.) resulted in the following maximal changes: PEP by 25%, LVET by 10%, and total electromechanical systole by 25%; these changes were reported without serum levels. Reductions of total electromechanical systole by 10% found with AR-L 115 BS were comparatively lower.

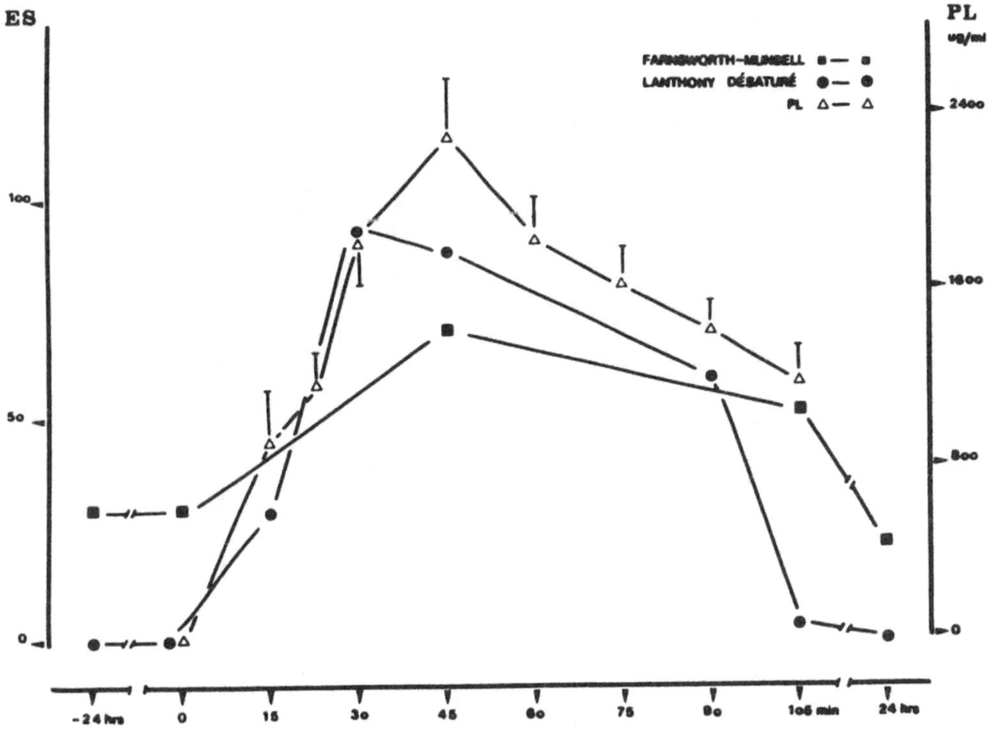

Fig. 6. Plasma levels (*PL*) of AR-L 115 BS in healthy volunteers. *ES*, error score

Under amrinone no side-effects were reported in the above mentioned investigations (1–6). With AR-L 115 BS patients complained of slight headaches and palpitations in connection with flush-like symptoms. These tolerable side-effects were noted in the 30th min, i.e., at maximal dosages; they had disappeared by the end of the investigation. Disturbances of light-darkness vision could be identified as disturbances in color discrimination, and were quantified by the specific tests mentioned (11); see Fig. 6. An error score increased from 0 to 93 and decreased to 2.5 at the end of the investigation. These side-effects were apparent only at serum levels where an additional therapeutic effect could no longer be obtained; they were tolerable and completely reversible in all instances, which was reflected also by test results 24 h later.

Congestive Cardiomyopathy

The aims of the study were:

1. To assess the effectiveness of AR-L 115 BS, both orally and by intravenous route, in patients with chronic congestive heart failure
2. To determine the therapeutic influence of chronic pretreatment with digitalis and diuretics
3. To assess tachyphylaxis, arrhythmogenicity, and side-effects of AR-L 115 BS.

Table 2. Clinical and hemodynamic characteristics of groups I and II of patients with congestive cardiomyopathy. *PAM*, mean pulmonary artery pressure; *CI*, cardiac index; *EF*, ejection fraction; *VCF*, mean circumferential fiber-shortening rate

Group I (ARL 3 mg/kg body wt. IV)	Age (years)	NYHA class	PAM (mmHg)	CI (l/min · m²)	EF (%)	VCF (circ/s)	Serum digitalis (μg/ml)	
n = 11	46	III–IV congestive cardiomyopathy	28	2.04	31	62	1.5	
Group II (ARL 200 mg orally)	**Age (years)**	**NYHA class**	**PAM (mmHg)**	**CI (l/min · m²)**	**EF (%)**	**VCF (circ/s)**	**Serum digoxin (μg/ml)**	**M-mode echocardiography**
n = 18	39	III–IV congestive cardiomyopathy	31	2.03	33	67	1.4	No assymmetr. contract. patterns

There were two patient groups: group I (n = 11) were given a 3 mg/kg body wt. intravenous bolus during routine heart catheterization; group II (n = 18) were given 200 mg orally every 8 hours for 4 days. The patients investigated were NYHA class III–IV (Table 2), clinically and by hemodynamic assessment. Coronary artery disease and primary valvular disease were excluded. Digoxin levels averaged 1.5 ng/ml, and furthermore, all patients were chronically pretreated with diuretics and nitrate preparations with no interruption of therapy during the investigation.

In patients of group I, AR-L 115 BS effects induced the following average changes (Fig. 7) in the 10th to 14th min after injection: Heart rate rose by 18%, aortic systolic pressure decreased slightly but not significantly, mean pulmonary artery pressure decreased by 25%, and cardiac index increased by 27%. In the 25th min after injection all parameters but aortic systolic pressure showed significant effects, somewhat lower than the listed maximal levels.

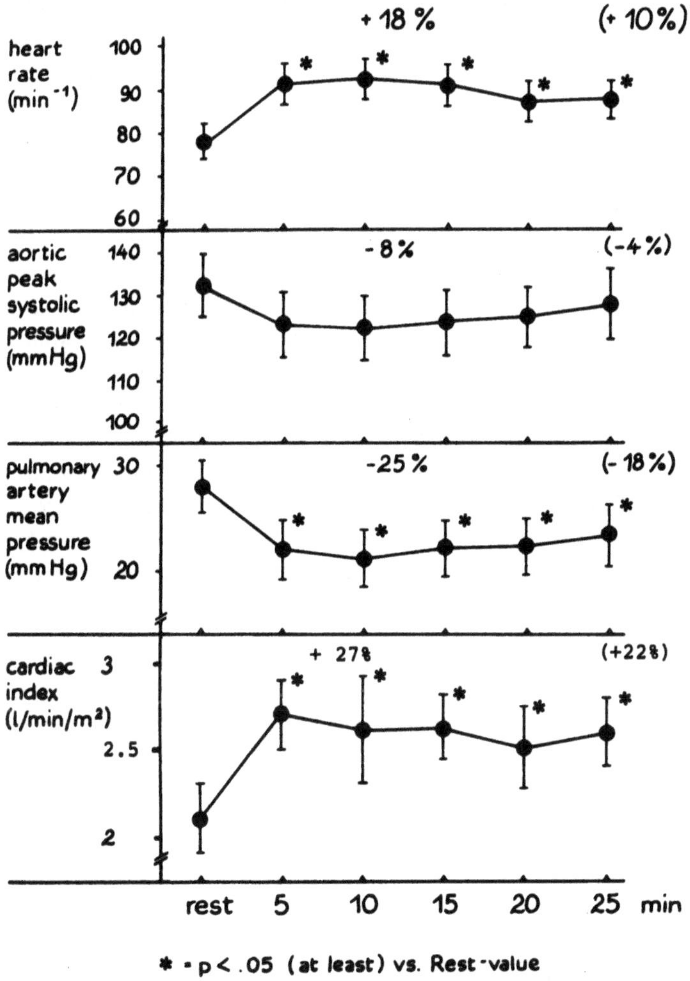

Fig. 7. Parameter changes with AR-L 115 BS (3 mg/kg body weight i.v.) in patients from group I of those with congestive cardiomyopathy

Continuous ECG recordings up to 2 h after the end of investigation showed no arrhythmic activities. All patients complained of transient disturbances in light-darkness vision during the injection phase only; some had palpitations.

An estimation of the tension time index suggests no significant increase in myocardial oxygen consumption. With respect to the decreased end-systolic left ventricular pressures and volumes, one can safely assume a decreased oxygen consumption. This would be in accordance with changes obtained after amrinone application, where in patients with congestive heart failure an increase of myocardial pump function was achieved with a decrease in myocardial oxygen consumption (5).

When AR-L 115 BS was applied orally (200 mg) an increase of left ventricular pump function was achieved, starting in the 40th min and with maximal values at the 120th min (Fig. 8); in the 3rd hour pretreatment levels were nearly reached and in the 5th hour no changes relative to control could be ascertained.

After AR-L 115 BS 200 mg administered orally the following average parameter changes were found in the 120th min (Fig. 9): Heart rate and blood pressure did not change, mean pulmonary artery pressure fell by 29%, cardiac output increased by 31% and ejection fraction by 35%, and mean circumferential fiber-shortening rate and fractional posterior wall shortening also increased significantly. A comparison of ejection fraction parameters before and 120 min after AR-L 115 BS over 4 days revealed no tachyphylaxic effects (Fig. 10). The percentage increase of ejection fraction was the same on each day of drug application. Twenty-four hours Holter monitoring did not reveal arrhythmogenic effects.

Oral application of amrinone reported in patients with congestive heart failure elicited similar effects (3, 4). Regarding the improvement in ventricular pump function with AR-L 115 BS, the results are comparable to those found after dobutamine application reported by Klein et al. (14). Our results with AR-L 115 BS are in accordance with those obtained by others using amrinone in patients with congestive heart failure due to different underlying myocardial disorders (1–6). In summary, the drug improves congestive heart failure by its positive inotropic effects and by the simultaneous reduction of pre- and afterload; this effect is also present in patients under the full regime with conventional therapeutic means. Finally, no arrhythmogenic activity could be ascertained, either during or following the intravenous application of AR-L 115 BS or during the 4-day period of oral treatment. Tachyphylaxis was not observed.

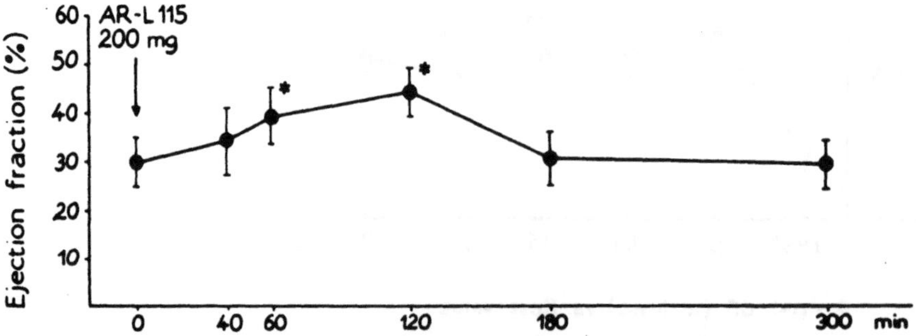

Fig. 8. Pilot study with orally administered AR-L 115 BS indicating maximum changes in ejection fraction at the 120th min and return to control levels at the 300th min

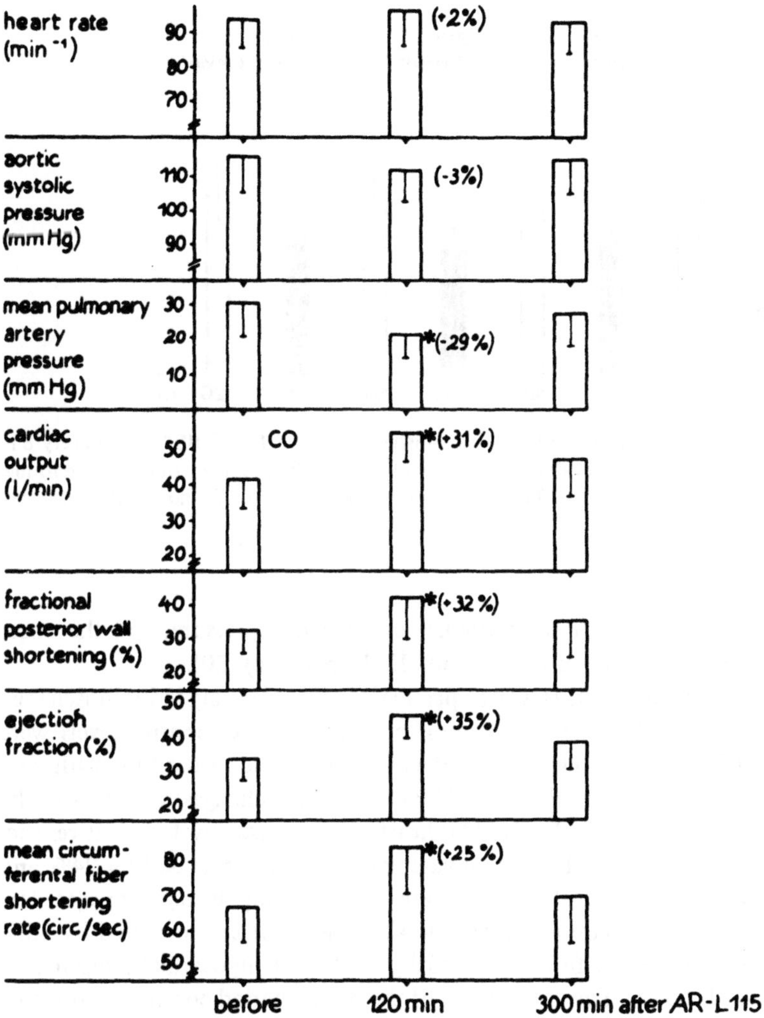

Fig. 9. Results of oral administration of AR-L 115 BS 200 mg every 8 h for 4 days (group II of patients with congestive cardiomyopathy)

Patients with Coronary Artery Disease

The aims of the study were to answer the following questions:

1. Does AR-L 115 BS induced increase of myocardial pump function cause myocardial ischemia?
2. Can the drug-induced reduction of pre- and afterload prevent ischemia in the presence of pacing-induced increased myocardial metabolic demand?

Seven patients were investigated in group I (Table 3) (8). All had significant 2–3 vessel disease according to coronarographic examination. During pacing (ventricular pacing of 170 beats/min over 1 min duration), all patients developed myocardial ischemia as evidenced by a change from lactate extraction to significant lactate pro-

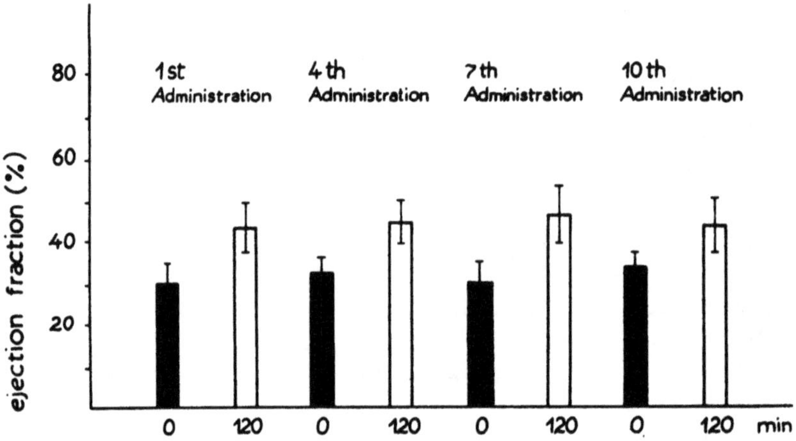

Fig. 10. Changes in ejection fraction with oral administration of AR-L 115 BS 200 mg every 8 h. There are no cumulative effects as evidenced by control values taken at a set time each morning. Neither are there any tachyphylactic effects, since percentage increase did not differ from day to day

duction, a pathological rise in left ventricular end-diastolic pressure, and the onset of angina. The same criteria were used for group II (17 patients) (20).

Control values were assessed, as well as parameters during pacing-induced ischemia (i.e., measured immediately after interruption of pacing). After control levels were reached, AR-L 115 BS was applied. Parameters were recorded in the 9th, 14th, and 19th min following injection of the drug. Maximal average changes occurring in the 14th min were as follows (Table 4, Fig. 11): heart rate increased by 19%, left ventricular peak pressure decreased by 13%, total vascular resistance decreased by 24%, and dp/dt_{max} increased by 27%. Most striking were the following average changes: left ventricular end-diastolic pressure decreased by 42%, cardiac index increased by 30%, thermodilution coronary sinus flow increased by 39%, and myocardial oxygen consumption increased by 34%. Toward the end of the pacing period coronary sinus flow and myocardial oxygen consumption were increased comparatively equally, and lactate production, indicating ischemia, was not present following the application of AR-L 115 BS. Since coronary flow and coronary resistance and myocardial oxygen consumption were increased to similar levels, it is safe to assume that metabolic adjustment had taken place. Obviously, due to the limitation of the thermodilution method used, myocardial oxygen consumption and coronary flow represent global left ventricular assessment only, and could account neither for regional changes nor for variations and redistribution in different myocardial layers. Rough estimations of the tension time index reveal no significant differences between values during resting conditions vs AR-L 115 BS, while there was an increase during pacing by nearly 90% vs control. For the sake of exact correlations coronary sinus flow was measured at the point of cessation of the pacing period, i.e., during the first sinus beats, when the ischemia-causing factor, the rate increase, had just disappeared.

In the 17 patients in group II the prevention of reproducible pacing-induced ischemia by AR-L 115 BS was studied. Following diagnostic cardiac catheterization including biplane left-ventricular angiography at rest and assessment of control values,

Table 3. Clinical characteristics of group I of patients with coronary artery disease at rest (*R*) and at pacing-induced anginal threshold. *EF*, ejection fraction; *CI*, cardiac index; *LVFP*, left ventricular filling pressure; *LAD*, left anterior descending; *CX*, circumflex artery; *RC* right coronary artery; *VT*, paced ventricular tachycardia

	Coronary disease		EF(%)	CI (l/min/m²)		LVFP (mmHg)		Lactate extraction rate (%)	
No. of patients	Infarct-ECG	Signif. coron. disease (> 75% stenosis)		(R)	(VT)	(R)	(VT)	(R)	(VT)
n = 7	3/7	LAD 7/7 CX 3/7 RC 2/7	62.6 ± 7	2.69 ±0.5	2.28 ±0.4	12 ± 4	25 ± 5	(5) ±7 (L extr.)	(−37) ±7 L production (= ischemia)
		p values vs (R)		<0.01		<0.001		<0.001	

Table 4. Maximal average changes occurring in the 14th min following injection of AR-L 115 BS in trials on patients with coronary heart disease. *R*, at rest; *VT*, at pacing-induced anginal threshold; *ARL*, at peak effectiveness of AR-L 115 BS 2 mg/kg body wt. IV; *HR*, heart rate; *LVSP*, left ventricular systolic pressure; dp/dt_{max}; *LVEDP*, left ventricular end-diastolic pressure; *CI*, cardiac index; *CSF*, coronary sinus flow; *CVR*, coronary vascular resistance; MVO_2, myocardial oxygen consumption; *LVSWI*, left ventricular stroke work index; *SVR*, stroke-volume ratio

HR (min^{-1})			LVSP (mmHg)			dp/dt_{max} (mmHg/s)			LVEDP (mmHg)		
R	VT	ARL	R	VT	ARL	R	VT	ARL	R	VT	ARL
74	170	88 (+19%)	142	110	123 (–13%)	1920	1970	2430 (+27%)	12	25	7 (–42%)
	< 0.001	< 0.001		< 0.001	< 0.05		N.S.	< 0.01		< 0.001	<0.05
CI (l/min · m^2)			**CSF (ml/min)**			**CVR (mmHg/ml/min)**			**MVO_2 (ml/min)**		
R	VT	ARL	R	VT	ARL	R	VT	ARL	R	VT	ARL
2.54	2.3	3.3 (+30%)	131	193	182 (+39%)	0.75	0.45	0.47 (–37%)	14.1	21.7	19 (+34%)
	<0.01	<0.05		< 0.05	< 0.05		<0.001	<0.05		< 0.05	< 0.05
LVSWI (mg/m^2)			**SVR (dyn · s · cm^{-5})**			**Lac. extr. ratio (100%)**					
R	VT	ARL	R	VT	ARL	R	VT	ARL			
51	15	49 (–4%)	1774	1563	1343 (–24%)	5	–37	15			
	< 0.001	N.S.		N.S.	< 0.05		< 0.001	N.S.			

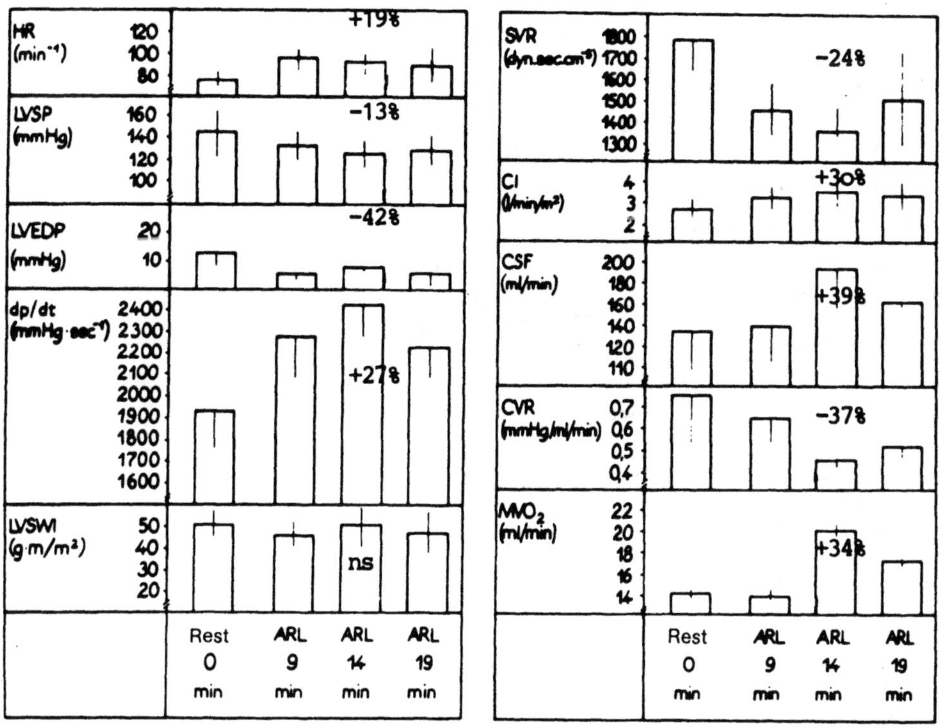

Fig. 11. Maximal average changes occurring in the 14th min following injection of AR-L 115 BS in trials on patients with coronary artery disease. *HR*, heart rate; *LVSP*, left ventricular systolic pressure; *LVEDP*, left ventricular end-diastolic pressure; *dp/dt*; *LVSWI*, left ventricular stroke work index; *SVR*, systemic vascular resistance; *CI*, cardiac index; *CSF*, coronary sinus flow; *CVR*, coronary vascular resistance; *MVO_2*, myocardial oxygen consumption

ventricular pacing at 170 beats/min was induced. Left ventricular end-diastolic pressure was measured and an ECG recording made at the anginal theshold and immediately after cessation of pacing. After the return to control values a second course of pacing was started, and a left ventricular angiography carried out at the anginal threshold immediately after interruption of pacing. Twenty min later AR-L 115 BS, 2 mg/kg body wt., was given intravenously, and in the 12th min with peak effect biplane angiography and pressure assessments were repeated. In the first two periods of pacing-induced angina there was an average increase of left ventricular end-diastolic pressure from 13 to 30 mmHg at the anginal threshold and it remained 13 mmHg during AR-L 115 BS. Ejection fraction decreased from 60% to 50% and increased to 65% during the influence of the drug plus pacing. Changes in wall motions as assessed angiographically are represented in the examples in Fig. 12. In a patient with significant stenoses of circumflex and right coronary arteries, left ventricular end-diastolic pressure increased to 30 mmHg during pacing-induced ischemia, and ejection fraction decreased from 53% to 38%. Apparently, there is a loss of regional wall motion in the basal area. Pacing during the influence of AR-L 115 BS caused no increase of left ventricular end-diastolic pressure; however, wall motion increased even compared to control conditions, and ejection fraction increased to 64%.

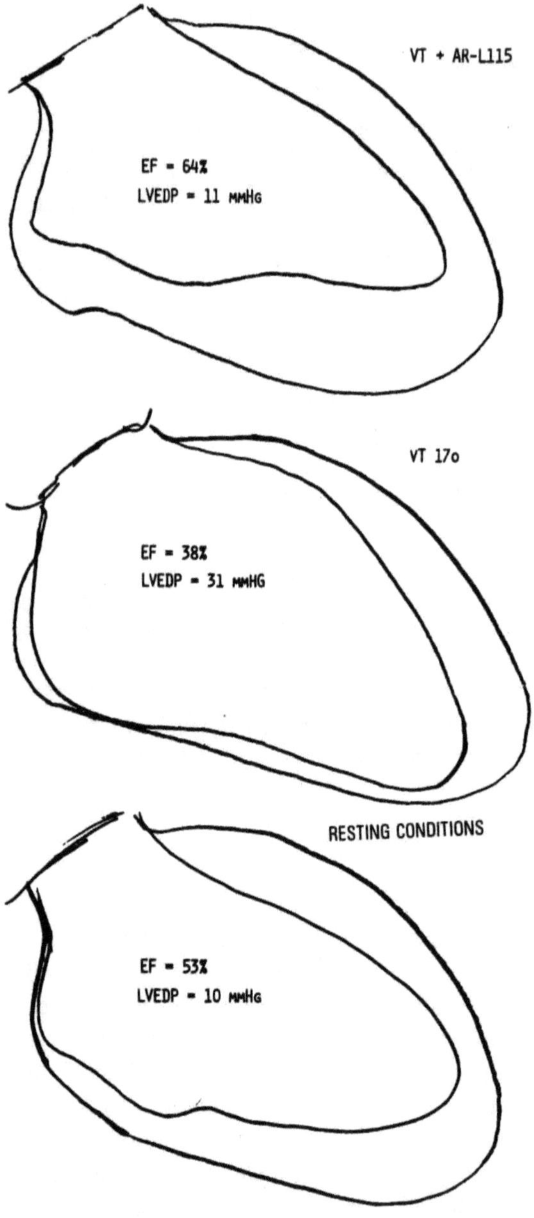

Fig. 12. Examples of changes in cardiac wall motions assessed angiographically following administration of AR-L 115 BS. *VT*, pacing-induced anginal threshold; *EF*, ejection fraction; *LVEDP*, left ventricular end-diastolic pressure

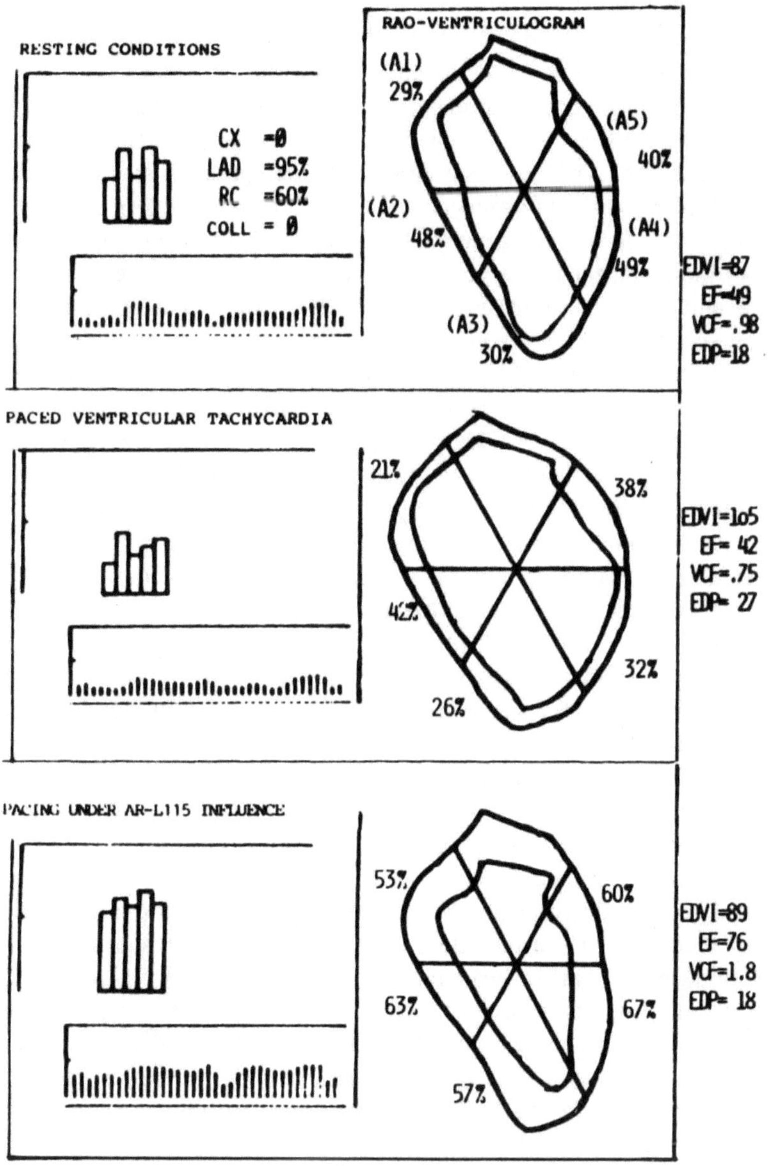

Fig. 13. Disturbances of regional wall motion in the anterior wall area due to left anterior descending stenosis. *EDVI*, end-diastolic volume index; *EF*, ejection fraction; *VCF*, mean circumferential fiber-shortening rate; *EDP*, end-diastolic pressure; *CX*, circumflex artery; *RC*, right coronary artery; *LAD*, left anterior descending artery; *COLL*, collateral coronary arteries

In a second patient (Fig. 13) disturbances of regional wall motion occurred primarily in the anterior wall area due to left anterior descending stenosis. Here again, during pacing-induced ischemia ejection fraction and mean circumferential fiber-shortening rate decreased, while left ventricular end-diastolic pressure rose and end-diastolic volume index increased from 89 to 105 ml/m^2. Decrease in regional wall motion involved the segments A_1, A_3, and A_4 according to the computer-assisted evaluation of 36 axes (in the biplane ventriculogram) corresponding to the perfusion area of the left coronary artery (15). After pacing plus AR-L 115 BS, myocardial pump function showed significant improvement in the presence of a low left ventricular end-diastolic pressure and low end-diastolic volume index; in addition, not only regional but also global contraction was improved.

Our findings differ from those of Pozenol (16), who did not find an improvement in stress tolerance after the application of AR-L 115 BS. Differences in the investigational protocol may be the reason. The dosage of 0.75 mg/kg body wt. was well below ours, and in addition, ergometric stress was started only 2 min after injection, i.e., at a time when AR-L 115 BS effects are not optimal. Drug-induced increases in myocardial oxygen consumption as found in our investigations are surprising only at first glance. Hellige et al. (17) found no increase in myocardial oxygen consumption in animal experiments with low dosages, since oxygen consumption of the left ventricle, according to Bretschneider's equation (18), does not undergo changes due to the inverse relation of development of tension and maintenance of tension; however, with higher dosages of AR-L 115 BS myocardial oxygen consumption rose in a dose-dependent manner. Furthermore, in patients with coronary artery disease, not only did contractility improve by 27%, but heart rate increased under the influence of the drug, so that an increase also of E_1 (energy for activation, de- and repolarization of membrane) can be assumed. Our results cannot without restriction be compared with those found under amrinone treatment (5), where congested ventricles were investigated with a comparatively greater volume decrease than in those of patients with normal-sized ventricles. In contrast, the results of Oeff et al. (19) showed neither an AR-L 115 BS induced increase of heart rate and cardiac output nor a decrease of arterial pressure. In addition, myocardial oxygen consumption measured by the argon method represents the mean value of a period of measurement, and therefore cannot account for relatively short-lasting changes. The abolition of myocardial ischemia by the drug should not be judged merely from the point of view of myocardial oxygen consumption, which, when measured clinically, represents more or less an approximation. The depression of left ventricular end-diastolic pressure, or rather the prevention of a rise, in itself entails an improvement of subendocardial perfusion and thereby may prevent ischemia. The beneficial effects of AR-L 115 BS on left ventricular pump function are certainly not achieved by inappropriate oxygen consumption.

References

1. Benotti JR, Grossman W, Braunwald E, Davolos DD, Alousi AA (1978) Hemodynamic assessment of amrinone. A new inotropic agent. N Engl J Med 299:1373–1377
2. Lejemtel TH, Keung E, Sonnenblick E, Ribner HS, Matsumoto M, Davis R, Schwartz W, Alousi AA, Davolis D (1979) Amrinone: A new non-glycosidic, non-adrenergic, cardiotonic agent effective in the treatment of intractable myocardial failure in man. Circulation 19: 1098–1104
3. Wynne J, Malacoff RF, Benotti JR, Curfman GD, Grossman W, Holman BL, Smith TW, Braunwald E (1980) Oral amrinone in refractory congestive heart failure. Am J Cardiol 45:1245
4. Lejemtel TH, Keung E, Ribner HS, Davis R, Wexler J, Blaufox MD, Sonnenblick EH (1980) Sustained beneficial effects of oral amrinone on cardiac and renal function in patients with severe congestive heart failure. Am J Cardiol 45:123
5. Benotti JR, Grossman W, Braunwald E, Carabello BA (1980) Effects of amrinone on myocardial energy metabolism and hemodynamics in patients with severe congestive heart failure due to coronary artery disease. Circulation 62:28
6. Hayes SL, Lasseter KC, Davolos DD, Palmer RF (1980) Amrinone, a novel cardiac inotrope: effects after oral application in normal man. Clin Res 56:524A
7. Thormann J, Kramer W, Schlepper M (1980) AR-L 115: A new non-glycosidic, non-adrenergic cardiotonic agent; hemodynamic proof for its effectiveness both intravenously and by oral application. Circulation [Suppl III] 62:III–233
8. Thormann J, Kramer W, Schlepper M (to be published) Changes of hemodynamics, coronary flow and myocardial energy metabolism in patients with coronary artery disease after the application of the new cardiotonic agent AR-L 115. Am Heart J
9. Weissler AM, Harris WS, Schoenfeld CD (1968) Systolic time intervals in heart failure in man. Circulation 37:149
10. Weissler AM, Schoenfeld CD (1970) Effect of digitalis on systolic time intervals in heart failure. Am J Med Sci 259:4–20
11. Farnworths D (1943) The Farnworths-Munsell 100-hue and dichotomous tests for color vision. J Opt Soc Am 33:568–603
12. Hauf G, Bubenheimer P, Roskamm H (1981) The acute effect of a new positive inotropic agent (AR-L 115 BS) on cardiac hemodynamics and contractility in patients with severe chronic congestive heart failure. Arzneim Forsch (Drug Res) 31:253–256
13. Zsotér TT (1980) Calcium antagonists. Am Heart J 99:805–810
14. Klein G, Sauer E, Bauer R, Wirtzfeld A, Sebening H (1981) The comparison of the effects of AR-L 115 BS and dobutamine in patients with severe cardiac failure. Arzneim Forsch (Drug Res) 31:257–259
15. Gottwik M, Frimmel W, Stämmler G, Winkler B, Schlepper M (1980) Computergestützte Beurteilung der regionalen Wandfunktion des linken Ventrikels. Abstr Commun Z Kardiol 69:692
16. Pozenel H (1981) Ergometric and hemodynamic studies on a new cardiotonic agent AR-L 115 BS. Arzneim Forsch (Drug Res) 31:263
17. Hellige G, Baller D, Ensink FBM, Zipfel J, Woplers HG, Bretschneider HJ (1981) Effects of AR-L 115 on hemodynamics; myodardial oxygen consumption in cardiac pulmonary pumping efficiency. Arzneim Forsch (Drug Res) 31:155
18. Schaper W (1979) Regulation of coronary blood flow. In: Schaper W (ed) The pathophysiology of myocardial perfusion. Elsevier/North-Holland Biomedical Press, Amsterdam New York Oxford, p 174
19. Oeff M, Prokein E, Schröder R (1981) Wirkung des Imidazol-Derivates (AR-115) auf Hämodynamik, Kontraktilität und myokardialen Sauerstoffverbrauch. Z Kardiol 70:271
20. Thormann J, Schlepper M, Kramer W, Gottwik M (1982) Effects of AR-L 115 BS in coronary artery disease and its mechanism: improved LV function and LV wall motion without inducing angina. Am J Cardiol 49:1039

Efficacy of AR-L 115 BS on Hemodynamics, Contractility, and Myocardial Oxygen Consumption in Man

M. Oeff, R. Schröder

In search of new drugs with positive inotropic and vasodilatory effects for treating chronic congestive heart failure, the imidazole derivative AR-L 115 BS (2-[(2-methoxy--4-methylsulfinyl)-phenyl]-1H-imidazo [4, 5-b]pyridine) was developed. It combines the following effects: high direct cardiac efficacy, a distinctly higher therapeutic range than digitalis glycosides, no substantial increase in the heart rate, long biological half life, and good intestinal reabsorption.

There are no structural analogies with digitalis glycosides, catecholamines, or other positive inotropic substances like glucagon or amrinone, but, as regards structure and effect, to some extent similarities to theophylline exist (5).

In animal experiments, after AR-L 115 BS a marked dose-dependent increase in the contractility parameters was found, reinforcing the increase resulting from digitalis and catecholamines (4, 5, 8, 10, 15). There was a distinctly relaxing effect on the tone of the vessels, in venous vessels more marked than in the arterial resistance vessels (13), but a blood-pressure-lowering effect was observed with doses markedly higher than those necessary to induce a positive inotropic effect.

The myocardial oxygen consumption was found to be nearly constant at doses of 2 mg/kg body wt, due to an opposite effect on the energy demand of increased contraction velocity and reduced ventricular volume and ejection period (8). Only doses of 4 mg/kg body wt lead to an essential increase in myocardial oxygen consumption (3, 8, 10). The cellular mode of action has not yet definitely been clarified; it seems, however, as though the flux of calcium ions from the cell is restrained or the slow calcium influx raised, and additionally an acute release of calcium ions from intracellular compartments takes place (1, 9, 15, 23). Accordingly, the effect can be neutralized by verapamil (14) but not by β-blockers (22). A direct effect on the contractile structures of the heart, like that of tropomyosin, may also be possible (1, 9).

Findings of clinical studies available so far with doses between 0.7 and 3 mg/kg body wt. in the acute test demonstrated a dose-dependent enhanced left ventricular contraction measured by dp/dt_{max} (18, 20, 22), fractional shortening by means of echocardiography (6, 21), or the ejection fraction by means of radionuclide ventriculography (11) and contrast injection (18).

With pathological increase in the preload, a distinct decrease ensues with consecutive increase of cardiac output. The increase in heart rate is small, and the arterial pressure remains constant. Except for the investigations of Thormann et al. (22), nothing has been published concerning myocardial oxygen consumption in patients with coronary heart disease. Hence the objectives of our study were to investigate the effects of AR-L 115 BS on coronary blood flow, myocardial oxygen consumption, and lactate extraction, relating the data to concurrent changes in hemodynamic determinants in patients with chronic coronary artery disease.

Method

Thirteen patients with coronary artery disease (two women, 11 men; 37–65 years of age, mean age 49.1 years) undergoing diagnostically required heart catheterization were chosen after informed consent. Measurements of hemodynamic and metabolic parameters at rest were done before and 10 min after administration of AR-L 115 BS. 1.3 mg/kg body wt. (0.3 mg/kg as a bolus and 1 mg/kg as short-time infusion for 10 min) were given. Contractility parameters were measured by means of a tip-manometer. Determination of coronary blood flow was done by the argon method (19). For statistical evaluation, the Wilcoxon-test was used.

Results

AR-L 115 BS did not change heart rate and mean arterial pressure. Right atrial pressure decreased significantly from 3.3 ± 1.9 to 2 ± 1.8 mmHg ($p < 0.01$), mean pulmonary artery pressure from 12.1 ± 4.5 to 10.1 ± 3.6 mmHg ($p < 0.005$) (Fig. 1).

Left ventricular end-diastolic pressure decreased significantly from 9 ± 5.6 to 4.8 ± 2.6 mmHg ($p < 0.005$); cardiac index remained constant. Contractility parameters dp/dt_{max} and $dp/dt_{max}/p$ significantly increased by approximately 25% each,

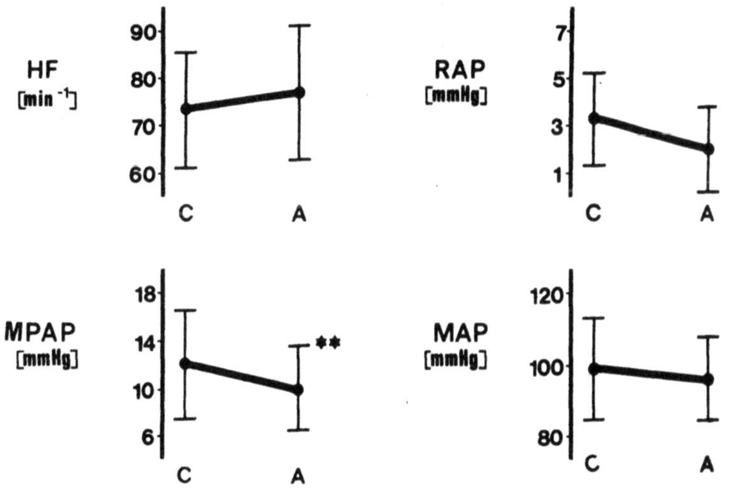

Fig. 1. Effect on heart rate (*HF*), right atrial pressure (*RAP*), mean pulmonary artery pressure (m*PAP*) and mean arterial pressure (*MAP*). *C*, control values; *A*, after AR-L 115 BS; ** $p < 0.01$

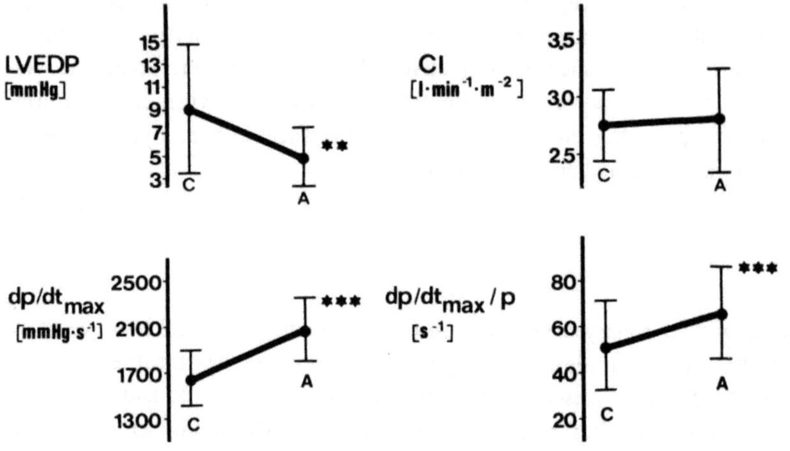

Fig. 2. Effect on left ventricular end-diastolic pressure (*LVEDP*), cardiac index (*CI*), dp/dt_{max}, and $dp/dt_{max}/p$, *C*, control values; *A*, after AR-L 115 BS; ** $p < 0.01$; *** $p < 0.001$

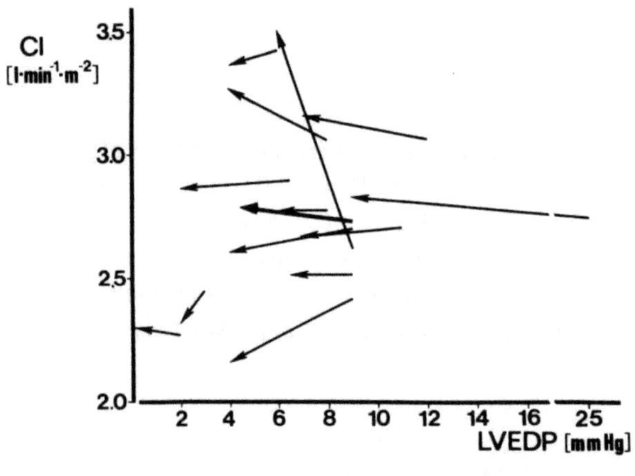

Fig. 3. Effect on cardiac index (*CI*) in relation to the left ventricular end-diastolic pressure (*LVEDP*)

from 1167 ± 238 to 209 ± 277 mmHg/s and from 51 ± 19 to 66 ± 20/s ($p < 0.005$) (Fig. 2).

As a result of a large decline in the mostly normal basic values of left ventricular end-diastolic pressure, the ventricular function curves were shifted mainly to the left but not upward (Fig. 3).

Discussion

In our investigations, we were able to confirm a decrease in right atrial pressure, mean pulmonary arterial pressure, and left ventricular end-diastolic pressure. This can be explained by an augmentation of the contraction of the left ventricle and by the venodilatory effect of AR-L 115 BS. With the dosage used, there was no effect on the mean arterial pressure. The contractility parameters dp/dt_{max} and $dp/dt_{max}/p$ increased significantly. However, there was no consistent increase in cardiac output. In our patients without left ventricular failure, the left ventricular end-diastolic pressure declined to an average of 4.8 mmHg. The lack of increase of cardiac output is explained by this low left ventricular filling pressure. The left ventricular function curve was shifted to the left only.

The myocardial oxygen consumption decreased after administration of AR-L 115 BS (Fig. 4). This decrease despite increased contractility may have been caused by the reduced wall tension due to the lowered left ventricular end-diastolic pressure and volume. Oxygen consumption of the myocardium is the primary factor regulating coronary flood flow (17). In general, coronary blood flow changes according to myocardial oxygen requirements, but arterial-coronary-venous oxygen content difference remains constant (2, 7). Therefore, the lowering of the arterial-coronary-venous oxygen difference which we have found, although statistically just not significant, probably indicates some direct dilating effect of AR-L 115 BS on the coronary resistance vessels. Thus although the oxygen consumption decreased in our patients, the coronary blood flow did not decline accordingly. There is a disagreement with the results of Thormann et al. (22), who found an increase of myocardial oxygen consumption of 34% after 2 mg/kg body wt. AR-L 115 BS. It may be that the ventricular stimulation, they had carried out, had influenced these results. Additionally, there is a great interindividual variation among their seven patients. The local thermodilution technique used to measure the coronary sinus flow may not be sufficient for accurate evaluation under clinical conditions, as shown by Sigmund-Duchanova et al. (16).

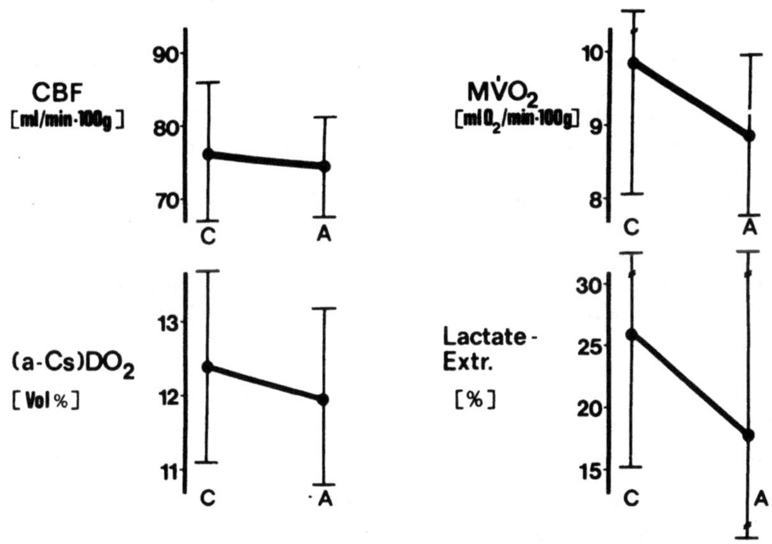

Fig. 4. Effect on the coronary blood flow (CBF), myocardial oxygen consumption (MVO_2), arterio-coronary-venous oxygen content difference ($(a\text{-}Cs)\ DO_2$), and lactate extraction. *C*, control values; *A*, after AR-L 115 BS

Summary

Intravenous administration of AR-L 115 BS leads to a drop in the left ventricular filling pressure (47%), in the mean pulmonary artery pressure (17%), and in the right atrial pressure, thus proving the venodilatory effect of this substance. The increase in the contractility parameters of approximately 25% reveals the positive inotropic effect. The decrease in myocardial oxygen consumption of 10% which nevertheless occurs may be due to the reduced wall tension of the left ventricle. Additionally, and independently of this, the reduced arterio-coronary-venous oxygen content difference has to be explained by a coronary dilatory effect.

References

1. Achenbach C, Hauswirth O, Roskothen R, Ziskoven R (1981) Effects of AR-L 115 BS on the electrophysiological and contractile properties of sheep cardiac fibres. Drug Res 31/1:171
2. Chatterjee K, Matloff JM, Swan HJC, Ganz W, Kaushik VS, Magnusson P, Henis MM, Forrester JS (1975) Abnormal regional metabolism and mechanical function in patients with ischemic heart disease. Circulation 52:390
3. Dämmgen J, Kadatz R, Diederen W (1981) Action of AR-L 115 BS on myocardial oxygen consumption, cardiac performance and vascular resistance and capacitance. Drug Res 31/1:151
4. Diederen W, Kadatz R (1981) Effects of AR-L 115 BS, a new cardiotonic compound, on cardiac contractility, heart rate and blood pressure in anaesthetized and conscious animals. Drug Res 31/1:146
5. Diederen W, Weisenberger H (1981) Studies on the mechanism of the positive-inotropic action of AR-L 115 BS, a new cardiotonic drug. Drug Res 31/1:177
6. Hauf GF, Bubenheimer P, Roskamm H (1981) The acute effect of a new positive inotropic agent (AR-L 115 BS) on cardiac hemodynamics and contractility in patients with severe chronic congestive heart failure. Drug Res 31/1:253
7. Heiss HW, Barmeyer J, Wink K, Huber G, Hagemann G, Beiter G, Keul J, Reindell H (1975) Durchblutung und Substratumsatz des gesunden menschlichen Herzens in Abhängigkeit vom Trainingszustand. Verh Dtsch Ges Kreislaufforsch 41:247
8. Hellige G, Baller D, Ensink FBM, Zipfel J, Wolpers HG, Bretschneider HJ (1981) Effects of AR-L 115 on hemodynamics, myocardial oxygen consumption and cardiac pumping efficiency. Drug Res 31/1:155
9. Herzig JW, Feile K, Rüegg JC (1981) Activating effects of AR-L 115 BS on the Ca^{++} sensitive force, stiffness and unloaded shortening velocity (V_{max}) in isolated contractile structures from mammalian heart muscle. Drug Res 31/1:188
10. Hockerts TH, Höcht B, Broll R, Trenkel K, Sieber J, Kilian A, Brendle B (1981) Die Wirkung von AR-L 115 BS auf die Kontraktionsdynamik des isolierten metabolisch unterstützten Hundeherzens. Drug Res 31/1:157
11. Klein G, Sauer E, Bauer R, Wirtzfeld A, Sebening H (1981) The comparison of the effects of AR-L 115 BS and dobutamine in patients with severe cardiac failure. Drug Res 31/1:257
12. Kramer W, Thormann J, Bitter C, Schlepper M (to be published) Aktivitätsprofil von AR-L 115 bei Herzgesunden: Wirkungsbeeinflussung durch Ca^{++}-Antagonisten (Verapamil). Verh Dtsch Ges Inn Med 69
13. Nguyen-Duong H, Rüdel R, Voss V (1981) Effects of AR-L 115 BS on arterial and venous muscle tone in vitro. Drug Res 31/1:197
14. Puschmann S, Kirkeeide RL, Nienaber C, Schaper W (1981) Wirkung von AR-L 115 auf die Pumpfunktion des linken Ventrikels mit und ohne Beta- und Calcium-Blockade (Abstr). Z Kardiol 70:271

15. Schrader J (1981) Interaction of AR-L 115 BS with Isoprenaline on hemodynamic parameters in the isolated guinea pig heart. Drug Res 31/1:183
16. Sigmund-Duchanova H, Baller D, Bretschneider HJ, Prennschütz-Schützenau H, Piesker KH, Vennebusch H, Zipfel J, Hellige G (1979) Experimental validation of methods for the measurement of coronary sinus blood flow in man. Basic Res Cardiol 74:277
16. Sonnenblick EH, Ross J, Braunwald E (1968) Oxygen consumption of the heart. Newer concepts of its multifactoral determination. Am J Cardiol 22:328
18. Stauch M, Nechwatal W (1981) Effects of AR-L 115 BS on hemodynamics and contractility of the heart in cardiac patients. Drug Res 31/1:267
19. Tauchert M, Kochsiek K, Heiss HW, Rau G, Bretschneider HJ (1971) Technik der Organdurchblutungsmessung mit der Argon-Methode. Kreislaufforsch 60:871
20. Thormann J, Kramer W, Schlepper M (1980) AR-L 115: A new non-glycosidic, non-adrenergic cardiotonic agent, hemodynamic proof for its effectiveness both intravenously and by oral application. Circulation [Suppl III]62:III–233
21. Thormann J, Kramer W, Schlepper M (1981) A new non-glycoside, non-adrenergic cardiotonic agent AR-L 115 BS. Drug Res 31/1:273
22. Thormann J, Kramer W, Schlepper S (1981) AR-L 115 induzierte Änderungen der Hämodynamik, des koronaren Blutflusses und des Energiemetabolismus bei Patienten mit koronarer Herzerkrankung. Z Kardiol 70:385
23. Trube G, Trautwein W (1981) Experiments with AR-L 115 BS on skinned cardiac fibers. Drug Res 31/1:185

Treatment of Cardiac Failure with AR-L 115 BS

W. Schulz, R. Kirsten, H.J. Hermann, G. Kober

In the present study, AR-L 115 BS was infused for 30 min in two groups of patients. Group I consisted of eight patients with severe congestive heart failure (class III and IV NYHA), and group II of eight patients with moderate heart failure (class I and II NYHA). The characteristics and pretreatment of the patients are given in Fig. 1, and the study protocol is shown in Fig. 2.

GROUP I

7 m, 1 w; 54 ± 13 ys.

6 pts class III, 2 pts class IV NYHA

Pretreatment:

	Digitalis	Diuretics	Nitrates
4 pts:	+	+	+
1 pt:	+	+	-
1 pt:	+	-	-
2 pt:	-	-	-

GROUP II

8 m; 50 ± 7 ys.

6 pts class I, 2 pts. class II NYHA

Pretreatment:

	Digitalis	Diuretics	Nitrates
1 pt:	+	+	+
1 pt:	+	+	-
1 pt:	+	-	-
1 pt:	-	-	+
4 pts:	-	-	-

Fig. 1. Characteristics and pretreatment of the patients. Age shown as mean ± SD

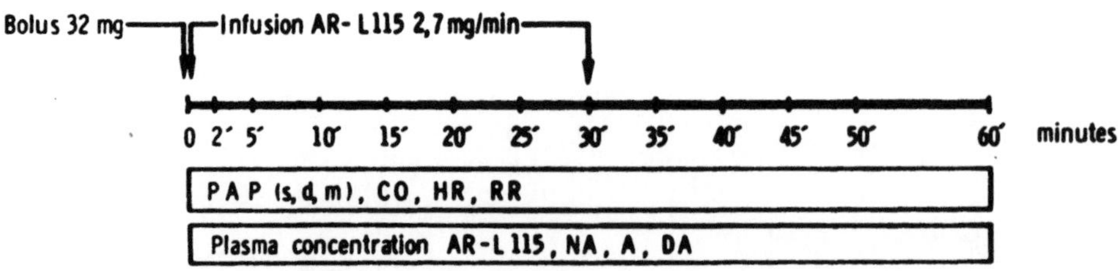

Fig. 2. Study protocol. *PAP*, pulmonary artery pressure; (*s*, systolic; *d*, diastolic; *m*, mean); *CO*, cardiac output; *HR*, heart rate; *RR*, blood pressure; *NA*, Noradrenalin; *A*, Adrenalin; *DA*, Dopamin

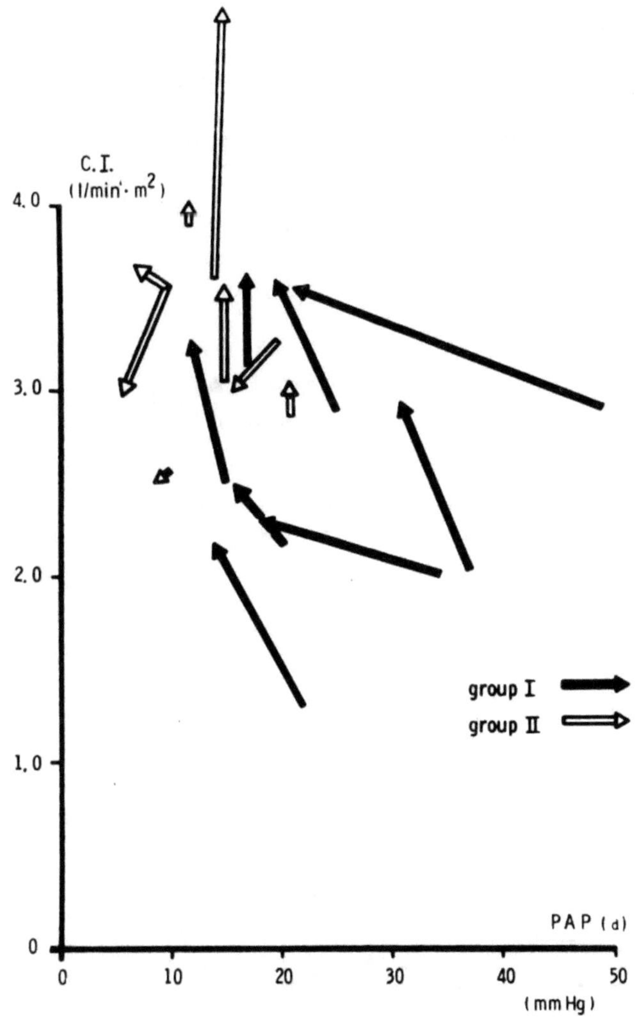

Fig. 3. Cardiac index and diastolic pulmonary pressure

Results

1. Fig. 3 shows are cardiac index (CI) on the ordinate and the diastolic pulmonary pressure on the abscissa. CI increased in all patients of group I, the enddiastolic pulmonary pressure decreasing markedly in seven patients and remaining unchanged in one patient. In group II, no significant changes of CI or pulmonary pressures could be observed.
2. Fig. 4 shows the reaction of heart rate (HR). In group II, the initial values were normal, increasing significantly from 79 to 94 beats/min by the end of infusion ($P < 0.01$). In group I, however, the initial HR was elevated and did not increase significantly during infusion of AR-L 115.

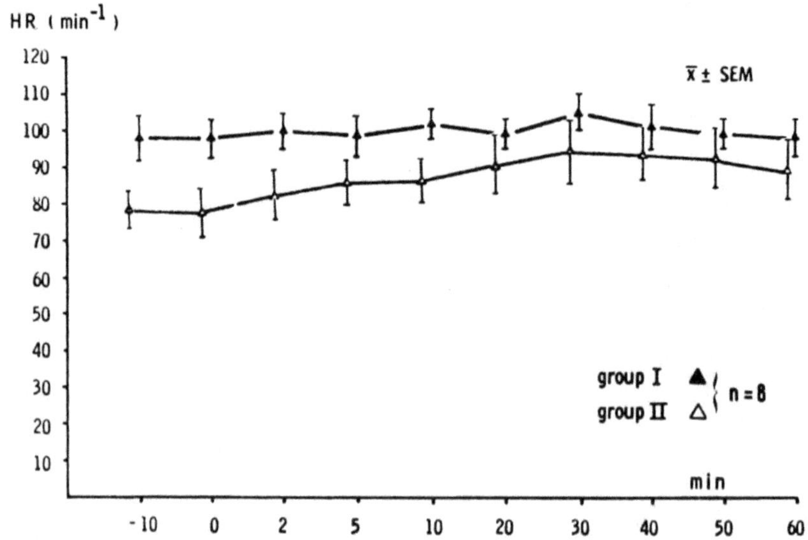

Fig. 4. Reaction of heart rate

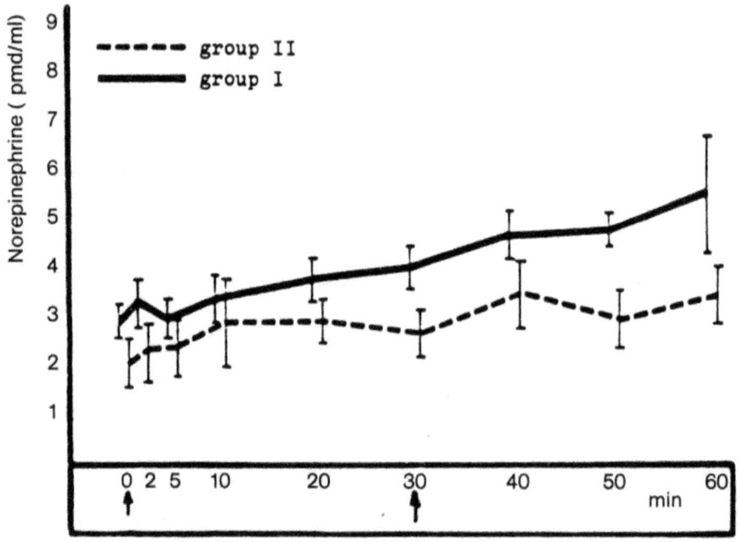

Fig. 5. Concentration of norepinephrine in venous blood during infusion of AR-L 115

3. Blood pressure, which was measured by the cuff method, did not change significantly in either group, despite marked increases in cardiac output.
4. The concentration of norepinephrine in central venous blood increased markedly in group I and slightly in group II (Fig. 5).

Discussion

The results show that AR-L 115 BS improves myocardial performance in patients with severe heart failure, but not in those with moderate heart failure. The hemodynamic mechanism is a positive inotropic effect on the myocardium and a vasodilatation in the peripheral arterial bed.

The biochemical mode of action is unknown. It is said not to be digitalis-like or catecholamine-like, but we found increasing levels of norepinephrine in the central venous blood, so we could argue that there is some indirect sympathomimetic effect. Some effect may be attributable to the inhibition of phosphodiesterase, so that the second messenger of intracellular beta-adrenergic stimulation is increased. There are also some reports of the inhibition of intracellular calcium sequestration into the sarcoplasmic reticulum. But mere regulation of calcium fluxes does not fit with augmented striated muscle contraction on one hand and diminished smooth muscle contraction on the other hand.

The clinical relevance of this new drug is in its beneficial effect on class III and IV patients already undergoing full conventional therapy.

Discussion (zu den Seiten 185–214)

Schulz, Frankfurt: Theophylline releases catecholamines as well. Is this release of catecholamines due to the pharmacodynamic changes or is it due to augmented release from the adrenal gland?

Schlepper, Bad Nauheim: We do not know exactly, and I think it is too early to speculate about it; but we should obtain blood samples out of the coronary sinus blood in order to measure catecholamines.

Auditorium: First, did you find augmented stroke volume index, as you demonstrated an increase in cardiac index but also in heart rate? Secondly, dp/dt_{max} is dependent upon preload, which limits its usefulness in determining the basic contractility. Therefore, increased dp/dt_{max} may be due to the decrease of preload by the drug, without affecting contractility!

Oeff, Berlin: We did not observe changes of stroke volume index: indeed, we measured a pronounced decrease in left ventricular end-diastolic pressure.

Schlepper, Bad Nauheim: This drug exerts positive inotropic effects, there is no doubt about it. Increased contractility due to AR-L 115 BS can be shown as follows:

1. Under conditions where preload and afterload are constant, e.g., in an isolated cardiac muscle
2. By studying indices that are independent of preload and afterload, e.g., total electromechanical systole adjusted for heart rate, which is shortened
3. By the fact that additional vasodilation caused by verapamil does not enhance the positive effects, but causes a drastic reduction of cardiac performance in humans and in experimental animals.

Bussmann, Frankfurt: Is it possible to assume that AR-L 115 BS produces effects similar to nitroglycerin when given to patients with coronary artery disease?

Schlepper, Bad Nauheim: Yes, but in addition inotropic effects are observed.

Auditorium: Nevertheless, nitroglycerin reduces oxygen consumption. You have shown oxygen consumption to be augmented by AR-L 115 BS. In addition, there is a certain inconsistency with the results of Bretschneider.

Schlepper, Bad Nauheim: There are no differences. Nitroglycerin reduces preload, which does not necessarily increase left ventricular cardiac output; this mechanism does not increase oxygen consumption. However, AR-L 115 BS produces an augmentation of cardiac output of about 30%. According to the results of Bretschneider, who also noticed that the extent of oxygen consumption depends on the dosages used, at low dosages AR-L 115 BS shows a distinct positive inotropic effect and reduces end-systolic volume without increase in myocardial oxygen consumption, and at higher doses there is an increase in oxygen consumption.

Schulz, Frankfurt: I wonder why AR-L 115 BS does not increase heart rate. Is there augmentation of heart rate when AR-L 115 BS is used in isolated preparations? Did you observe CNS stimulation or are such observations published, and is this stimulation bothering for the patients?

Schlepper, Bad Nauheim: First, I think in view of studies in humans we can assume that heart rate does not increase, probably due to reflex mechanisms following decrease in pulmonary arterial pressure. Furthermore, no essential increase in heart rate could be observed in animal studies and experiments using isolated organs.

Secondly, the CNS effect of the drug might be compared to that of two cups of coffee. Apart from transient brightness of vision, I do not know of any adverse CNS reactions. High dosages of theophylline, for example, would be needed to effect a comparable decrease in pulmonary artery and wedge pressure.

Schröder, Berlin: I did not understand your experiment with ventricular stimulation!

Schlepper, Bad Nauheim: We measured oxygen demand at rest and at ventricular tachycardia, and again following AR-L 115 BS administration using the same protocol, in order to answer two questions. First, does AR-L 115 BS provoke ischemia in patients with coronary artery disease? Secondly, might ischemia be prevented by AR-L 115 BS?

Auditorium: What about the blocking effect of calcium antagonists? There is a study by Schaper and co-workers, who did not find a complete inhibition, as far as I know.

Schlepper, Bad Nauheim: Depending on the dosage used, Schaper et al. observed the same complete inhibition using verapamil. It is a question of the dosages used. We used relatively high dosages, therefore we observed almost complete inhibition.

Lang, Villingen: Did you observe any influence on impulse automaticity, impulse conduction, or development of ectopic depolarization?

Schlepper, Bad Nauheim: We did not detect arrhythmias, although we carried out 24-h ambulatory ECG in all these patients.

IV. Conception and Planning of Future Studies

Can Vasodilator Treatment Improve Life Expectancy and Physical Performance Heart Failure?*

Jay N. Cohn

The acute hemodynamic response to vasodilator drugs in patients with congestive heart failure is well known. In response to administration of drugs that dilate both arterial and venous vascular beds there is an increase in cardiac output, a reduction in right atrial and left atrial pressure, a reduction in pulmonary arterial pressure, a modest reduction in arterial pressure, and little change in heart rate. This hemodynamic response has been reported during infusion of sodium nitroprusside (1), and after oral administration of hydralazine and isosorbide dinitrate (2), prazosin (3), trimazosin (4), or captopril (5). Despite the salutary acute hemodynamic response to these drugs, the major question is whether chronic administration of any of these agents will result in improved exercise performance, reduced morbidity, and longer life expectancy.

Exercise Capacity

Despite the improved left ventricular performance measured at rest after acute administration of a variety of vasodilator drugs, exercise tolerance does not appear to be improved. After single doses of hydralazine, isosorbide dinitrate, or prazosin no demonstrable increase in exercise capacity or maximal oxygen consumption can be demonstrated (6–8). After chronic administration of these drugs, however, an improvement in exercise tolerance has been noted in most controlled and uncontrolled studies (9–12). The magnitude of this increased exercise tolerance varies, but has been sizable enough to result in reduced clinical severity of the disease as classified by the New York Heart Association.

The mechanism by which exercise tolerance is unaffected acutely but improved chronically has not been established. The chronic response may reflect progressive improvement of left ventricular performance as a result of a sustained reduction of impedance and preload, or it may in part reflect improved peripheral utilization of oxygen because of "retraining" of the skeletal muscle bed (13). The likelihood that the latter mechanism is important is supported by observations that the increased

* Supported in part by PHS Grant HL 22977 from the National Heart, Lung and Blood Institute

exercise cardiac output after chronic therapy with these drugs appears to be less than the increase in total body oxygen consumption (7).

Most controlled trials with vasodilator drugs have been carried out for limited periods of time, usually not exceeding 3 months. Therefore, the long-term benefits of this therapy on exercise performance have not been addressed. Furthermore, exercise testing has a very subjective end point when exercise duration is used a criterion for improvement. If the patient and physician are aware of the therapy that the patient is receiving, bias may also significantly influence the overall results. Therefore, a double-blind study is necessary and objective measurement of exercise capacity, such as maximum oxygen consumption, is highly desirable.

Furthermore, consideration must be given to whether maximum exercise capacity in a test with progressively increasing load, or total duration of exercise capacity at a submaximal load, is a more appropriate means of assessing the response to therapy. A cooperative study involving ten Veterans Administration Hospitals in the United States is currently under way, comparing the effect of addition of a hydralazine- and nitrate combination or prazosin to a standard medical regimen for treatment of congestive heart failure. This vasodilator heart failure trial (V-HEFT) will assess both the maximal exercise capacity and the submaximal exercise duration, utilizing oxygen consumption data as an objective end point. These studies are planned to be continued for a period of 5 years, and should thus provide an answer to the question of whether such therapy has a sustained effect on exercise capacity.

Life Expectancy

The mortality rate in patients with congestive heart failure who continue to exhibit symptoms while taking digitalis and diuretic drugs is surprisingly high. In patients studied in our laboratory the mortality rate in 1 year exceeds 50%, despite administration of what is viewed as optimal therapy, including vasodilators. It is clear that patients who reach this stage of the clinical syndrome already have far advanced heart disease and might not prove to have a meaningfully prolonged life expectancy even if effective therapy were instituted. Nonetheless, this patient population with a high mortality provides an opportunity to study the effect of vasodilator drugs on mortality rate. These data will be accumulated as part of the V-HEFT and, indeed, mortality rate is the primary end point of that study.

Although initial studies of vasodilator drugs have been carried out primarily in patients with advanced heart failure, it is likely that the major impact of an effective therapy for heart failure would be in preventing progression of the diasease rather than in forestalling death in the terminal stages of the disease. The natural history of congestive heart failure suggests that it may often primarily reflect a positive feedback system in which cardiac failure begets more cardiac failure (Fig. 1). The left ventricular dilatation that is characteristic of cardiomyopathy may progress independently of the pathological process in the myocardium, and this progression may in part be a manifestation of heightened preload and heightened outflow resistance on the performance of the left ventricle. Any drug that could reduce wall stress and improve performance of the left ventricle might be effective in forestalling the vicious circle

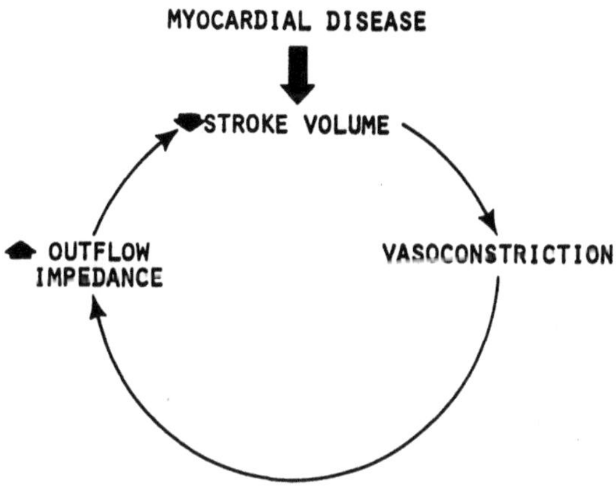

Fig. 1. Theoretical vicious cycle that may lead to progressive deterioration of cardiac function in congestive heart failure

that may precipitate end-stage heart failure. Neither diuretics nor inotropic agents have been subjected to objective evaluation of their effectiveness in altering the course of the disease. The vasodilator drugs are attractive agents to utilize for this purpose, since activation of the sympathetic nervous system and the renin-angiotensin system seems to be an important component of the neurohumoral response to congestive heart failure, and the resultant vasoconstriction might be a vital factor in the progressive left ventricular dysfunction. Thus evidence of prolongation of life when these drugs are instituted in the late stages of the disease would encourage a more extensive and rewarding trial instituted in the earlier stages.

Despite evidence supporting the deleterious effects of vasoconstriction on the performance of the left ventricle and the salutary effects of vasodilator drugs on hemodynamics, rational recommendations for chronic use of these agents must still await more extensive and definitive trials. Until these trials are completed, the physician has at his disposal a potent group of drugs which seem to be effective in improving the performance of the left ventricle. Patients whose clinical course has been unsatisfactory while they were taking more traditional drugs for the treatment of heart failure should certainly be considered for a trial of vasodilator drugs. Which patients would be most likely to benefit and those in whom the risk of therapy might be too great remains to be determined.

References

1. Guiha NH, Cohn JN, Mikulic E, Franciosa JA, Limas CJ (1974) Treatment of refractory heart failure with infusion of nitroprusside. N Engl J Med 291:587
2. Pierpont GL, Cohn JN, Franciosa JA (1978) Combined oral hydralazine-nitrate therapy in left ventricular failure. Hemodynamic equivalency to sodium nitroprusside. Chest 73:8
3. Miller RR, Awan NA, Maxwell KS, Mason DT (1977) Sustained reduction of cardiac impedance and preload in congestive heart failure with the antihypertensive vasodilator, prazosin. N Engl J Med 297:303

4. Franciosa JA, Cohn JN (1978) Hemodynamic effects of trimazosin in patients with left ventricular failure. Clin Pharmacol Ther 23:11
5. Levine TB, Franciosa JA, Cohn JN (1980) Acute and long-term response to an oral converting-enzyme inhibitor, captopril, in congestive heart failure. Circulation 62:35–41
6. Franciosa JA, Cohn JN (1979) Immediate effects of hydralazine-isosorbide dinitrate combination on exercise capacity and exercise hemodynamics in patients with left ventricular failure. Circulation 59:1085–1091
7. Franciosa JA, Goldsmith SR, Cohn JN (1980) Contrasting immediate and long-term effects of isosorbide dinitrate on exercise capacity in congestive heart failure. Am J Med 69:559–566
8. Rubin SA, Chatterjee K, Gelberg HJ, Ports TA, Brundage BH, Parmley WW (1979) Paradox of improved exercise but not resting hemodynamics with short-term prazosin in chronic heart failure. Am J Cardiol 43:810
9. Aronow W, Greenfield RS, Alimadadian H, Danahy DT (1977) Effect of the vasodilator trimazosin versus placebo on exercise performance in left ventricular failure. Am J Cardiol 40: 789
10. Franciosa JA, Nordstrom LA, Cohn JN (1978) Nitrate therapy for congestive heart failure. JAMA 240:433
11. Weber KT, Kinasewitz GT, West JS, Janicki JS, Reichek N, Fishman AP (1980) Long-term vasodilator therapy with trimazosin in chronic cardiac failure. N Engl J Med 303:242–249
12. Colucci WS, Wynne J, Holman BL, Braunwald E (1980) Long-term therapy of heart failure with prazosin, a randomized double-blind trial. Am J Cardiol 45:337–343
13. Holloszy JO (1976) Adaptations of muscular tissue to training. Prog Cardiovasc Dis 18:445

Discussion

Sigwart, Lausanne: Among patients we studied there was a total of seven who increased their ejection fraction angiocardiographically during exercise. I think this is a significant point, because the sympathetic tone plays a very important role in certain patients with a low ejection fraction. This situation is observed at rest and is easily reversible by exercise, that is, by increasing sympathetic tone. We call this "lazy left ventricle", and have presented material on this unusual and interesting syndrome. At first glance, the poor performance of the left ventricle at rest seems absurd, but the reversibility of this state with even minor degrees of exercise immediately clarifies the problem, which in all likelihood is related to the sympathetic innervation of the heart.

Cohn, Minneapolis: This is an important clarification. We have not performed exercise ventriculography, and so cannot exclude exercise ventriculography. We cannot, therefore, exclude the syndrome of lazy left ventricle in our patient.

Taylor, Leeds: I should like to take up two further points for emphasis. First, exercise studies are most valuable in evaluating the response to drugs in patients with congestive heart failure. It has to be realized, however, that with submaximal exercise a rather poorly defined end point is utilized. Oxygen debt may be only minimal or may be pronounced. Maximum or peak exercise may be a more valuable point in time for measurement. We have to realize that hemodynamic measurements at rest, including such parameters as ejection fraction, cannot be used to predict the response to exercise in an individual patient.

The second point is the very important relationship between the death rate and the level of sympathetic activity measured by the plasma norepinephrine concentration. An equally reliable indicator of risk in congestive heart failure may be the plasma renin activity or the serum sodium concentration.

Cohn, Minneapolis: The dissociation between resting function of the heart and the response to exercise is, of course, a prominent and well-known feature. Yet many clinicians still do not take full account of this fact. The plasma norepinephrine content is indeed a reliable indicator of risk in congestive heart failure. The plasma renin activity, however, did not correlate well in our studies with survival or mortality. It was only weakly correlated with the plasma norepinephrine content when we tested the relationship using multivariance analysis. Serum sodium did not correlate either, which means that in our hands this parameter could not be used as a risk indicator. By further analysis of the predictive power of plasma norepinephrine content we were able to show that this correlation does not exist for the cases of sudden death. It was the group of patients with the most severe degree of congestive heart failure who showed the highest plasma norepinephrine content, and in these patients the incidence of death in progressive heart failure. We have to conclude, therefore, that today we still do not possess reliable predictors of sudden death in heart failure patients, although we are able to identify patients at risk of dying in progressive heart failure.

Jahrmärker, München: Cohn presented three groups with plasma norepinephrine levels in relation to prognosis. Can the range of "normal" and of levels predictive of a poor prognosis be defined in numerical terms? It would further be of interest to know whether the blood samples for norepinephrine determination are to be obtained under special conditions.

Cohn, Minneapolis: In our studies all samples for norepinephrine determination were taken after a 1 h rest in a supine position. They were always taken in the morning in fasting patients and from a central venous line. Low values were defined as those up to 400 pg/ml, the middle group included valves of 400–800 pg/ml, and the high group values above 800 pg/ml. These plasma levels tended to be sustained when determinations were made on successive days.

Vasodilators in Chronic Heart Failure: Problems of Design and Analysis of Controlled Clinical Trials

L. Horbach

Introduction

The present state of our knowledge of the long-term therapeutic efficacy of vasodilators chronic heart failure calls for controlled clinical trials. In spite of a great many pharmacological findings and clinical trials, therapeutic decisions taken in the future should be based on systematic empirical observations following a well-defined statistical design. The object of this chapter is to outline and highlight the essential elements of the design and implementation of controlled clinical trials involving the administration of vasodilators in patients suffering from chronic heart failure.

Therapeutic Hypothesis

The first step in designing a controlled clinical trial is the formulation of a clear-cut therapeutic question to be answered by the study. Such a question should be capable of being transformed into a statistical hypothesis corresponding to a mathematical model of evaluation appropriate to the therapeutic problem; the testing of the therapeutic hypothesis by the aid of this model is the aim of the study.

This methodological procedure underlying a controlled clinical trial, although apparently simple, is often difficult to carry out. The therapeutic problem to be dealt with may be very complex, with multiple prerequisites for its solution. In the interests of a clear interpretation of the results, the formulation of a singular cardinal question is to be preferred to a number of questions, each of which may be only of partial interest.

In a first approach, physicians, pharmacologists, and statisticians have to gather and analyze all available information from personal experience or published literature, to give a broad and sound basis for defining the hypothesis of a study. With regard to vasodilators, there are many parameters, but these need only briefly be outlined here:

1. Pharmacological and toxicological examinations in at least two species of animals
   a) Examination of special effects on smooth muscular fibers of veins and/or arterioles
   b) Examination of toxicity, especially with regard to carcinogenic, teratogenic, and mutagenic effects, fertility, and perinatal mortality

2. Clinical examinations of phase I type including
   a) Tolerance of the drug in healthy volunteers
   b) Pharmacokinetics
   c) Pharmacodynamics
   d) Interaction with other drugs
3. Clinical examination of phase II type: observations of the course of disease after administration of vasodilators in patients suffering from chronic heart failure.

From these pharmacological and clinical findings and observations is to be derived a worthwile therapeutic advantage in a clinically important situation. The chief question raised by the contributors to this study is that of the efficacy of long-term therapy with vasodilators in chronic heart failure. Efficacy in the sense of the recent German law on permission for new drugs means the combination of all effects influencing the development of a disease in a beneficial way. The total efficacy can be more than the sum of singular effects. By definition, the notion of efficacy demands the use of criteria based on the essential pattern of evolution of the disease. In severe illness like chronic heart failure, the survival time is the essential variable that may be influenced by the therapy. other factors may contribute to this, but are of secondary importance in the final evaluation of observations. The central question is: Do vasodilators prolong survival in patients with chronic heart failure?

Long-term therapy with vasodilators requires continuing medical care for these patients, so that hemodynamic and clinical variables (e.g., states of heart failure as defined by NYHA) can be observed and documented for further statistical analysis.

Demarcation of the Therapeutic Indication for Vasodilators

According to the rules of inductive statistical reasoning (3, 4), the population of patients to whom the drug is to be administered must be determined. The patients have to be classified on the basis of clinical findings. The category of chronic heart disease certainly represents a heterogeneous group which can roughly be stratified into three according to cardiac pathology:
1. Coronary heart disease
2. Cardiomyopathies with other etiology
3. Acquired valvular defects.

Such definitions often delineate the basic characteristics of the disease but are not sufficient for a clear-cut demarcation. In order to come to an unequivocal decision as to whether a certain patient should be included in a study or not, a list must be drawn up of all possible grounds for the exclusion of the patient from the study, e.g., the presence of an acquired valvular defect or unsuitability of age.

In the interests of avoiding bias, all patients treated in a clinic who fulfill the criteria for inclusion in the study and who present no grounds for exclusion should be considered.

According to the rules of medical ethics the patients considered should be informed of the modalities and aims of the controlled clinical trial, and must give their agreement to participation. It is generally assumed that patients refusing to take part do not represent a selection as far as the randomness of the survey is concerned.

Elements of a Comparative Design

Therapeutic Measures To Be Compared

The essence of statistical methodology is comparison. Every test of a therapeutic hypothesis demands a comparison, e.g., of the drug in question with another one or with a placebo. For ethical reasons, as far as is known at the beginning of the trial the drug used for comparison must be as effective as the drug to be tested, so in chronic heart failure there is no question of a placebo. But it is a matter of good therapeutic strategy that therapeutic trials be carried out early, before unfounded therapeutic prejudices are established. In view of the large number of vasodilators available, there is a certain temptation to try to carry out multiple comparisons of some of these drugs simultaneously. In order to establish a sound inductive basis, however, one should select one drug with an appropriate indication according to previous findings and compare it to standard therapy as hitherto applied without vasodilators. A very critical question is that of the so-called basic therapy, i.e., drugs given at the same time as those of which the effects are to be compared, e.g., the administration of digitalis and diuretic agents in chronic heart failure. Personal experiences in a therapeutic study concerning myocardial infarction (6) have revealed the importance of formulating rules concerning the application of drugs other than those to be compared: there were drugs with competitive effects and it was impossible to analyze the interactions of different drugs in view of the multitude of drugs administered. In a multicenter clinical trial the therapeutic usages customary in the different hospitals should be simplified and standardized. Otherwise, different therapeutic habits particular to a clinic must be grounds for stratification of overall results by hospitals.

Comparability of Observations

The assessment of clinical findings is often subject to the individual judgment of the physician. But even if data from different laboratories, e.g., chemical or hemodynamic measurements, are regarded as criteria for the evaluation of therapeutic effects, there is a need for standardization, especially if different laboratories are involved in a multicenter study. Technical or chemical methods, as well as modifications of such methods, differ from one laboratory to the next. Hardly comparable distributions of values of certain enzymes between different clinics turned out in a multicenter therapeutic study of myocardial infarction (5). The comparability of observed and measured data becomes more difficult to achieve as the number of clinics involved rises. Even the assessment of survival time must be made with the same carefulness and accuracy in all participating clinics.

As in the study of vasodilators, survival time and hemodynamic variables are the important criteria and subjective medical judgments are of minor importance; there is no need of a double-blind study, which in some cases is necessary to avoid the observer bias. Nevertheless, special standardization of observation rules has to be worked out for observing the occurrence of adverse effects. The accuracy of observation has to be the same in the compared therapeutic series. A prescribed documentation helps to achieve this aim. Expected adverse effects are to be questioned in a defined way.

The occurrence of hitherto unknown effects can only be ascertained by the notation of extraordinary observations in the treated patients: a higher frequency of complications of a certain type in one of the series implies an adverse effect of this therapy.

Adequate Patient Totals

In the first phase of planning, the classic question, put to the statistician involved in a study is: What should be the size of the study, i.e., the number of patients to be included in order to give a good chance of discovering really beneficial or adverse effects? Many prerequisites are necessary for a reliable answer. The final statistical model designed to test the therapeutic hypothesis must first be defined. The first factor determining the model to be taken into consideration is the sort of cardinal variable which characterizes the therapeutic efficacy, i.e., quantitative variables lead to other models than qualitative variables. Generally the following factors influence the model and the power of the statistical test:

1. Sort of variable to be tested (e.g., measurements, frequencies)
2. Homogeneity of the observed patient group
3. Size of the study
4. Real difference in efficacy of the compared therapeutic agents (unknown).

Knowledge of these factors, perhaps the results of pilot studies, is useful as a basis for planning.

If frequencies are considered, e.g., the mortality in the two series within a certain period of time from the beginning of therapy, the first question to be put may be: what is the mortality rate with standard therapy? What amelioration by a vasodilator would be therapeutically relevant? The comparison of frequencies generally demands large observed series.

In testing hypotheses, there are two types of errors which can be made: a type I error is the rejection of a hypothesis which is true, a type II error is the nonrejection of a hypothesis which is false.

When designing a clinical trial to test a hypothesis, it is desirable to minimize the probabilities of making these errors. Type I errors may be designated $\alpha \leqslant 0.05$ (α is generally referred to as the significance level of the test) and type II errors $\beta \leqslant 0.10$ ($1\text{-}\beta = 0.9$ is known as the power of the test). Then the differences between the following mortalities in series A (symbolized by p_A) and in series B (symbolized by p_B) are detected by the test with the indicated numbers of patients in each of the series $n_A = n_B$:

p_A (%)	p_B (%)	$n_A = n_B$
20	80	10
30	70	25
40	60	105
45	55	426
49	51	10704

p_A (%)	p_B (%)	$n_A = n_B$
20	40	88
10	20	217
5	10	473
2	4	1284
1	2	2564

These figures calculated by the aid of the operating characteristic function of the frequency test show that frequencies near 50% demand the biggest sample sizes. The detection of differences in the order of 2% generally exceeds the realistic possibilities of the size of a study, even if a multicenter study is carried out. The chance of getting significant results if mortality is the cardinal criterion can be augmented by the consideration of survival times, which leads to tests with more power. There are other models of statistical tests for comparing measured variables which generally demand smaller numbers of observed patients.

The problem of deciding upon a fixed sample size in advance can sometimes be avoided by the application of sequential tests of hypotheses. By this procedure economy can be achieved by means of the possibility of terminating sampling as soon as it is possible to reach a decision. However, given the necessary information, e.g., statements concerning the allowable risks associated with incorrect decisions, the required maximum sample size may be specified to avoid an endless sequential procedure.

Random Allocation

Random allocation is a fundamental procedure which confers on controlled clinical trials an experimental character. The ethical aspects of randomization are currently being widely discussed. Today, the question of randomization has to be judged by ethical committees according to the principles of the Helsinki agreement. Patients have to be informed if they are participating in a clinical trial; they can refuse to take part. If all available information and empirical experience do not demonstrate the superiority of one of the compared therapeutic measures or agents, the carrying out of a controlled clinical trial using the technique of randomization is fully ethical. Randomization is the only unbiased method of establishing treatment series of structural similarity that can be compared statistically.

The technique of randomization today is based on the use of random numbers. In practice, in order of admission of patients to be included in the study there are closed envelopes with the number of the patient in the study outside and his allocation inside. Special care has to be taken that the procedure of random allocation is strictly observed. In the case of heterogeneous a simple random allocation can lead – at a limited size – according to the rules of binomial distribution series, to a certain accumulation of rare types of patients in one series. In this situation the application of a stratified random allocation offers relevant advantages and can strengthen the power of the final test.

Plan of Observation: Documentation

The plan of observation is prescribed by documentation sheets for each patient in a form adapted to computerization. It includes items of the following types:

1. Identification of the patient
2. Variables characterizing secondary conditions of the disease, e.g., sex, age, and other chronic diseases, such as diabetes
3. Cardinal variable characterizing the efficacy of therapy (survival time defined by date of onset of therapy and date of death)
4. Laboratory variables complementing the characterization of the course of disease, possibly repeatedly observed
5. Signs of adverse effects or extraordinary observations
6. Administered drugs with dosage
7. Compliance.

A special problem is the timing of repeated observations, which should give reliable information about the evolutional pattern of the disease under therapy. This timing has to be adapted to the dynamics of the disease.

Manual

Informed and dedicated investigators are needed if a controlled clinical trial is really to follow the rules of the statistical design. For reasons of the advance documentation of the essential therapeutic hypothesis and the information of all participating personnel, a manual has to be worked out at the beginning of the study. It should include all elements of design hitherto pointed out, especially guidelines of how to follow the different procedures for making and recording observations. This working protocol will be a detailed reference document and must be available to all participating investigators.

Statistical Analysis of Data

The records describing the course of disease in individuals under therapy have to be linked and stored on the computer destined for calculations. Controls of completeness and plausibility and formal controls precede statistical analysis. In long-term studies data from different places and different times must be linked with the aid of the computer during the course of the study. An appropriate data bank is very useful.

In most studies criteria have to be defined in advance in case of augmented adverse effects in one series. For this reason, sequential tests regarding these criteria are very usefull. They have to take into account every new observation made during a study relevant for a possible premature stop end to the study.

A descriptive statistical presentation of the results serves to control the comparability of the distribution of variables in the two samples after random allocation and to show the structure of these samples regarding essential concomitant variables such as sex, age, and stage of disease.

The cardinal variable, survival time after the onset of the special therapy, is tested by a log rank test (1). The testing of further variables is of secondary importance but may support the result concerning survival time.

Further Explorative Statistical Analysis

Given a rather high variability of state of chronic heart failure even in the patients who were given the superior therapy, there will be more or less convincing therapeutic efficacy. The important question is raised: How does the distribution of results of the administered therapy differ from the pattern of medical findings at the outset? The answer to this question can give us a key to better differential therapeutic decisions. Possibly one drug gives good results in certain types of disease, the other one is superior in another subgroup of patients.

One approach to the solution of this problem uses an appropriate discriminant function. This method was initiated by Hughes et al. (7) and by Norris et al. (9). Relevant findings observed at the beginning of therapy figure as K variables in individual vectors of two groups of patients: those

$$X = (x_1, x_2, \ldots x_K)$$

with poor results, e.g., those alive and those who died within a certain interval of time after the beginning of therapy. Under the assumptions of multivariate normal distributions and equality of covariance matrices, the linear discriminant function ascertains a

$$D(X) = a_1 x_1 + a_2 x_2 + \ldots t a_K x_K$$

maximal separation of the two groups. The individual values of the function D (X) of the two groups have generally overlapping but distinctly differing distributions. If the whole range of the two distributions is divided into classes of values D (X), groups of more or less different prognoses can be discerned. The multivariate pattern of disease giving good results can be found out.

Cox (2) has presented a model responding to the requirement for calculation of prognostic indices as a function of time, e.g., expected individual survival curves under given circumstances of disease pattern and therapy. All available information on the follow-up of patients is taken into account. The estimated function of the survival time P (A) has the form

$$P(t) = e^{-e^{\beta_1 x_1 + \beta_2 x_2 + \ldots \beta_n x_n} \Lambda(t)}$$

The unknown parameters $\beta_1, \beta_2, \ldots, \beta_n$ and the unknown cumulative hazard function Λ, which characterizes the intensity of the probability of dying as a function of time, are estimated on the basis of the individual observation vectors at the outset of therapy and the documented survival time. Procedures of calculation have been presented by Cox (2) and by Kalbfleisch and Prentice (8). With the aid of this model, life expectancy curves for future patients under different therapy can be calculated, and the most effective drug can be chosen.

References

1. Breslow NE (1970) A generalized Kruskal-Wallis test for comparing K samples subject to unequal pattern of censorship. Biometrika 57:579–594
2. Cox DR (1972) Regression models and life tables. JR Stat Soc [B] 34:187–220
3. Fisher RA (1925) Statistical methods for research workers. Oliver and Boyd, Edinburgh London
4. Fisher RA (1935) The design of experiments. Oliver and Boyd, Edinburgh London
5. Horbaoh L, Gunselmann W, Just H, Schicketanz K-H, Schmidt W (1976) Verlaufsindizes bei Herzinfarkten. Klinisch-statistische Forschung, Ber. 19. Jahrestag. GMDS. Mainz 1974. Schattauer, Stuttgart, pp 271–287
6. Horbach L, Just H (1979) Klinisch-therapeutische Studie: Trasylol bei Herzinfarkt. Intensivmedizin 16:338–360
7. Hughes WL, Kalbfleisch JM, Brandt EN, Costiloc JP (1963) Myocardial infarction prognosis by discriminant analysis. Arch Intern Med 3:338–360
8. Kalbfleisch JD, Prentice RL (1973) Marginal likelihoods based on Cox's regression and life model. Biometrika 60:267–278
9. Norris RM, Brandt PWT, Caughey DE, Lee AJ, Scott PJ (1967) A new coronary prognostic index. Lancet I:274–278

Discussion

Schröder, Berlin: The definition of the problem to be studied and the questions to be answered have to be drawn up according to the principles of biostatistics. I can only underline the importance of careful statistical planning of each clinical study. A study should be planned in a prospective way. Retrospective analyses are of limited value. Prospective studies allow the formulation of hypotheses, but these have to be proved by a prospective study.

Clinical studies should be controlled, which means that each treatment group should be compared with a control group selected according to statistical principles. This claim also applies in the case of difficult patient populations, such as those with congestive heart failure. Cohn has demonstrated conclusively that even in a placebo group the physical performance of patients in chronic congestive heart failure may be improved. Without this control group a false positive effect of the treatment with vasodilators would have to be concluded.

A study should be randomized. This may seem self-evident. Unfortunately, however many a clinical study published nowadays does not fulfill this criterion. Criteria for inclusion and for exclusion must be defined with sufficient clarity to avoid subjective influence on the selection. It is also of the greatest importance that each patient who has been randomly included in a study remain within his group. This also holds if, for example, he does not continue to take the drug.

Pharmacological studies should always be designed in a double-blind fashion. This is not always feasible. It must be pointed out, however, that the less conclusive the criteria of judgment are, the more important the double-blind design is. With death as the primary end point, a double-blind design is obviously not needed. Other means of judging, for example, physical performance in heart failure, demand double-blind design. It should also be observed that the evaluation of certain clinical data, including angiocardiograms or ventriculograms, electrocardiograms or echocardiograms, should always be performed in a blinded fashion.

Jahrmärker, München: What are the criteria for determining the size of study groups and for the inclusion or exclusion of randomized patients?

Horbach, Erlangen: This question is not easily answered. The number of patients needed to answer a certain question requires knowledge about the mortality in the placebo group or in the control group. You will also need to know the present influence of your intervention on the criterion to be judged, for example, mortality. Furthermore, the dropout rate has to be estimated, and a whole host of other factors will enter into the calculations of the statistician who determines the sample size needed.

In the case of a well-known β-blocker study, the question may be raised why a patient should remain in the group he has been randomized into, in spite of the fact that he is not taking the drug any longer. These patients are not to be regarded as dropouts. They are being followed up, even though they are not taking the drug. At the onset of a study it has to be realized that it is both the "intention to treat," and the "intention to treat with placebo," that determine the process of randomization. There will be patients who will not continue to take the drug, or the placebo, in each such study. This will have to be taken into account in additional processes of analysis, notwithstanding the fact that the rule of randomization must be obeyed in the strictest fashion.

Auditorium: What about the case of a study on acute myocardial infarction? It is desirable to treat the patients as early as possible. I begin the treatment at a point in time when I cannot be certain about the diagnosis, whether the patient had an acute infarct or not. How does one decide in this situation? Will patients remain in the group they have been entered into through randomization, even if the diagnosis turns out to be incorrect?

Schröder, Berlin: These patients will have to be considered separately. The main thing is that they are randomized into the study. The study will have to be continued until a sufficient number of patients exhibiting the problem to be studied have been accumulated in the study and the control groups.

Horbach, Erlangen: The basic principal is to avoid selection, because this will inevitably render the results erroneous. If patients with misdiagnosed acute myocardial infarction enter the study group, they will also be found in the control group in a quantitatively similar fashion, if the process of randomization has been organized in a proper way. Therefore, no problems will arise if this group of patients with a false diagnosis is excluded later on in the process of evaluation of the study.

Just, Freiburg: In the study of vasodilating drugs and their effect on patients with heart failure, patients with widely differing severity of disease have to be considered. It will be essential for studies in this field that the groups to be studied be stratified according to the severity of heart failure. This applies not only to studies concerned with mortality or longevity, but also to any study in this group of patients concerned with clinical or hemodynamic data.

Cohn, Minneapolis: The question of qualification of a patient for a study is always difficult to answer. Of course, cirteria will have to be selected that can be handled in a clinical study; as far as possible, it will have to be made certain that the degree of severity at the time of entrance into the study will be maintained approximately over the time of the study. If this can be assumed, a stratification according to the degree of severity of heart failure is possible and advisable. On entering into a study one should always take time over the process of patient selection and randomization. We always prefer a fixed number of patients in a study and a control group. We tend to make the size of the sample slightly larger in order to maintain a sufficient number of patients for a statistically meaningful analysis in each group.

Löllgen, Freiburg: Can errors and disadvantages of retrospective studies be minimized or abolished if larger groups of patients are studied?

Schröder, Berlin: The main problem with retrospective studies rests with selection. Earlier cases may not be as well documented, the design of observation not be as complete, etc. One will always have incomplete data in retrospective studies. Here, even large numbers remain useless, because the systematic error remains and may even become more pronounced.

Drexler, Freiburg: How can a study be handled double-blind, randomized, and with variable doses of the drug to be studied? Should not patients with heart failure generally be treated with fixed doses of one drug in order to avoid the difficulty of a complex study fulfilling all three of the mentioned criteria.

Schröder, Berlin: I believe that a double-blind design is unimportant for some studies. Studies with variable doses will hardly ever support a double-blind design.

Auditorium: In practice, double-blind studies are quite frequently begun but cannot be continued to the end point because the drug effect is evident and cannot be overlooked. In studies concerned with highly active drugs, a double-blind design will hardly ever be perfect.

Taylor, Leeds: I want to suggest that considerations of statistical planning in this kind of disease are irrelevant. I believe that we are applying vertical thinking to a problem that requires horizontal thinking. We are attempting to utilize statistical models, which are not compatible with a process the biology of which we do not even understand. We cannot utilize the results of clinical trials for statistical set-ups in heart failure, because the requirements are totally different from those in other heart diseases. We have to take into account that we do not know all the effects involved and that we cannot even design a model. The models that I have seen are the conventional ones. It is not evident that they apply to what we know of clinical trial situations in heart failure. Where doses are adjusted in different conditions, multiple effects will be seen, which obviates the possibility of any meaning ful statistical analysis. This is not just a question of lowering the blood pressure. The complexity of the disease to be dealt with is so tremendous that in my beliefe statistical means of planning are imperfect from the beginning.

Schröder, Berlin: I would tend to accept your comment; however, the statistical means of planning and evaluating a study will, even at this stage, have their usefulness and applicability. Their value in questions of detail is undebatable. I do agree, however, that the problem is exceedingly complex. I hope that our knowledge will improve and that our means of measuring functional alterations will improve as well. We might then reach a point where we could apply statistical methods to the basic question of whether function and survival can be improved. I am afraid that otherwise we would be lost very easily.

Just, Freiburg: The nature of the problem of congestive heart failure is extraordinarily complex: we all realize this fact.

All the more, then, are we called upon to search for ways and means of clarification. This applies to the pathophysiology and pharmacology as well as to the means of evaluation, that is the methods of statistics. The value of statistical methods in clinical studies cannot be overemphasized.

It is my firm belief that only through statistically well-planned studies, such as Horbach has outlined, will we be able to shed light upon the effects of vasodilators in congestive heart failure in particular, and to improve our understanding of disease and our means of treatment. If foresee real danger if we should ever depart from the lines of sound statistical planning and evaluation of clinical studies.

Subject Index

MIX
Papier aus verantwortungsvollen Quellen
Paper from responsible sources
FSC® C105338
www.fsc.org

If you have any concerns about our products,
you can contact us on
ProductSafety@springernature.com

In case Publisher is established outside the EU,
the EU authorized representative is:
Springer Nature Customer Service Center GmbH
Europaplatz 3, 69115 Heidelberg, Germany

Printed by Libri Plureos GmbH
in Hamburg, Germany